Human Reproduction

This book is outdated, inaccurate, and male-biased.

Human Reproduction
BIOLOGY AND SOCIAL CHANGE

HAROLD D. SWANSON
Drake University

New York
OXFORD UNIVERSITY PRESS
London Toronto 1974

Original drawings by Gene A. Lucas, rendered by Vantage Art, Inc.
Copyright © 1974 by Oxford University Press, Inc.
Library of Congress Catalogue Card Number: 73-92868
Printed in the United States of America

Preface

Human reproduction is a subject having universal appeal. Each of us is a sexual being with special sexual needs, desires, problems, and satisfactions. In addition to this, each of us is a *unique* human being with the potential for helping to produce another *unique* human being. Everyone should understand as much as possible about this process—so basic to all human life.

Many books have been written about this intriguing subject. Some of them have been useful in shedding new light on a complex subject; others have been full of misinformation or partial truths, and therefore misleading.

It is the purpose of this book to look at human reproduction from the point of view of a biologist, viewing human problems and concerns in light of our evolutionary origin and our kinship with all living things. I am convinced that biological observation and theory, gained from the study of many kinds of living things, can be usefully applied to our concern with human reproduction. The importance of biological understanding will increase rapidly as new reproductive technology progressively alters the manner of origin of new human individuals. It is my intention here to establish a biological understanding of human reproduction for clear thought and intelligent action.

Most of the material in the book is based on standard biological information from a multitude of disciplines, and from sources usually too technical for the layman to read, even if he could find them. The book touches on medical, psychological, sociological, and ethical aspects, without concentrating on any one area. Information is included on the biological topics of genetics, cell biology, embryology, pregnancy and birth, conception, sexual physiology, evolution, and behavior. This information is applied to the consideration of such subjects as contraception, population problems, abortion, genetic counseling, race differences, origin of human nature, styles of parenthood, and human reproduction as it may be practiced in the future. I have included considerable material from several popular scientific books in those areas which are most distant from my personal professional competence. I have also included a number of new ideas of my own.

Those reading *Human Reproduction* are assumed to have a basic acquaintance with the terminology and assumptions of modern biology, at about the level of high school biology. Thus the book can be read with profit by almost anyone interested in the subject. It was actually written for use as the text in a college course in the biology of human reproduction for nonbiologists, and would also serve easily as one of the texts in a more broadly directed course.

The reader of this book may start with only a high school level of understanding about human reproduction, but I hope he does not remain at that level. I have intended to include all the biological fact and theory that can profitably be brought to bear on individual understanding of topics of concern in human reproduction. Material that is normally scattered throughout many different advanced college texts in biology is collected here and applied to the reader's understanding and experience. The necessary vocabulary is defined when first used; these words and some additional terms that might appear in other reading are included in the Glossary.

There are numerous popular accounts, some very good, of individual topics: human inheritance, reproductive hormones, sexual responses, experiences of pregnancy, fertilization and its prevention, problems of our growing population, new developments in reproductive technology. I am not aware, though, of any other book that includes all these topics. The outline of formative events in embryonic development and the predicted timetable of future innovations in this field may not be available in any popular works.

Humans have long exercised a unique level of control over their own reproduction. Within our lifetime, the process of human reproduction has begun to come under direct human control more than ever before. New powers to control not only the timing, but also the description of our children may have tragic results if we are not prepared to cope with these new powers and their products. A biological understanding of the reproductive process is useful now in making wise decisions among the choices which are already available to us. But without such an understanding, the potentially greater control of human reproduction in the future can lead only to disaster. I firmly believe that the only protection against the hazards of dangerous knowledge is more knowledge and better understanding, as well as a thorough discussion of the uses of this knowledge. I hope this book contributes to that end.

Des Moines, Iowa H.D.S.
January 1974

Contents

1. Understanding human reproduction: a basis for making decisions, 3
2. Gametes, genes, and gonads, 16
3. Hormonal and nervous control of reproductive behavior, 70
4. Conception and contraception, 110
5. The early embryo, 144
6. Embryonic development: a commonplace miracle, 179
7. Pregnancy and birth, 236
8. From child to adult: growing individuality, 269
9. Revolution in reproduction: double-edged swords, 318

Glossary, 367

Suggested readings, 385

Index, 389

Human Reproduction

I | *Understanding human reproduction: a basis for making decisions*

Building a computer or a robot is an expensive undertaking that requires advanced technology and high competence. In contrast, two ordinary people can put together a tiny cell that is capable of growing into the most wonderful and complicated system known, a human being. The tiny cell that grows into an infant needs only to be given the ordinary support that is usually available even if the parents live in a poor and primitive society.

Understanding how a tiny cell can contain the patterns necessary for growth into a person, and how this development proceeds, is difficult but also rewarding. Any reader of a book about human reproduction is probably interested in three things about this process:

1. Understanding the entire fascinating process and seeing how the individual's experience fits into the pattern.
2. Being able to predict such things as what pregnancy is like and what a newborn baby might look like. Also, knowing what kinds of variations are common so as to avoid needless worry about harmless differences, but to be able to spot conditions that should be corrected.

3. Acquiring the necessary background to make decisions related to reproduction. Each year brings a further increase in our power to decide whether or not to have children, to detect defective embryos before birth, and to correct problems that plague us. Thoughtful individuals want enough information to make wise decisions about whether to exercise these new powers and if so, in what way.

I hope this book can help its readers understand reproduction clearly, predict events accurately, and choose their actions wisely.

I have tried throughout the book to emphasize the biological topics on which I can speak with some authority and with special understanding. In contrast, I have put much less emphasis on the ethical, psychological, and social ramifications of reproduction. These are at least as important as the biological aspects, but they are not within my area of special competence. Like anyone else, I have strong personal opinions in many of these areas, but in writing the book my intention has been to include opinions only where my biological point of view could contribute to an improved understanding of the issue.

The usual popular book related to human health gives a number of interesting and generally accurate facts and answers many of the specific questions that people commonly ask. Most such books, however, do not attempt to help the reader gain enough understanding of the subject to draw his own conclusions about new problems as they arise. Scientific books on the subject are usually intent on presenting theories or philosophical concepts that explain the data gathered by research workers in the field. They usually answer only a few of the questions the average person might ask about the subject. Furthermore, the discussion of the theory is often aimed at producing further technical research rather than at its application to common human problems.

I have included in this book both facts and theory—enough facts to answer the questions that most people have, and enough theory to provide responses to new problems as they come up. I have tried to restate principles or theories so that they may be applied by the nonspecialist, and to demonstrate with examples just how the theories might be applied.

There is no scientific field called "reproduction." Instead, the description of reproduction and the thinking about it are done by workers in the various biological fields of evolution, anatomy, physiology, genetics, embryology, and behavior, and in the medical specialities of gynecology, obstetrics, and psychiatry. These fields have not been taken up separately in this book; instead, materials from each field are introduced at the places in the discussion where they apply. The sequence of topics in the book follows the sequence of events in the reproductive cycle, from gamete to infant to parent to gamete again. To describe any one part of the reproductive cycle, it is usually necessary first to understand something that has happened previously in the cycle. This is precisely because the process *is* a cycle that goes on repeating itself generation after generation. Thus the biologist's old joke: "Which came first, the chicken or the egg?"

Summary of the reproductive sequence

In this section an overview of the reproductive cycle will be given (Fig. 1-1). The various processes involved will be explained in later chapters.

Mature men and women produce special reproductive cells, or gametes; these are called sperm (or spermatozoan) in the male and egg (or ovum) in the female. During sexual intercourse, the sperm from the man is deposited in the body of the woman. There the gametes join to form the fertilized egg (or zygote). The zygote begins immediately to divide into a mass of smaller cells, which can now collectively be called an embryo. The embryo attaches to the wall of the mother's womb (or uterus) and grows and develops there. Part of the embryo fuses with part of the uterine wall to form the placenta, through which materials necessary for growth and regulation are exchanged between the blood

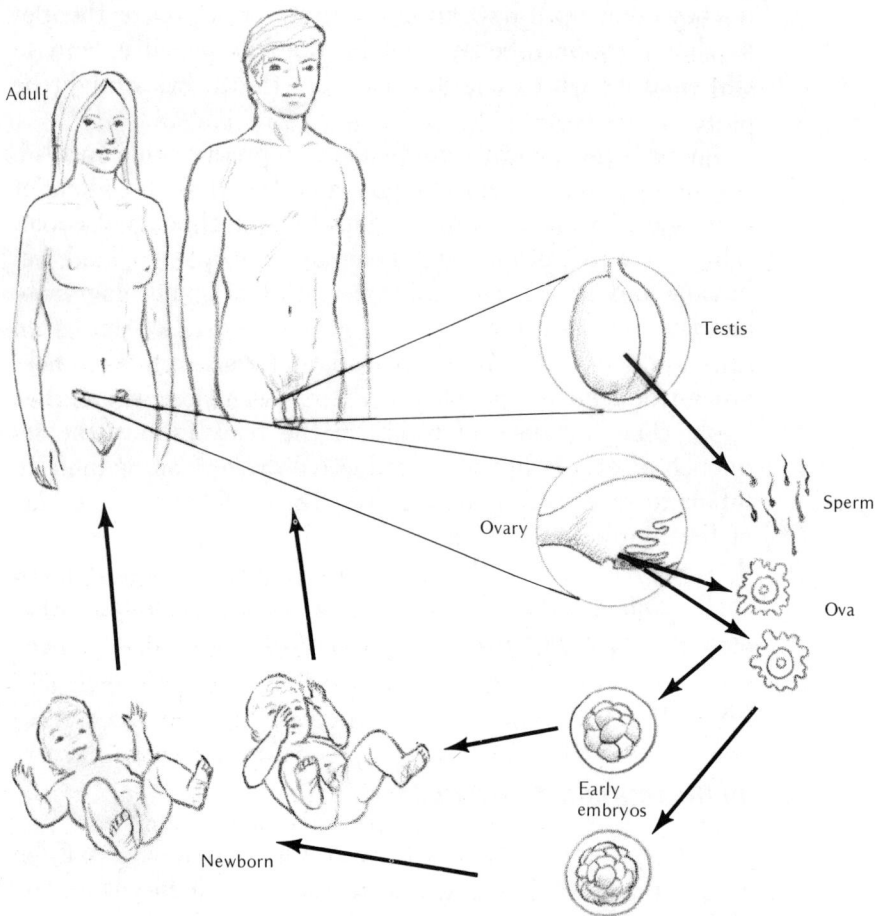

Fig. 1–1: The circle of life. Stages in the human reproductive cycle.

systems of the mother and the embryo. The alterations going on in the embryo, which change it from a shapeless mass to a human being, are fascinating but not easy to understand. The structures and processes involved include many unfamiliar ones that do not have common names. After 9 months, when the embryo has passed through a progressive series of profound changes, it is expelled to the outside world as a newborn infant. The infant continues to change as it grows through childhood. Then, sometime after the child reaches the age of 10, the gonads (testes in the male, ovaries in the female) enlarge, and they begin to produce sex hormones and then gametes as well. These hormones stimulate other parts of the body to be ready for reproduction—for transferring the sperm and for nurturing the zygote that may be formed. The same hormones also stimulate special behavioral patterns—the sexual drives. Sexual desire, acting through socially learned and controlled patterns, culminates in sexual intercourse: The sperm are deposited through the penis into the vagina of the woman and swim to where the ovum is approaching the uterus. One sperm unites with the ovum, forming a zygote. Thus the cycle has been completed and continues.

The importance of sexual reproduction

In any plant or animal species the process of reproduction is necessary simply to provide replacements for those members disabled or lost by accident or death. In modern human society this replacement has an additional importance: Old people have trouble adjusting to changing times, so we depend on the young people coming along to respond to new problems with new solutions.

The reproductive process would be much simpler if offspring were produced from a single parent, as is common among bacteria and protozoa. Instead of utilizing this simpler process, however, almost all higher organisms depend on the much more complicated process of sexual reproduction. Sexual reproduction requires that two different kinds of adults come together when each is in a special state of readiness—a very complicated requirement. As might be deduced from these facts, there is a compelling reason for the nearly universal occurrence of sexual (two-parent) reproduction instead of the simpler asexual (single-parent) process. Sexual reproduction produces a valuable reshuffling of the elements of inheritance, so that each zygote receives a new combination containing some elements from each parent. These offspring of two parents, each offspring genetically unique, provide the chance that a few may be capable of adapting to gradual, periodic or even catastrophic changes in the environment. If the offspring live in a world even slightly different from the one in which their parents lived, they need to be somewhat different. Since the genetic reshuffling is random, most of them would be no better suited to the new conditions than their parents. But some of the offspring might be better fitted to live under changed conditions, and these would survive to reproduce. On the other hand, carbon copies of previously successful types (offspring produced from a single parent) would never be suited to cope with new problems.

It is this sexual reshuffling, coupled with the survival of those individuals having the best combination of traits for changing conditions, that makes up the process of evolution. Evolution has produced all the different kinds of plants and animals that have ever existed, including man. Evolutionary selection has, meanwhile, ensured that plants and animals continue to reproduce sexually, since sexual reproduction helps to make them better adapted to changing environments and therefore more successful at living and reproducing.

The few exceptions to sexual reproduction among higher animals help to illustrate how important this sexual recombination is. The few insects and many worms that reproduce from one parent generally do this at a stage in their life cycle when they are isolated from other members of the species (for example, flightless aphids on a tall plant and parasitic worms inside a snail). They may have a sexual phase of some kind at another stage of their life cycles. The animals most closely related to humans that utilize asexual reproduction belong to a few species of desert lizards (Fig. 1-2). These species, made up of females only, live in very sparse desert locations that will support so few lizards per unit area that the chances of finding a mate at breeding time are apparently too low for survival. In these regions the females laying eggs that will develop and hatch without fertilization will be the only lizards to perpetuate themselves. These female-only strains are at a disadvantage in slightly less harsh areas that can support denser populations of lizards. There, lizards that

Fig. 1–2: Lizard species, *Cnemidophorus neomexicanus*, that has no males. The entire reproductive process is carried out by females. (Courtesy of James Christiansen, Drake University.)

reproduce sexually are more successful at surviving and reproducing than are the solitary females.

The retention of sexual reproduction in almost all animals, including man, has certainly been the result of its genetic advantages. But in the human species, sexual behavior has acquired a much wider importance. Sexual attractions cement the bond between the parents so that both adults will be available to aid in raising the helpless babies. Beyond that necessity, sexuality has enriched the emotional lives of humans in ways not even approached in other species. Sex will continue to be important even in cultures where society rather than the parents takes the major responsibility for rearing the children. An awareness of the evolutionary basis of human reproductive patterns provides a helpful framework for understanding why and how these patterns are important in so many aspects of our lives.

Applications of knowledge

Knowledge about human reproduction has many applications to our thoughts and actions. The ideas that will be introduced in this section will be dealt with in greater detail in later chapters.

The most basic and important reproductive process going on in adults is the development of the sex cells, or gametes—the sperm and the egg, or ovum. The gametes develop in the primary reproductive organs—the testis of the man, the ovary of the woman. These organs (called the gonads) produce, in addition, hormones that interact with still other hormones and organs to influence many other parts of the body—all the parts that operate in a specifically male or female manner. Hormones synchronize the production and release of gametes (gametogenesis) and coordinate the release with other bodily changes. Study of all these hormones made possible the development of the most effective contraceptive yet found, "the pill." For the first time in history, reliable contraception allows sexual intercourse, with its special human importance, functions, and meaning, to be securely separated from reproduction at times when the production of children is not desired. Knowledge of the mechanisms of gametogenesis and fertilization can be of help in choosing a contraceptive.

Whereas sex and conception can now be separated, the physiology of sex cannot be separated from its social role. It is impossible to divide the physiological mechanisms that allow intercourse to take place from the personal attitudes and interpersonal relationships of the people involved. When problems arise in sexual relations, they may appear as physical ones, but they usually have emotional origins. Knowing the mechanism of sexual intercourse and understanding some of the things that can influence it may be of help in choosing a satisfying approach to love. An understanding of the evolutionary origin and the present functions of our sexual behavior may help us even further to make wise choices, whether of contraceptives or of conduct.

If, without or in spite of contraception, fertilization does occur, the new cell formed—the zygote—becomes imbedded in the wall of the uterus, beginning the wonderful, fascinating and mysterious process that culminates in a baby, a unique new person. The changing structures of the embryo have been described in some detail for a good many years. But our understanding is still very sketchy of why these things happen and why they occasionally go tragically wrong. Our understanding is limited both because the process is so complicated and changes so rapidly, and because the manner of thinking required in embryology had not been needed and developed in the older sciences. In recent years, scientific understanding of the embryological process has been improving rapidly.

A flood of recent research has contributed to our knowledge of how very small changes (called mutations) in the chemical constitution of the gametes of the parents can be passed on to the embryo. There these changes persist so that the embryo may be unable to manufacture some chemical necessary for life or health. As a result, it may die early in development or may suffer at birth from some defect such as phenylketonuria (PKU), albinism, or hemophilia. The principles of inheritance that Mendel worked out for garden peas 100 years ago, and that have been developed in greater detail for vinegar flies during the last 50 years, can be used to predict the chances that such a defective embryo will occur, or will occur again in the next pregnancy. It is now possible, and useful when the risk of such a defect is great, to get a sample of some cast-off cells from an embryo in an early stage of its development and find out from these cells whether or not the embryo has certain of the defects. It is then also possible to terminate the pregnancy if the law and the beliefs of the parents allow this action. Thinking out such beliefs requires an understanding of what kind of an object an embryo is. Is it alive? Is it human? Such questions can be considered intelligently only by someone with knowledge of embryonic development.

Most defects appearing at birth are not so simply explained as are single genetic defects, but we can at least see how, in the pattern of structural development, certain defects should appear instead of others—why a cleft palate rather than a cleft cheek, for instance. Further, when we understand a little of the complicated interaction of forces that shape the embryo, we can see why much of the old argument about which is more important in the determination of human traits, nature or nurture, is really not useful. A look at the process of embryonic development makes it clear that influences of nature and of nurture both are present and important for the embryo. They reinforce or diminish each other over and over again in the course of development, so that their effects cannot be fully distinguished.

Much more is known about developments before birth from the outside, so to speak, than from the inside, since the experiences of pregnant women are many, well described, and well discussed. But the medical advice given to expectant mothers has changed considerably in recent years, and probably still contains a generous dose of folklore mixed with sound advice. Consequently, there are still sharp controversies about the best diet and activities for a pregnant woman, the advisability of using pain killers and of having the father present at birth, and the importance of breast feeding the baby. Knowing the reasons for the generally accepted rules of conduct in pregnancy, and the reasons for the disagreements on other points, may help expectant parents to accept the necessary rules more happily and still to have confidence in their medical advisor when he tells them something different from what their neighbor was told.

The birth of a child is one of the most dramatic of human events. This is true for several reasons. Childbirth is the end of a long period of anticipation, which has involved many changes in the physical state of the mother and the life styles of both parents. The process is unfamiliar or at least nonroutine to most parents. There is some risk in childbirth and always considerable uncertainty as to when it will happen and how long it will take. But what makes it most dramatic is the question: "What will the baby be like?" The parents are concerned first about the baby's health and soundness and then about other variables. Is it a boy or a girl? How big is it? What color hair does it have? Later still, the other variables become more apparent: family resemblances, growth rate, disposition, intelligence, special talents. It is now possible, through the laws of genetics, to predict the chances for appearance of a number of defects and of many normal variations. It is also possible to prevent a few of the defects by counselling before conception. Less is known about the major variations—size, intelligence, longevity, talents—than is known about trivia such as eye color and earlobe shape, since major traits are determined by many genes inherited independently but acting together. More is being learned about the differences, both significant and trivial, between populations of people in different parts of the world.

Any increase in our power to solve old problems is likely to lead to new problems, usually including some unexpected ones. This is because both the natural world and society are exceedingly complex; we don't understand even the predictable matters very well, and people have a way of acting in ways that we haven't anticipated. The new problems may not be as bad as the ones that the new development solved, but we are faced, all the same, with the need to handle problems that we haven't previously dealt with. One of our greatest present world problems, and several of our future ones, are side effects of scientific and technical advancements. The worldwide population explosion, which threatens the world with famine, resource depletion, and despair, is the clear result of advances in medicine and public health that prevent premature death, especially of children. This elimination of childhood death is the clearest advance of human good that modern scientific civilization has brought —a mother no longer has to prepare herself for the probability that half her children will die before they are grown. This same advance is also responsible for one of our greatest perils, the population explosion. Solving this problem will require basic changes in our attitudes toward reproduction and in our behavior, and these changes will undoubtedly produce further unforeseen consequences. The recent and still anticipated additions to our ability to control and predict events related to reproduction (some of which have been developed because of the population problem) will require that we make some decisions we would rather avoid, as well as some that we are glad to have the power to carry out. The moral, emotional, and social consequences that we can already foresee probably mean that there are other significant consequences still unforeseen.

It is already theoretically possible to control the sex of the unborn child by selective abortion of the "wrong kind" of embryo early in development. Would anyone want to do that? Probably a few. But much more acceptable methods will surely soon be available, and being able to make just this simple choice will almost certainly upset the nearly exact balance of the sexes, with consequent loneliness and social disruption. Even more significant choices await us. Already certain terrible hereditary defects are detectable early in development of the fetus and can be eliminated by abortion. This will become standard procedure for severe defects except among those opposed to abortion under all circumstances. But then, as the list of detectable defects grows, we will have the question of just how serious a defect needs to be for us to want to eliminate it. Feeblemindedness? Diabetes? Harelip? Color blindness? We will have to decide. It is probably possible now, and will certainly be tried in the near future if it hasn't been already, to "adopt" a zygote produced by another couple and deposit it in the uterus of an adoptive mother, to develop there. But if this is done, shouldn't the ambitious parents adopt the embryo of the most admirable biological parents they can find? Or would that make them expect too much of their superchild? We now have a few ways, and will soon have more, to eliminate harmful genes, and improve the general condition of the next generation. But do we run a risk of eliminating some useful trait that is genetically connected with an unpleasant one? What is an ideal person, anyway? The measures necessary to bring about these changes will interfere with people's habits. Wouldn't that be dictatorial? Would it be better to improve people by taking better care of children and making sure they get the right kind of upbringing? Or would this interfere with an individual's rights even more?

Such painful questions have made many people think that we should control scientific progress and prohibit its application to new human problems. But if we do that, we are choosing to have all the misery and suffering that we could prevent by further scientific progress. We have no choice but to make choices. We had better understand and be ready for them.

2 | Gametes, genes, and gonads

Characteristics of gametes

Special cells called **gametes** combine, following sexual intercourse, to form a **zygote**, which can develop into a unique new individual. Gametes must meet several difficult requirements if they are to fulfill this function. They must be bulky enough to carry sufficient food to support the first growth of the embryo and still be agile enough to move together and unite. They need to lose the characteristics that they have had as rather inactive cells in the adult, and also to acquire the capability of dividing rapidly as a separate organism after they have united. Though the gametes must contain everything necessary to begin a life of rapid growth and development, this development must not begin until sperm and egg have united to become a zygote. Finally, each gamete must contain just enough hereditary information so that the zygote will have a mixture of some of the characteristics of each of its parents, without being overstocked with all the genetic information of both. This chapter will describe how these conditions are met, and some of the practical consequences resulting from the nature of gametes and of their formation.

Division of labor

The first two requirements mentioned for gametes—containing a large supply of nutrients and being able to move together to unite—obviously conflict with each other, since a

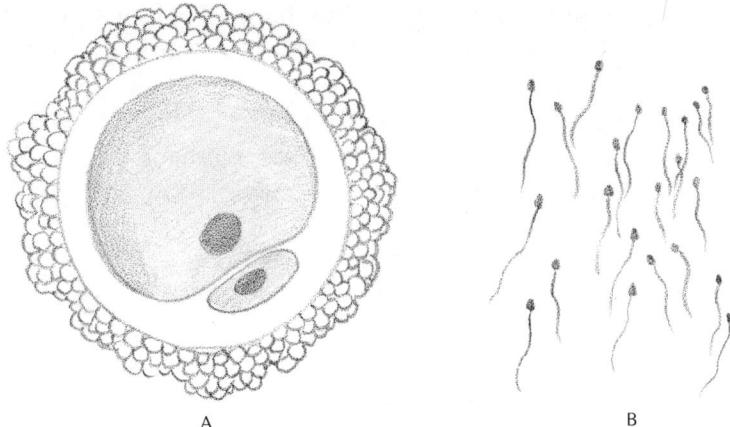

Fig. 2–1: Comparison of female and male human gametes. A: Ovum with attached polar body, surrounded by a layer of follicle cells. (Technically, this is a secondary oocyte.) B: Spermatozoa shown at approximately the same magnification as the ovum.

cell that is swollen with a load of nutrients can move only very slowly. This problem is solved by a division of labor (Fig. 2-1). One gamete (the ovum) is large and food-filled, and does not move by itself. The other gamete (the sperm) has shed most of its bulk and seems little more than a swimming speck of genetic information. Division of labor also solves the problem of starting the rapid cell division of development, since the ovum contains all the machinery for rapid cell division, but normally does not divide until stimulated by the entry of the sperm.

The ovum

The human **ovum** is 0.14 mm in diameter—smaller than a pinhead and just barely visible to the naked eye under good conditions. Yet it is a giant among body cells, since it has 10 times the diameter and 1000 times the volume of an average cell of the human body. As ova go, any mammalian ovum is rather small. The comparatively large size of bird eggs and reptile eggs is due to the presence of large amounts of yolk. This substance, familiar to us in the hen's egg, is a mixture of stored foodstuffs suitable for supporting the growth of the embryo. Since mammalian embryos will receive a supply of food directly from their mothers, the yolk supply needs only to be great enough to nourish the embryo briefly—until a good connection is established between the embryo and the maternal circulation. In birds and reptiles the embryo depends on the yolk stored within the egg to supply it with food throughout the whole period of embryonic development. Thus the familiar egg of a reptile or bird, though technically a single cell, bears little resemblance to an ordinary cell. These huge eggs are large masses of stored yolk with just a tiny spot of active cellular material at one point on the surface.

The mammalian ovum, except for its size and the presence of yolk, has the appearance of a rather ordinary cell. It is nearly spherical, like most cells that are not crowded together. It has no distinctive projections or visible internal complications. It is not known to perform any obvious or unusual actions, such as moving or emitting light. Yet the ovum is more than an inert food cache. Under exceptional circumstances mammalian eggs and bird eggs have developed, without fertilization, into embryos and then into mature organisms. The ovum must possess, in addition to one set of the genetic instructions found in all the cells of the body, some internal organization (so far undiscovered) that directs the first cell divisions and controls the modifications that produce the changing, developing embryo.

The sperm

The human **sperm** (or spermatozoan), looking and acting a bit like a tadpole, is only 0.06 mm long. The bulkiest part is called the head; it is essentially a much condensed nucleus or mass of chromosomes. In front of the head is a kind of cap called the **acrosome**, thought to contain materials or structures important for entering the ovum and stimulating it to divide. Behind the head are the midpiece and tail. (This tail is a slightly larger version of the flagella that propel some kinds of protozoa through the water; flagella are all built on a plan like that of the cilia found in various parts of the body, particularly on some of the cells lining the respiratory tract.) The midpiece contains condensed versions of several standard intracellular structures and is mainly concerned with supplying energy for the beating of the tail. It is still unknown whether or not these organelles of the sperm contribute to the nature of the zygote that is to be formed from the two gametes.

Whereas mammalian ova all have much the same featureless appearance from one species to another, sperm show considerable variation of the basic pattern. In the rat, for instance, the acrosome is a hook-like structure. No one knows whether these differences are important, or why they appear. Human sperm happen to be pretty close to the basic design, without noticeable frills. Sperm of some organisms, probably including humans, have some kind of guidance system to point them in the right direction. They certainly swim upstream against the slow current of fluid coming out of the female passages. In addition, sperm probably have a more specific guidance mechanism, perhaps an attraction to some chemical released by the ovum. This would make them swim toward an ovum in the vicinity and thus bring them close enough to unite with it. Their maximum unassisted rate of travel is about 3 mm per hour.

Defective gametes

Due to rare combinations of inherited factors, or to injury from disease, poisons, or radiation, gametes may be defective and unable to fulfill their functions. A good deal is known about the possibility of defective sperm, since sperm are produced in very large numbers and are easily obtained for study. Sperm of some men are misshapen, broken in two, or unable to swim (Fig. 2-2). If this is true of a large fraction of a man's sperm, he will usually be sterile. Perhaps even the normal-looking sperm of such a sterile man may have invisible defects that make them unable to operate. There may

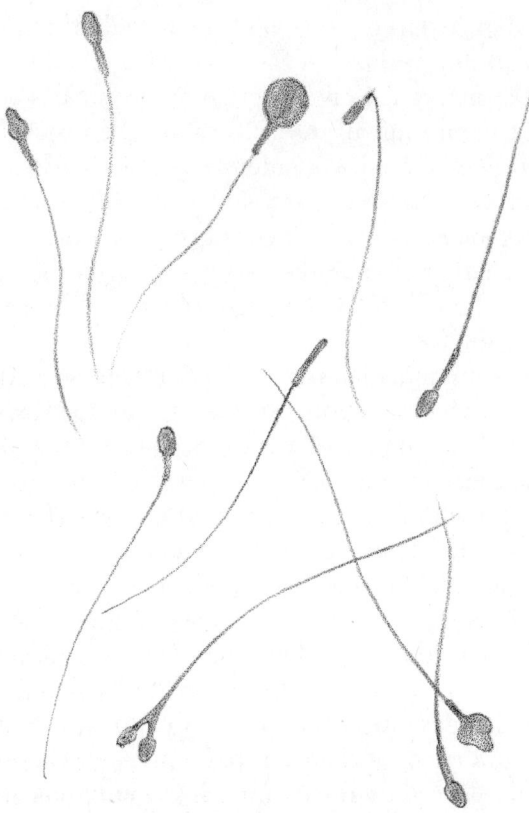

Fig. 2–2: Normal and abnormal sperm. Normal male fertility depends upon adequate numbers of normal spermatozoa. Here, among several normal sperm, are abnormal ones with giant head, dwarf head, two heads, misshapen head, and bent tail. These reduce fertility.

be as many things that can go wrong with the ova, but we know less about them. Eggs are released in the human only once a month, and it is difficult to remove and find them. Only with very special techniques can ova be obtained for study before they begin to break down. An ovum is thought to be capable of being fertilized for a day or so after it is released from the ovary. The sperm, in contrast, may be stored in the male tracts for weeks, but they retain their effectiveness for only a few hours after being deposited in the vagina.

Retraining to meet new job requirements

In order for the zygote that is formed from the union of sperm and egg to be ready to develop as an embryo, it must have cast off the limitations that an ordinary cell of the body has. This does not involve a lack of instructions or information. Each cell of the body, no matter what its function happens to be, normally contains the same set of genetic instructions. These instructions are sufficient to direct the activities of any kind of cell in the body. But in each of the 100 or so different kinds of cells in the human body, probably only a small fraction of the genes present in the nucleus of the cell are active. Those genes not in use in any particular cell type are under the influence of some inactivating chemical, and so are not available to participate in the biochemistry of the cell.

The hereditary instructions of each cell are contained in giant molecules, referred to in the biochemist's shorthand as DNA. These molecules are part of the chromosomes, which are located in the nucleus. (The way DNA functions will be discussed in Chapter 5.)

The cells whose descendants will be gametes grow slowly and divide infrequently. The two gametes, ova and sperm, when combined, produce a zygote that will divide rapidly into a mass of cells. These cells will then start to change from a collection of similar cells into a variety of different kinds of embryonic cells. During the formation of gametes, those genes that had been suitable for the quiescent precursor cell must be put out of action, and other genes, suitable for fast-dividing zygotes, must take over. The cell fluid now contains an excess of whatever agent brings about this change.

Nuclear transplantation

If a nucleus from a partially differentiated cell (perhaps an intestinal cell) of a frog embryo is transplanted into a zygote (from the same or another frog) from which the nucleus has been removed, the composite zygote will develop into a normal tadpole and frog. The transplanted nucleus, which was already well on the way to becoming the nucleus of an intestinal cell, apparently is changed by its new environment in such a way that it now gives instructions for the development of an entire embryo, rather than just for intestinal tissue, as in the original embryo. Frogs produced by this nuclear transplantation process have the genetic characteristics of the embryo contributing the nucleus, rather than those of the host zygote. Such experiments make it clear that the nucleus of any kind of cell still contains the information for a whole organism. But the nuclear transplantation technique is probably too difficult for routine use in domestic mammals or in humans.

Reproduction with only one parent

The requirement that the cells giving rise to new organisms be unrestricted, zygote-like cells rather than specialized cells must be met even for asexual reproduction in the plants and animals for which this is part of the life cycle. A small twig of a geranium plant or of a willow tree, will, if placed in good, well-watered soil, routinely give rise to a complete plant. This works so readily because plant tissues usually contain some unspecialized cells that are able to develop into any of the various kinds of plant cells. Similarly, in a parasitic flatworm such as the liver fluke, certain cells within the body of a larval stage of the fluke that lives within a snail are able to become like embryonic cells, and develop into replicas of the larval fluke. This process of asexual reproduction is probably related to the ability of many animals less complicated than mammals to regenerate lost parts of the body to an extent far beyond the simple wound-healing with which we are familiar. It is not understood at all why mammals, reptiles, and birds have lost so much of the ability to regenerate, as well as losing the ability to reproduce asexually. In any case, it now looks likely that we will "recover" those lost abilities through technology.

Mass-produced people

Many embryologists believe that in a number of years it will become possible to develop some solution of chemicals that will change an ordinary body cell into a zygote capable of development, just as the cellular fluids of the zygote can modify a specialized nucleus to become suitable for directing embryonic development. It will then be possible to take an ordinary body cell from an adult person, treat it with the solution to transform it into a zygote, insert this zygote into a woman's uterus, and have the normal development of a genetic twin of the original person. If this technical development occurs, we will be able to produce any number we might want of people having the same genetic makeup as some admired person, just as all the Golden Delicious apple trees in the world are derived from cuttings of one original superior tree found growing in Iowa. How much alike these carbon-copy people would turn out to be no one knows, since they would have slightly different environments during embryonic life and very different childhood experiences. Equally unresolved is whether this would be a blessing or a curse. But in either case, I think it will become possible. Maybe you could "adopt" a zygote developed from the cell of a famous athlete, actor, scientist, or statesman, nourish it as any other embryo, and raise the baby as your own child. Would you expect too much and be disappointed if your copy of Mark Spitz didn't win many swimming races, or your twin of Dame Margot Fonteyn didn't want to work at her ballet lessons? Less controversial, but more distant, is the possibility that we may eventually be able to take a cell from your body and use it to grow an arm or leg to replace one lost in an accident. The technical obstacles would be many, but this dream may come true within your lifetime.

Women's lib with a vengeance

By the time the ovum is mature, it must contain a full food supply and be ready for the beginning of embryonic development. It ordinarily does not develop until it is fertilized by a sperm. If no sperm comes, it will slowly disintegrate. Experimenters found as long as 50 years ago that unfertilized frog eggs, if pricked carefully with a pin, or if treated with certain chemical solutions, could occasionally be activated so that they would proceed with development as though they had been fertilized. Since then, other workers have done the same thing with eggs of turkeys and ova of rabbits. In the laying-chicken industry, where only the female chicks are of any value for selling to farmers, this procedure might have become routine long ago, except for the fact that in birds half of such artificially stimulated eggs fail to develop, while the other half produce cockerels. In most mammals, any offspring from an unfertilized egg is female. The legendary tribe of the Amazons, which the ancient Greeks said contained only women, could have existed and perpetuated itself if they had had this technology. This is essentially what has happened naturally among the desert lizards mentioned earlier, which have only females in the species. Apparently the blockage to development that is ordinarily built into the ovum and needs to be unblocked by a sperm does not operate in this kind of lizard. Eggs laid without fertilization develop into females (like mammals instead of like birds). The phenomenon of development by unfertilized eggs is called **parthenogenesis**, from the Greek words for virgin and reproduction; it occurs also in several species of insects as a normal part of the life cycle.

A baby without a father?

Since parthenogenesis is normal in several different kinds of animals and can be induced experimentally in many other kinds, it seems possible that parthenogenesis may occur spontaneously in people, once in a long while. Somewhere there may be a few pregnant unmarried girls who are telling the truth when they deny violating the social code. With this idea in mind, a British doctor investigated about 100 cases where the unwed mother denied having had intercourse. Some of these claims seemed unlikely from the history of the case, and in other instances blood tests of the mother and her baby indicated that the baby must have inherited some genes for blood type from someone in addition to its mother. In one case, the genetic tests were favorable to the idea of parthenogenesis. The history of the case showed that the mother had been under intensive care in a well-run hospital when she became pregnant, and she had not been unconscious. She would not have been raped without her knowledge, nor would she have had any reason to shield her daughter's father, if one existed. Who knows?

Weak Amazons

The legendary Amazons could, fancifully, have reproduced by parthenogenesis, but they probably could not have been very strong or healthy if that were the case. Individuals produced parthenogenically have a special kind of unmixed heredity, in which any existing hereditary weakness is intensified. Such a method might be desirable in some species of small animal where perfect types could be rigidly selected for several generations, but it would probably only lead to an increase in hereditary defects if tried in people.

Genetic segregation

It seems to have been commonly assumed by most people in history, without much thought, that the genetic contributions of the parents were "blended" in each offspring, so that what the child passed on to his own children was a kind of compromise between the traits his parents had bequeathed to him. From the vantage point of modern understanding it is possible to see that the common assumption was wrong, even on the evidence that was always available. If our heredity were an even blend of the genetic make-up of all our ancestors, everybody in an isolated village, for instance, would be almost exactly alike, since they would have practically the same ancestors and the same blend of heredity. What is actually seen, even in isolated communities where everyone is some kind of cousin of everyone else, is a considerable variety of humans, with many strikingly individual traits both important and trivial. This understanding was skillfully expounded just over 100 years ago by the founder of modern genetics, Gregor Mendel. Working with common garden peas, he found that even if an ancestral trait was not visible in a particular plant, it might sometimes show up unblended and undiminished in a later descendant of that plant. Such an occurrence has always been recognized by animal breeders as a "throwback"—usually supposed to be an exceptional occurrence due to an unusually influential ancestor. Mendel showed that this was not exceptional but was a standard pattern of inheritance. He found from his study of peas a pattern of inheritance that has since been found to apply to all sexually reproducing organisms. For any one characteristic (such as seed texture), each individual pea plant has two genetic instructions, one derived from its male parent and one from its female parent. The information inherited from the male parent might be for smooth seed, while the information from the female parent might be for wrinkled seed. The pea plant in question will pass on, through any one of its gametes, either the information for smooth seed or for wrinkled seed, but not information for "slightly wrinkled" seed. The inherited information is preserved in unblended form and then **segregated** from other instructions as it is passed on to the next generation. Some gametes may get one kind of instructions, some the other; none gets a blended message.

Mendel's contemporaries did not understand what he was talking about, so they failed to appreciate the importance of his work. This neglect was in part because no one then understood how the cell could go about a process of shuffling some durable, distinct hereditary factors—it seemed easier to believe that the characteristics of gametes sort of oozed together. In the last decades of the 19th century, however, much more was discovered about the elaborate events that take place during the formation of gametes. In 1900, three biologists each independently found Mendel's published account and realized its importance, because they saw from the new information about gametes how genetic segregation might work.

Gametes ensure a new deal

The most important reason for the formation of special reproductive cells—gametes—is that it makes possible the process of sexual reproduction, specifically the orderly reshuffling of the genetic inheritance. This provides genetic variety in the next generation. Since two gametes combine to form a zygote, the amount of hereditary information contained in a gamete must be *half* that contained in a regular cell. This keeps the amount of genetic material the same from generation to generation.

Genes and chromosomes

The units of inheritance, commonly called **genes**, are small structural elements that are arranged in a sequence, almost as though they were beads on a string. This string of genes, much coiled and folded, is called a **chromosome**, and it is located in the nucleus of a cell. It seems likely that chromosomes have evolved as devices to handle the parcelling out of many genes at once, to avoid the confusion and complicated machinery that would be required for separating the thousands of pairs of (maternally and paternally derived) genes in any cell. Most species of higher organisms have many chromosomes, presumably because a single chromosome long enough to contain all the necessary genes would not separate efficiently either; it would get tangled up in itself.

Human chromosomes

The chromosomes, when visible early in cell division, are rod-shaped bodies of various lengths, some with constrictions. The number, sizes, and shapes of the chromosomes are characteristic for each species of plant and animals. In fact, when two populations of similar-looking organisms are being studied to find out how closely related they are, a discovery of different chromosome numbers indicates that they are of different species. On the other hand, two distinct species may happen to have such similar chromosomes that they are not easily distinguished in this manner. Human cells each contain 46 chromosomes, assorted in 23 pairs. With one important exception, to be discussed later, the two members of each such pair of similar chromosomes (called **homologous chromosomes**) are identical in size, shape, and general appearance. In each pair of homologous chromosomes, one chromosome was derived from the mother, the other from the father.

The locus of the gene

Each gene is a bit of the structure of the chromosome at a certain place, on a certain size and shape of chromosome. This location of a gene is called the **locus** of that gene. For instance, the blood type of a person is determined by the genetic information contained at a certain locus—a certain position on one size and shape of chromosome. This locus of genes for blood type is called, naturally, the "blood type locus." The information at the blood type locus may give instructions for any of several different blood types. Many thousands of loci for other characteristics are found at other locations on the same kind of chromosome and on all the other kinds of chromosomes.

Why bother to learn about mitosis and meiosis?

When ordinary body (somatic) cells divide, each new cell must have all the genetic information, all the chromosomes, of the original cell. Higher organisms have evolved an elaborate process to ensure that this genetic information is accurately passed on. This process of somatic cell division is called **mitosis**. The term refers to the movements of chromosomes during somatic cell division. When gametes are being formed, however, the process of cell division goes on a little differently, and this process is called **meiosis**.

It is unfortunate that the two words, mitosis and meiosis, are so similar and sound so much alike. But there are no common substitutes so it is just necessary to notice carefully which word is being used.

Mitosis and meiosis are rather complicated processes to understand. Still, it is worthwhile to make the effort, since once a person is over this hurdle, most of the study of inheritance is fairly easy. For any one pattern of inheritance, it would be possible to learn a rule that would suffice without knowing anything about meiosis and mitosis. But understanding meiosis enables one to work out the rules for many different patterns of inheritance, and to understand abnormalities such as extra chromosomes. This is useful and powerful knowledge, leading to an understanding of many otherwise puzzling phenomena.

Getting ready for mitosis

The nucleus of a cell that is getting ready to divide undergoes a change in its internal structure so that the chromosomes in it become visible by means of a special microscope or a special stain and an ordinary microscope. A close study of a human cell in this stage of mitosis reveals that each of the 46 chromosomes has split longitudinally into two strands (Fig. 2-3). It is not as though a sausage had been split with a knife, however, because each of the chromosome strands is as thick as the original chromosome. (As the chromosome splits, each of its parts is reconstructed, to make a complete chromosome identical to the original.) These chromosome strands, when completely separated, will be new chromosomes each containing all the original genes. During the early part of mitosis the split between the chromosome strands is not complete; the two parts are still attached at a point called the **centromere**. (The centromere is also the point of attachment to the structures that will later pull the chromosomes apart.) There are 46 such pairs of chromosome strands visible in a microscopic preparation of a human cell preparing for division, making a total of 92 such strands in each cell. In mitosis (taking place in an ordinary cell, not one that is becoming a gamete), each split-chromosome pair of strands is separate from the 45 other such pairs and does not join with any others.

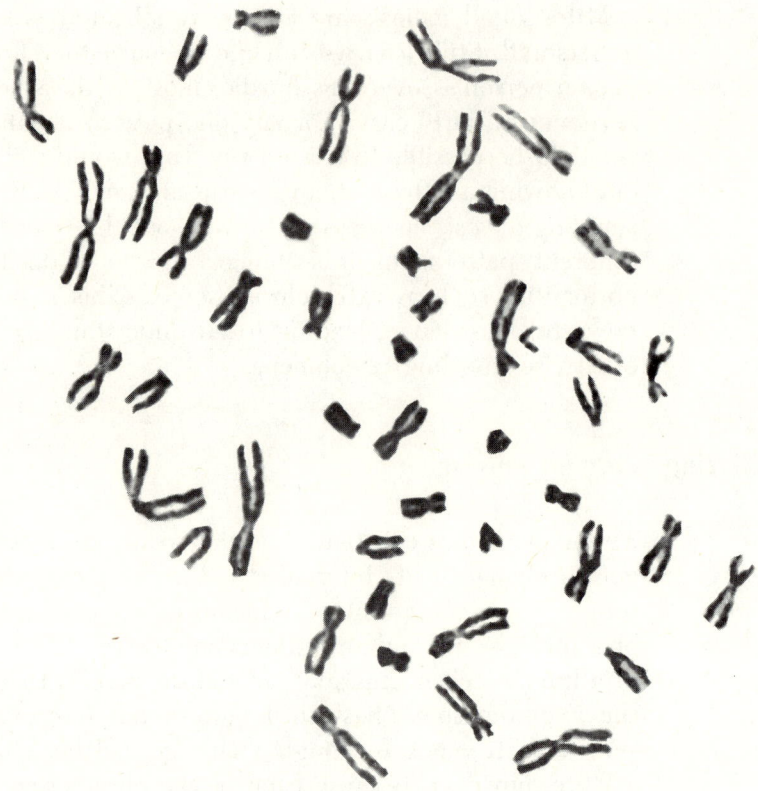

Fig. 2–3: Human chromosomes early in mitosis. Note that each chromosome is split into two strands held together at one point, the centromere. (Courtesy of Kurt Hirschhorn, Mt. Sinai School of Medicine.)

Mitotic cell division

As the process of mitotic division begins (Fig. 2-4), over a period of minutes or hours the chromosomes become more distinctly visible and more clearly duplicated. All of them line up in the center of the cell just like runners lining up for a race. Each is attached, by its centromere, to a fibrous structure called the **spindle**. When the chromosomes are all aligned at the center of the cell, the nuclear membrane disappears and the nucleus as such cannot be distinguished for the rest of the division process. At this point the centromere of each chromosome splits to complete the separation of the chromosome strands into individual new chromosomes. The two new chromosomes of each of the 46 pairs are pulled apart from each other and moved toward opposite ends or **poles** of the cell.

The chromosomes are pulled to the poles by an elongation of the spindle. Chromosomes trail behind their centromeres, each chromosome forming a V shape if the centromere is near the middle of the chromosome, something more like a "J" shape if the attachment is somewhat off-center, and a straight line if the attachment is at one end. This characteristic shape during movement to the poles is one of the ways in which the various kinds of chromosomes can be recognized.

Fig. 2–4: Chromosome duplication and behavoir in mitosis (ordinary cell division). Diagram includes only two pairs—pair 1 and pair 13 —of 23 normally present in human nuclei. M 1, maternal 1; M 13, maternal 13; P 1, paternal 1; P 13, paternal 13. See Fig. 2–9 for the standard number of chromosomes.

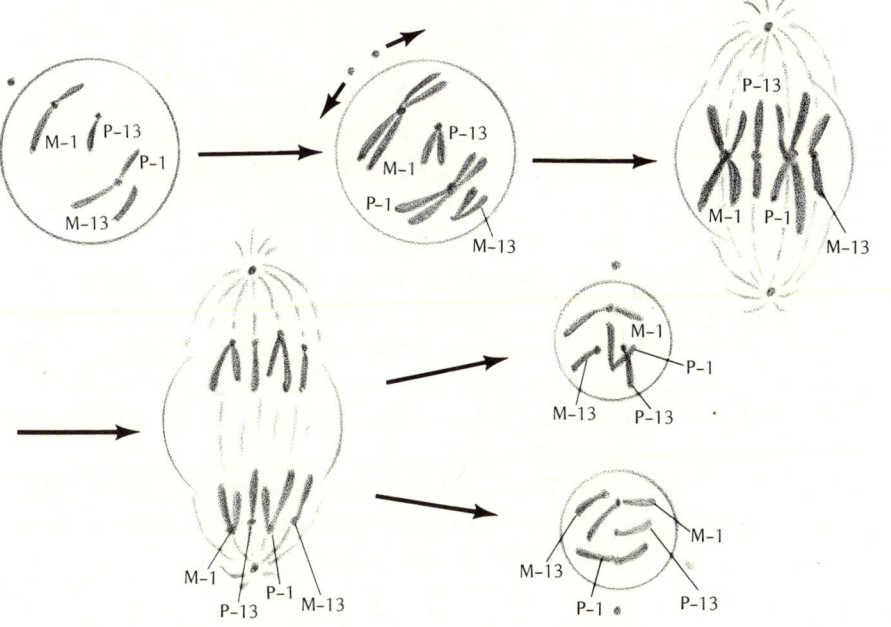

31

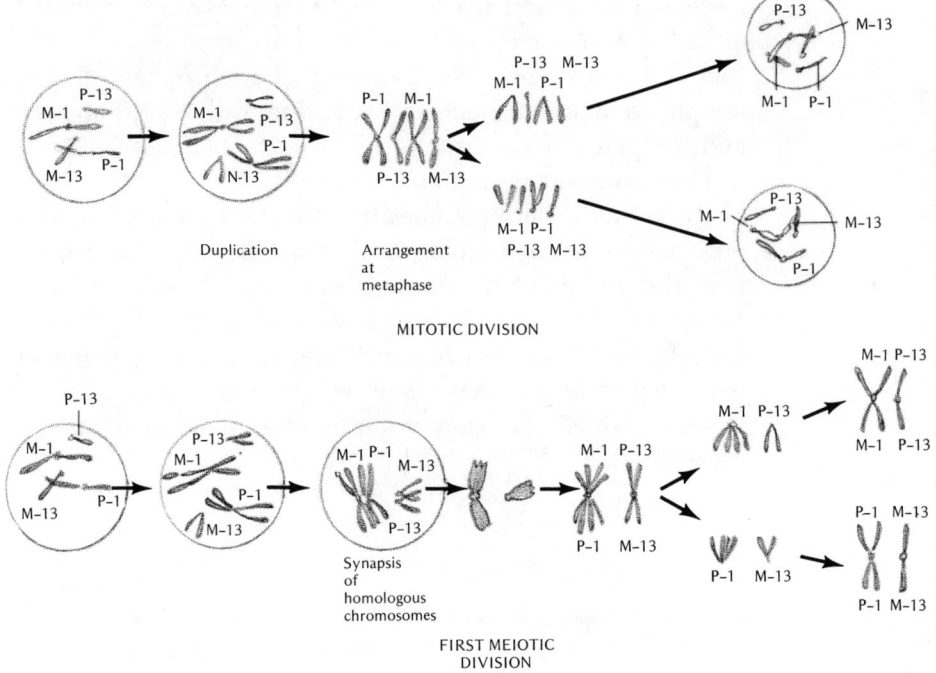

MITOTIC DIVISION

FIRST MEIOTIC DIVISION

The final state of mitosis

When the cluster of 46 new chromosomes arrives at each pole, the condensed, distinct chromosomes start to expand and fade out of sight, nuclear membranes form again, and the rest of the cell, called the **cytoplasm**, also divides, making two cells where there had been only one.

Mitosis can be seen to be a very conservative process; each new cell is assured of having an exact copy of all the chromosomes and all the genes contained in the original cell, no more and no less. This elaborate process goes on in each of the cell divisions that produce the trillions of cells in the adult human body.

In ordinary cell division (mitosis) each chromosome splits once and the cell divides once, producing two cells, each with the original number of chromosomes.

Fig. 2–5: Comparison of mitosis (ordinary cell division) and meiosis (gamete formation). For demonstration, only two chromosome pairs are shown.

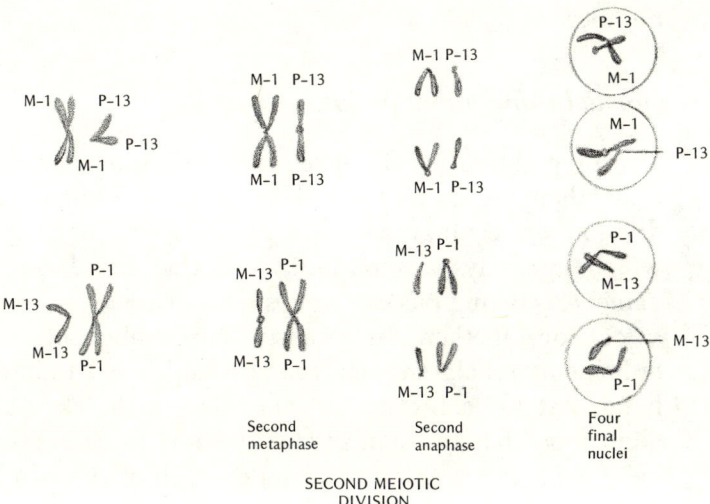

Second metaphase   Second anaphase   Four final nuclei

SECOND MEIOTIC DIVISION

Cell division in gamete formation: meiosis

In gamete-producing cell division (meiosis) the chromosomes split once, but the cell divides *twice*, producing four cells, each with *half* the original number of chromosomes (Fig. 2-5).

When the sperm fuses with the ovum to form the zygote, this new cell contains the genetic messages from both gametes. If ordinary cells from each parent, rather than gametes, fused into a zygote, each generation would have double the bulk of chromosomes that its parents had, and soon there would be no room for other structures in the cells. Therefore, a critical step in the formation of gametes is meiosis, the special kind of cell division by which the number of chromosomes is reduced to half the number found in ordinary body cells—one chromosome of each kind.

This special kind of cell division, meiosis, can best be understood as a variation of ordinary cell division. Meiosis can be thought of as a modified sequence of two mitotic divisions in which the chromosomes split up only once, so that the final cells each end up with half the original chromosome number.

The first unusual thing about meiosis: synapsis

The preparation for the first meiotic division is more elaborate than the corresponding stage in an ordinary mitotic division and contains a unique phenomenon called **synapsis**, which is always a sure indication that the formation of gametes is taking place. "Synapsis" is a Greek word meaning a "coming together." In synapsis, the members of each pair of homologous chromosomes come together in a tight bundle, lying next to each other so that they touch. They are also aligned so that the similar loci are next to each other; the loci for blood type, for instance, on each of the two homologous chromosomes where they are found, are next to each other or even touching, and the same is true for every other locus. During this time, just as in ordinary mitosis, each chromosome is split into two strands held together by the centromere. Since each pair of homologous chromosomes is synapsed, there are 23 bundles, each of four strands, representing the 23 kinds of chromosomes, two of each, each split into two strands. (Just as in mitosis, there are 92 strands in the cell, but they are now arranged in 23 bundles of four instead of 46 bundles of two.)

When the chromosomes begin to pull apart, there is another deviation from the events of ordinary mitosis. Each centromere does *not* split and carry its strands to opposite poles; in first meiotic division the centromeres of each homologous pair of chromosomes seem to repel each other, so that each pulls its bundle of two strands (half the bundle of four) toward opposite poles. The pattern of direction seems random, so that each group of chromosomes that collects at a pole usually contains some maternal and some paternal centromeres and chromosomes.

Second meiotic division

As meiosis continues, the cells divide again, but the chromosomes do not split. However, each of the centromeres (now 23) divides, so that single strands, as single chromosomes, are carried to opposite poles. The chromosomes, now reduced in number to 23, unfold and are lost from the view of an observer looking through a microscope. Though each cell now contains only 23 chromosomes, the 23 include one of each homologous pair—that is, one of each size and shape of chromosome.

Why meiosis works

Each of the ways in which meiosis differs from standard division—mitosis—is necessary to produce some required characteristic of the gametes. First, having the chromosomes split only once while the cells divide twice helps to ensure that the end product, the gamete, will have only half the number of chromosomes of the original cell. Thus the zygote, formed by the union of two gametes, will have the original number of chromosomes. (No one knows why gamete formation couldn't have been simpler, with one cell division and no chromosome splitting, but it just isn't that way.) The second difference, synapsis—coming together of homologous chromosomes—along with the pattern of separation of centromeres, ensures that as a result of the meiotic cell divisions, one chromosome of each kind will end up in each of the new cells. The result of meiosis is not just that the chromosomes are reduced in number in the gamete. In addition, the process occurs precisely, so that one and only one of each of the 23 kinds of chromosomes is represented in each gamete.

Simply by knowing and understanding the process of meiosis just described, it is possible to deduce directly many of the genetic ideas that were discovered from lengthy research by Mendel and his successors. First, however, it is necessary to understand about **alleles**, as well.

Alternative genes: alleles

Any specific gene locus always contains information for one particular trait, such as blood type or eye color. But the instructions are not always the same. The two or more alternative instructions for a given trait that may be found at a specific locus are called **alleles**. There may be two or many different alleles known for a locus; some may be common in a population of people, some may be very rare. Each person generally has two examples of each locus, one on each member of the pair of homologous chromosomes containing the locus. For instance, each person has two genes giving instructions for eye color, one on each of the two homologous chromosomes containing the eye-color locus. A person may have the same instructions (perhaps the allele for brown eyes) at both examples of the locus; in that case he is said to be "**homozygous** for brown eyes," or "homozygous for brown at the eye-color locus." He may instead be homozygous for another allele, most commonly the allele for blue eyes. On the other hand, a person may have two different alleles (one for brown eyes, say, and one for blue eyes) at the two examples of some locus; he is then said to be **heterozygous** at that locus. Note that there may be any number (tens, maybe hundreds) of possible alleles at any one locus, but a person can normally carry a maximum of two of these alleles, one at each example of the locus for that series of alleles. For example, there are at least three common alleles for ABO blood type; a person may have two examples of one of these in each cell, or he may have any two of the three (but the same two throughout the body).

Behavior of alleles during meiosis

During meiosis in the formation of gametes, the stem cells that are dividing to give rise to gametes contain two examples, ordinarily, of each locus, like any ordinary cell in the body. In the meiotic process, the two examples of each locus are separated into individual cells (eventually gametes) because the homologous chromosomes carrying that locus are separated from each other and go into different gametes. If the person is homozygous for brown eyes, for instance, each gamete gets one example of the allele for brown eyes. If a person is heterozygous at the eye color locus, with one allele for brown eyes, one for blue eyes, half of the gametes will carry the allele for brown eyes, half for blue eyes. It is a general rule, that *with regard to alleles at any one locus*, all of a person's gametes will be alike if he is homozygous for some allele at that locus; all gametes will carry one copy of that allele. If the person is heterozygous, it is a general rule that there are two kinds of gametes with respect to that locus, and that these two kinds are found in equal numbers; half of the gametes carry one allele, half the other.

Recombination of genes in the offspring

When two gametes, a sperm and an egg, unite to form a zygote—a new individual—this zygote will again have two examples of each locus. If the two gametes happen to be carrying the same allele, the zygote will be homozygous for that allele (at that locus). If the sperm and egg happen to be carrying different alleles, the zygote is heterozygous at that locus. It makes no difference which gamete (sperm or egg) carries which of two different alleles—the result to the zygote is the same.

It is obvious that the behavior of alleles just described corresponds exactly to the pattern of inheritance found by Mendel over 100 years ago, before anyone had heard of meiosis. Different alleles that happen to be together in one individual are **segregated** from each other in meiosis. It does not matter that an allele for brown eyes has been existing in the same cell as an allele for blue eyes for 30 years; when they are separated at meiosis neither shows any effect of this long association. There has been no blending. The alleles of the parents, now contained in individual gametes, can **recombine** in the offspring, often forming new combinations. For instance, a man homozygous for the gene for blue eyes will produce sperm each of which carries one gene for blue eyes. If his wife happens to be homozygous, too, but for the gene for brown eyes, all her ova will carry the brown-eyes allele. All their offspring will be heterozygous, at the eye-color locus, for the allele for blue and the allele for brown. This is a combination not seen in the parents, who were both homozygous, but for different alleles.

Genotype and phenotype

If the alleles that a person carries for a certain trait are known, it is usually possible to give a description of the person with regard to that trait. This is not necessarily so, however. More than one combination of alleles may give persons with the same appearance, and people with the same combinations of alleles may sometimes have different appearances, because of different environmental influences. Such lack of correspondence complicates the study of heredity, and it is necessary to understand the complications to understand many patterns of inheritance.

The combination of genes that a person has is called his **genotype**. Depending on the problem being studied, the genotype may refer to the alleles carried at a single locus, or it may refer to the alleles at several loci. The genotype is represented by a set of letters or symbols. The same letter or set of letters is used for all the alleles possible at one locus, often with superscripts to distinguish the different alleles (for instance, I^A, I^B, and I^O are alleles for blood type). If a person is homozygous for one of the alleles, his genotype is represented with two identical symbols, such as $I^A I^A$. If he is heterozygous, it might be $I^A I^B$, or any of the other combinations. If the study of inheritance is concerned with more than one locus at a time, the genotype contains both kinds of information. The locus for eye color is often represented by the letter "b," with "B" the allele for brown eyes, "b" the allele for blue eyes. A person heterozygous at the eye-color locus and homozygous for I^A at the blood-type locus would have the genotype $BbI^A I^A$. There is no rule as to which locus is written first, but it avoids confusion if the same order is used throughout the working out of a problem.

The physical description of a person with respect to a

certain trait is called his **phenotype**. Thus a person may have the phenotype of blue eyes or of brown eyes, and his phenotype may include any of the four ABO blood types. As with genotype, the completeness of the phenotype—the number of traits mentioned—depends on what is being studied.

Dominance and recessiveness

Some sets of alleles for one locus exhibit the phenomenon of **dominance**. If one allele is **dominant** to a second allele, the second is said to be **recessive** to the first. This means that the influence of the recessive allele on the phenotype (physical appearance) cannot be observed in the presence of the dominant allele—that is, in a heterozygous individual. Thus a person would have the same phenotype whether he were homozygous for the dominant allele or only heterozygous for it. Conversely, a person would show phenotypic evidence of a recessive gene only if he were homozygous for that recessive allele. An example commonly given concerns eye color: The allele for brown eyes is dominant to the allele for blue eyes. (Recessive alleles, like that for blue eyes, are often written with a lower case letter, as distinguished from a capital letter for the dominant allele.) Since there is dominance, both of the genotypes BB and Bb (homozygous and heterozygous for the dominant gene) produce the same phenotype—brown eyes. Only the genotype bb, homozygous for the recessive allele for blue eyes, will produce the phenotype blue eyes. Since the heterozygous brown-eyed person has a gene for blue eyes, he will still contribute this recessive gene to half his gametes, even though the gene's effects are not visible in his own body. Thus if his wife is also heterozygous for those two eye-color alleles, or if she is homozygous recessive, some of their children may be homozygous recessive and have the blue-eye phenotype. Two blue-eyed parents, on the other hand, would each be homozygous recessive and could have only blue-eyed children, who would also be homozygous recessive.

(A note of caution needs to be injected here. The inheritance of eye color is a popular example in simple discussions of genetics such as this one, because in most families eye color seems to be inherited as described above, by means of a single locus with two possible alleles, one dominant to the other. This is certainly an oversimplification. There really are several alleles, not just two; some of them have effects that are difficult to distinguish from effects of other alleles. Eye color is also affected by modifying genes at other loci, so that the pattern of inheritance is made still more complicated. Some of this complication is evident in the existence of people with eyes of a color called green, gray, or hazel. Some authorities suggest that these colors are due to other alleles intermediate in effect between the alleles for brown and blue, while other authorities say the intermediate color is caused by effects of genes at other loci. In any case, the simple pattern indicated above is the one usually observed, but it does not always occur. No one should be disturbed if inheritance of eye color in his family follows a different pattern.)

What dominance is not

Almost everyone has heard something about dominant genes, but many people get the wrong impression about them. Dominance simply tells something about our ability to distinguish certain genotypes by the phenotypes they produce. A dominant gene is one that has the same observable effect whether it is in double or single dose—that is, whether it is homozygous or heterozygous and accompanied by an allele recessive to it.

Dominance does *not* mean that the dominant gene is more likely to be transmitted by the heterozygous person than the recessive gene is. A dominant allele is not necessarily more common in the population than its recessive allele, nor does it necessarily replace the recessive gene as evolution proceeds. Dominant genes are not necessarily better for us or worse for us than recessive ones—they just show their effects more readily.

There is, though, a little bit of truth in several of these misconceptions about dominant genes. Recessive alleles are likely to cause the absence of some substance that should be produced; so recessive alleles are often harmful in double dose and are in such cases likely to be comparatively rare. It doesn't always work that way, though; blue eyes are caused by a recessive gene, but this is a perfectly healthy condition, quite common in some populations.

What will the children be like?

If the genotypes of a set of parents are known, then the different kinds of gametes they will produce, with regard to the alleles they carry, and the proportion of each kind, can be predicted on the basis of what happens in meiosis. From the proportions of the gametes carrying different genes, it is possible to predict the chances for one or another of the possible genotypes in the offspring (Fig. 2-6).

For example, if both the father and mother are homozygous for the b gene for blue eyes, all gametes produced by each parent will contain the b gene, and all children will also be homozygous for the b gene (genotype bb) and will have blue eyes as their phenotype. Similarly, if both parents are homozygous for brown eyes (not normally something they could be sure about), all the children will also be homozygous for the brown-eye gene, and have brown eyes. If one parent is homozygous for the b gene for blue eyes, and the other is homozygous for the B gene for brown eyes, all the gametes of the first parent will carry b, while all the gametes of the second will carry B; all the children will have the genotype Bb. Since B is dominant, these children will all have brown eyes, even though one parent had blue eyes.

If one parent is heterozygous Bb (with the phenotype brown eyes), and the other is homozygous bb (phenotype blue eyes), all the gametes of the second parent will carry the b gene, while half of the gametes of the first parent will carry B and half b. Each of the offspring will receive a b gene from the homozygous parent; of these, half will receive another b gene from the heterozygous parent and have the genotype bb, phenotype blue eyes. The other half will re-

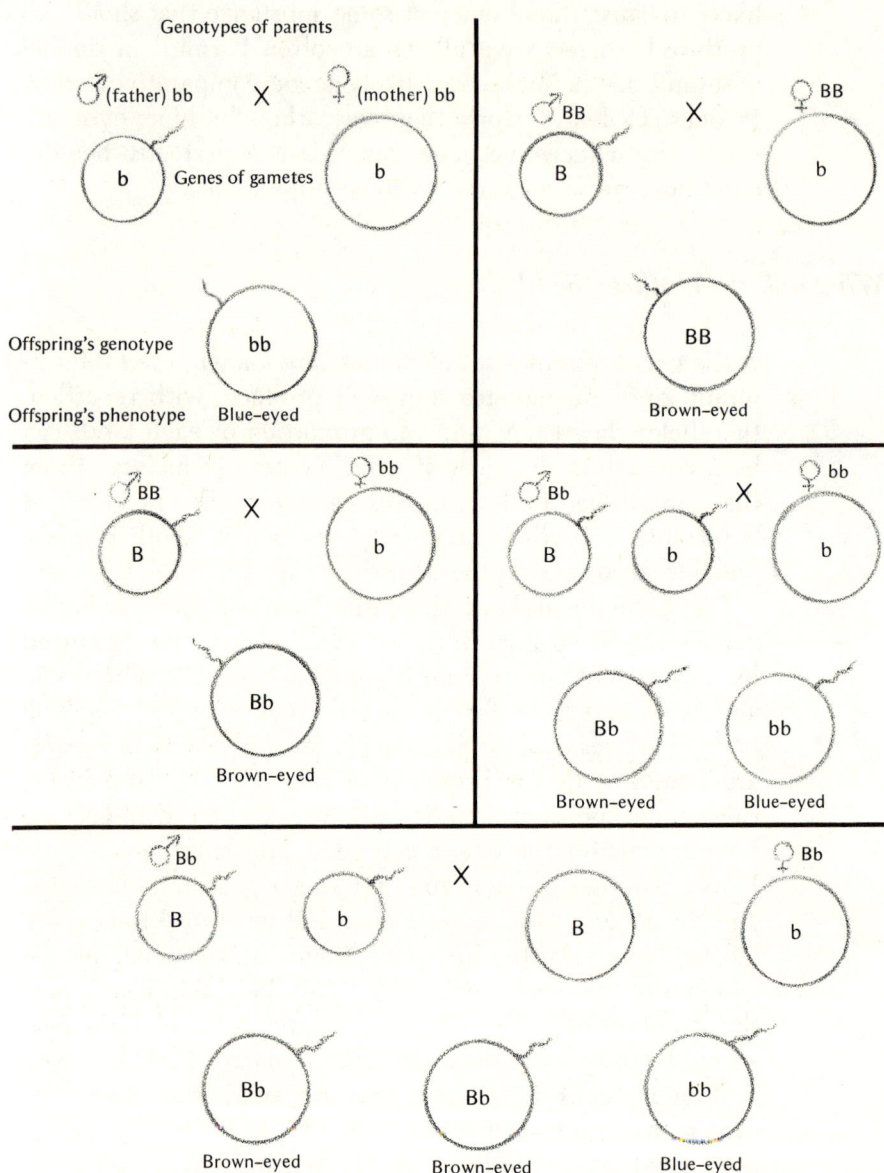

Fig. 2-6: Inheritance of eye color, showing the results of several possible combinations of parental genotypes. For each combination, the first row represents the parental genotypes and the second the gametes that can be produced from these genotypes. The third row gives the genotypes of the offspring that may result from combinations of the gametes, while the last row shows their phenotypes.

ceive a B gene from the heterozygous parent and have the genotype Bb, phenotype brown eyes.

The most complicated situation arises when both parents are heterozygous at the same locus. If the woman is Bb, half of her ova will carry B and half b. If the man is also Bb, half of his sperm will carry B too, and half b. Of all the zygotes that could be conceived by this couple, half would be from a B ovum. Of these, half (a quarter of the total number of zygotes) would be fertilized by a B sperm (producing BB) and half by a b sperm (producing Bb). Another half of the zygotes would come from b ova. Half of these would, as before, be fertilized by a B sperm (producing Bb), and half by a b sperm (producing bb). Summarizing, of all possible offspring:

1. One-quarter of the total would have the genotype BB, because they received a B gene from each parent.
2. One-quarter of the total would have the genotype bb, because they received a b gene from each parent.
3. One-half would have the genotype Bb; one-quarter because they received a b from the mother and a B from the father, one-quarter because they received a B from the mother and a b from the father. It would not make any difference which way they received this genotype.

Since there is dominance of B over b, the ratio of phenotypes will not be the same as the ratio of genotypes. Both genotypes BB and Bb will yield the phenotype brown eyes; only the genotype bb will yield blue eyes. Therefore, three-quarters of all possible offspring will have brown eyes and one-quarter blue eyes.

What do the ratios mean?

Remember, the ratios given above take into account all possible offspring of a particular couple. Actually, of course, families contain only a very small fraction of the total possible progeny. The actual distribution of phenotypes among the children whose parents have these genotypes may be quite

43

different, since each conception is a matter of chance, having no relation to what happened in the last conception. Thus, for example, two heterozygous brown-eyed parents could easily have four blue-eyed children. Such a family would be considered rather unlikely if the odds were considered ahead of time. But even if the first three children happened to be blue-eyed, the chances of the next one also being blue-eyed would still be one-quarter. However, in species that have hundreds or thousands of offspring (some fish, insects, chickens), the genetic ratios may describe the actual ratios of different kinds of progeny. In human families, we can only predict that a certain genotype is possible, with a certain probability.

Gene combinations without dominance

If there is no dominance, the probability of different phenotypes is the same as that of different genotypes. The next example illustrates both "no dominance" and "dominance."

All people can be classified in one of four blood groups, called A, B, AB, and O. (Group A can be divided into two subgroups, A_1 and A_2, but this complication will be ignored in the present discussion.) The blood group of a person is based on the presence of certain specific chemical substances on the surface of the red blood cells. These substances can cause fatal reactions if blood of the wrong group is transfused into the person. The substances that can cause trouble are called the A and B substances. Type A blood cells have A substance on them, type B cells have B substance, type AB cells have both. Type O blood cells have neither substance A nor substance B. Presence of these blood group substances is dependent on possession of one or two of the three allelic genes at the ABO blood group locus. These genes are designated I^A, I^B, and I^O. (The A and B substances are sometimes called "isoagglutinogens"; hence the use of the letter I.) Any cell that contains the I^A allele will have substance A on its surface. If it contains allele I^B, it will have substance B. If both I^A and I^B alleles are present, both A and B substances will be formed.

The allele I^O actually causes the production of a substance O, but this causes no transfusion problems and can be detected only by very complicated tests. Thus, ordinarily, it is impossible to distinguish, by observing the blood type, between a person who is homozygous for the I^A allele ($I^A I^O$) whose cells contain substance A and the undetectable substance O. So we say that the allele I^O is recessive to the allele I^A, which is dominant to I^O. There is the same kind of dominant and recessive relationship between I^B and I^O (Fig. 2-7).

Fig. 2–7: Inheritance of ABO blood types.

Genotype	Cell surface substances produced	Blood type (phenotype)
$I^A I^A$	A	A
$I^A I^O$	A, O	A
$I^B I^B$	B	B
$I^B I^O$	B, O	B
$I^A I^B$	A, B	AB
$I^O I^O$	O	O

To work out any problem of inheritance, put down the genotypes of the parents, the kinds of gametes they produce, the kinds of zygotes (genotypes) that can be formed by combining the gametes, and the phenotypes produced by these genotypes. Can you do it for the combination mentioned in the text?

AB X AB

Parental genotypes?

Gametes formed?

Zygote combinations?

Phenotypes?

The parental genotypes are both $I^A I^B$. Each parent produced gametes half of which carry I^A, half I^B. Of the offspring, $\frac{1}{4}$ are type A ($I^A I^A$), $\frac{1}{2}$ type AB ($I^A I^B$), $\frac{1}{4}$ type B ($I^B I^B$).

If two people of blood type AB marry (uncommon, since AB's are rare) the chances are 25 percent that any particular child will be group A, 50 percent that he will be AB, and 25 percent that he will be B (homozygous for the I^B gene). (If this is not obvious to you, work it out by figuring the kinds of gametes that are possible and the probabilities of each combination. Do this also for parents of other blood types.) Keep in mind constantly, in making these predictions, that we are dealing with probabilities. The probability of two AB's having a B type child is only 25 percent: 1 out of 4. It would be possible, however, for such parents (both AB) to have a family of four children all of whom were type B—for the same reason that a person flipping coins could flip a series of eight "heads" in a row. (In both cases the probability of the result—four B children or eight heads—is 1/256.) In both tossing coins and bearing children, what happens one time does not affect the probabilities for the next time. But prediction of probabilities is possible because over the long run, with many examples, heads will turn up about half the time when you flip coins, and about one-fourth of the children of parents heterozygous for a certain gene will be homozygous for the gene.

Assortments of different traits

The situation gets more complicated if we consider more than one characteristic at a time. Yet the principles are the same, and the probabilities can be worked out from knowing how the chromosomes are distributed during meiosis. Suppose a mother is heterozygous for sickle cell hemoglobin (Hb^sHb) and for blood groups (I^AI^B), while her husband is homozygous for normal hemoglobin (HbHb) and for blood group A (I^AI^A). Suppose further that the mother knew she had inherited both the I^B and the Hb^s gene from her own father. The couple in question has a child of blood group AB (easy to test for). Will this child also have a gene for abnormal hemoglobin? Will such a gene come along with the other gene from the same grandparent? The answer is normally "no," that the inheritance of hemoglobin type cannot

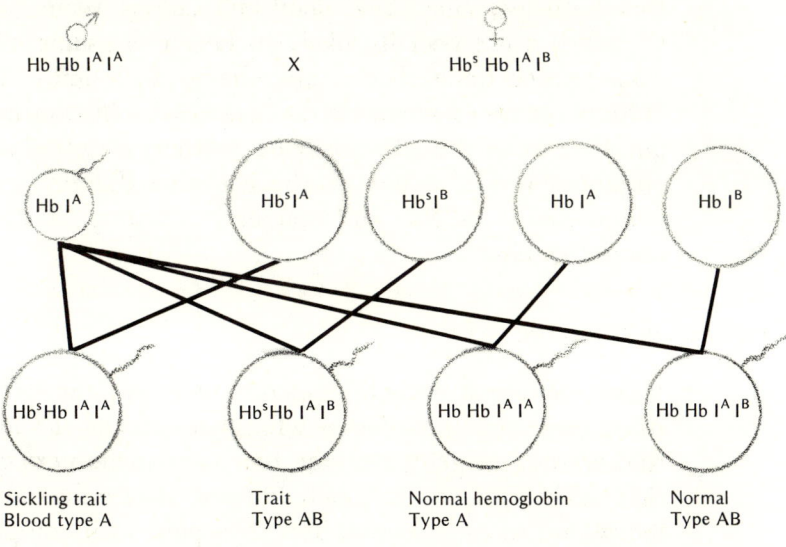

Fig. 2–8: Inheritance probabilities with two loci. In the first row, the parental genotypes; next, all possible gametes. In the third row, the genotypes of zygotes that can be formed, and finally the phenotypes of the possible offspring. The probability of each outcome is one-fourth.

generally be expected to tell us anything about what has happened to the blood type genes (even though both affect the red blood cell). We say that there is **independent assortment**: that the genes at different loci are sorted out independently of each other.

In calculating the probabilities in an example with two loci involved, each locus is calculated independently, and then the information is combined. In the present example, half the children will be homozygous for normal hemoglobin, and half will have the sickling trait (heterozygous). Similarly, half will have blood group AB, and half will have A. Putting this together, they will be ¼ AB, normal Hb; ¼ AB, sickling trait; ¼ A, normal Hb; ¼ A, sickling trait (Fig. 2-8).

Part of the reason for independent assortment is evident from what has already been said about meiosis. The chances are that the two loci are on different chromosomes (the fact that both loci affect the blood cell doesn't seem to mean that they are especially likely to be on the same chromosome.) Since for each size and shape of chromosome it is entirely random which pole the "maternal" chromosome (the one that came from the person's mother) and the paternal chromosome each move to, the genes on different pairs of chromosomes will be "independently assorted," just as the chromosomes are.

Linkage and crossing over

If two loci being studied happen to be on the same chromosome, there is a little tendency for genes on this chromosome that are once together to stay that way in the next generation. This tendency is called **linkage**; it is only rarely detected in human genetics, and then only with complicated statistical records of many families or with special laboratory techniques. It would probably never show up in casual observation of a person's own family. People will often notice that some combination of facial characteristics (like big nose, puffy cheeks, protruding lower lip) seems to be inherited as a unit over several generations of a family. Couldn't that be linkage? No, this is always a false explanation. The real reason is probably that what seem to be several facial traits inherited together are really due to a single gene acting so early in development that it affects several different structures.

The reason linkage is not more evident than it is lies in another phenomenon during meiosis that has not yet been discussed. When the pairs of homologous chromosomes synapse in first meiotic division, the association between them is so close that the two chromosomes of a pair may actually fuse and interchange parts. When they get sorted out in the gametes, one end of a chromosome originally received from the person's father may be joined to the other end of the homologous chromosome received originally from his mother. There may be several interchanges of parts (called **crossing over**) on each chromosome pair, so that each chromosome in the gametes has alternating sections derived from each of the homologous partner chromosomes. Thus the various loci on one chromosome get nearly as well mixed as if they were on separate chromosomes, and linkage doesn't concern us much.

Is it a boy or a girl?

There is one extremely important exception to the rule that each cell contains two examples of each locus. Among many animals, including all mammals, there is a pair of chromosomes called the **sex chromosomes**. A female mammal has a certain pair of medium-sized chromosomes, known as **X-chromosomes**. A male has an X chromosome and a smaller, Y chromosome (Fig. 2-9). (X and Y are just handy names to apply and have nothing to do with the shapes of the chromosomes.) The X and Y chromosomes share at least a few loci, and they synapse and divide as homologous chromosomes, despite their differing size and shape. Strictly speaking, they are homologous along only part of their length. The X chromosome contains a number of loci that have nothing specific to do with one sex or the other but are called **sex-linked** genes because they are on the X chromosome. As we shall see, their pattern of inheritance is unusual, and more is therefore known about what loci are on the X chromosome than about the loci on any other chromosome. The smaller Y chromosome has one or a few genes necessary for development into a male and possibly some few others unrelated to sex. The loci shared by X and Y are not usually recognized; their nature and pattern of inheritance are not unusual. Half of the sperm produced by meiosis will carry an X chromosome and will thus have an example of each possible gene locus (including, of course, those located on the X chromosome) with the exception of the few but important maleness loci on the Y chromosome. Half of the sperm will carry the Y chromosome. They lack any copy of the loci carried on the X. Even if the sperm were otherwise capable of developing into an embryo by itself, the Y-bearing ones would lack some genes essential for life. (We know this to be so because in birds it is the female that produces two kinds of gametes, one with and one without a major chromosome.

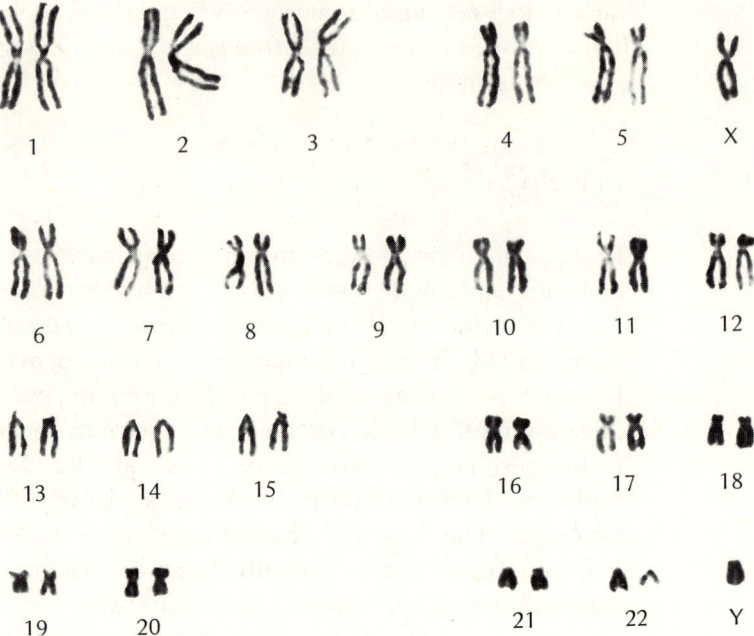

Fig. 2–9: The 23 pairs of human chromosomes, seen in Fig. 2–3, are arranged and numbered in order of size. Such an arrangement is known as a karyotype. (The chromosomes are also sometimes grouped as follows: Group A, 1, 2, and 3; Group B, 4, 5, and X; Group C, 6 through 12; Group D, 13, 14, and 15; Group E, 16, 17, and 18; Group F, 19 and 20; Group G, 21, 22, and Y.) (Courtesy of Kurt Hirschhorn, Mt. Sinai School of Medicine.)

It is sometimes possible to stimulate the egg of a turkey or chicken to divide and develop without fertilization, producing a chick with only one parent. Half of the eggs, those lacking the sex chromosome, will never hatch, however skillful the stimulation, since they lack essential genes.)

Sex-linkage

A significant difference from the ordinary inheritance pattern is found in sex-linked traits—those determined by genes at loci on the X chromosome. In the example given below, the notation X^c is used for gene C on the X chromosome; X^C

is the gene for normal color vision and X^c is the gene for color blindness:

$X^cX^c \times X^cY$ give offspring: ¼ X^cX^c, ¼ X^cX^c (girls)
  ¼ X^cY, ¼X^cY (boys)

In this example, the color blindness trait is recessive and sex-linked. It will show up phenotypically only in half the sons, since in them there is no dominant normal gene to cover its effects. A daughter can express such a trait phenotypically only if her father expresses it and her mother carries it, in either single or double dose.

When chromosomes get stuck

The process of meiosis does not always go smoothly. One important mishap is called **nondisjunction**. This means that a pair of chromosomes fails to separate during one of the meiotic divisions; both members of the pair go to one pole and none to the other pole. If the ovum lacks one of the 23 kinds of chromosomes, it will not develop completely when fertilized (with only one exception), apparently because the balance of instructions among the various kinds of chromosomes is uneven. The zygote will have two of every other size and shape of chromosome, but only one of the kind not contributed by the ovum. If, on the other hand, the ovum receives the two chromosomes that failed to separate, these will be joined at fertilization by a similar chromosome from the sperm, making a total of three. This abnormal condition is called **trisomy**, meaning "three bodies"; that is, three of one kind of chromosome. In most cases this also is fatal during early embryonic development. It is only a few of the smallest kinds of chromosome that the cell can tolerate in triple dose, and then the result is far from normal. Probably nondisjunction is just as common with large chromosomes as with small ones, but embryos containing them do not survive. It has been recently found that over 20 percent of the embryos that are spontaneously aborted in the first few months of pregnancy had trisomy of one or more chromosomes, or else **monosomy**, just one chromosome of some kind.

Fig. 2–10: A girl with trisomy 21, shown with her normal sister. (Courtesy of Gene Lucas, Drake University.)

Trisomy 21

The most familiar kind of trisomy (because it is not as likely to be fatal before birth) is that involving the chromosome pair number 21. Trisomy 21 produces a child with moderate to severe mental deficiency and many distinctive incidental traits: folded eyelids and dry, furrowed palms, together with a surprisingly good disposition (Fig. 2-10). The folded eyelids resemble, at least when observed by a European, the differently constructed epicanthic fold of most Asian peoples, and the condition was for some time called Mongolism, Mongoloid idiocy, or Mongolian idiocy. Some people even suggested that these unfortunate children were "throwbacks" to the Mongolian peoples who invaded Europe in medieval times. Because the common terms are so misleading, there has been a growing effort to substitute "Down's syndrome," derived from the name of the man who first described the abnormality. This name has the disadvantage that it lacks any clue that would help people remember what it means. Many now think that **trisomy 21** is the best designation. This has just one disadvantage: The same symptoms are found in another chromosome abnormality that is not strictly trisomy. It may sometimes happen that a small chromosome, or part of one, becomes permanently attached to the end of a larger chromosome. This is called **translocation**. Then an ovum may receive the normal number of chromosomes but one of them is longer than normal. In such a case, the person is effectively trisomic and shows the same symptoms (Fig. 2-11).

It is possible to have a case of Down's syndrome caused by a translocation of chromosome 21 onto chromosome 13, 14, or 15. Half the zygotes produced from such oocytes have Down's syndrome. In the other half, however, the ovum has not received the normal chromosome 21 in addition to the translocated one, and the chromosome constitution of the zygote is normal, except that one each of chromosome 21 and 14 are attached to each other. The person produced by such a zygote is said to have a **balanced translocation**. The person's offspring have one chance in four of carrying the balanced translocation without symptoms, one chance of being normal, and one of being trisomic with Down's syndrome. One-quarter of the zygotes will be **monosomic** for chromo-

Fig. 2–11: Chromosome arrangements that result in Down's syndrome.

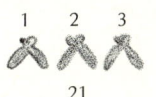

21

Trisomy 21. An extra chromosome 21 is present, making a total of three.

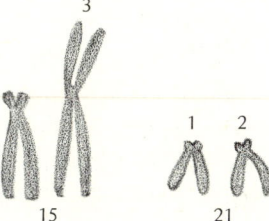

21-15 translocation. On the left is a normal chromosome 15 paired with an abnormal chromosome 15, having attached to it a substantial portion of a chromosome 21 (translocation). The result is the same as trisomy 21.

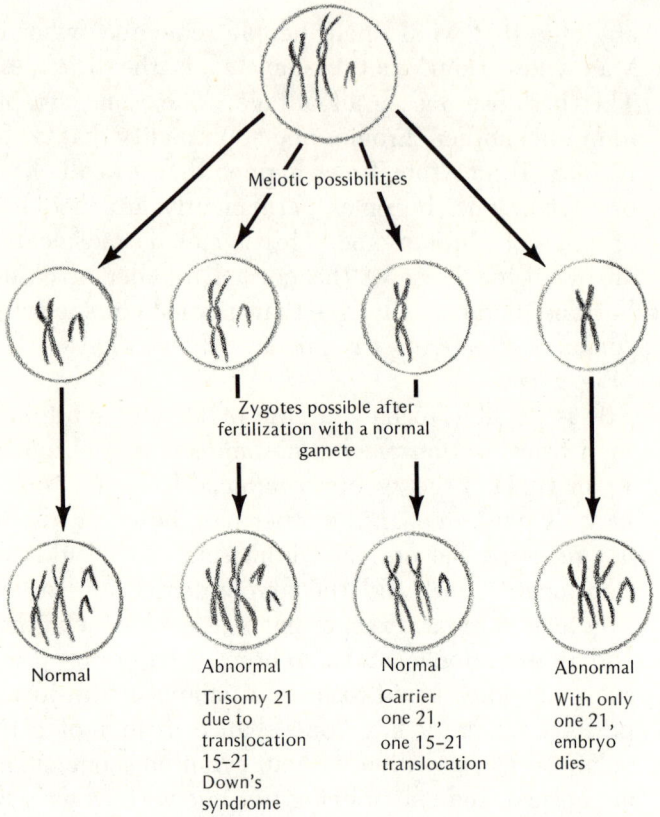

Fig. 2–12: A normal person may be a carrier for the rare translocation form of Down's syndrome if he has only one free chromosome 21 along with the chromosome 21 attached to another kind of chromosome.

some 21, and will fail to develop (Fig. 2-12). If a child with Down's syndrome is found to have the translocation form of trisomy, the parents are tested, too, since one of them may carry the balanced translocation. If so, they are advised that they have one chance in four, or one chance per three live births, of having another child with Down's syndrome.

The risk of a repeat of regular trisomy 21 is not noticeably different from the average risk for all parents of a given age. It is the age of the mother that seems to make the difference. Among women over 40 years old, the risk of trisomy is 40 times that for women between 20 and 30. Presumably the same unknown process that gradually destroys the oocytes during the reproductive period of a woman's life weakens some that are still able to develop, in such a way that the meiotic mechanism does not work normally. The age of the father does not seem to be important in trisomy; his gametes are produced from actively dividing cells in which large defects are usually eliminated by cell death.

Other trisomies

Two trisomies of small chromosomes other than number 21 occasionally persist to birth, but rarely much beyond. One of these is a trisomy of a chromosome of the E group: 17 or 18. The resulting condition, called the Edwards syndrome, includes multiple malformations and is usually fatal within the first 6 months of life. The other trisomy occasionally found, causing Patau's syndrome, involves a chromosome of the D group—13, 14, or 15. More severe even than the Edwards syndrome, it usually causes death within 3 months. Other autosomal trisomies do not survive to birth.

There is another group of fairly common trisomies—the abnormalities of the sex chromosomes. The X chromosome is medium-sized and the Y is small, but embryos having trisomies of either can survive with surprisingly mild afflictions. The lack of severe effect is due to a special peculiarity of these chromosomes and of their behavior during development.

Sex chromatin

The stained nucleus of cells of mammalian females has a dark spot not found in males, called **sex chromatin**. It is present in these female cells because one of the X chromosomes in adult cells is always inactivated. Something about the mechanism by which it is inactivated makes that particular X chromosome take up stain and become visible during the period of cell life when chromosomes are not normally visible in a slice of stained tissue. Most of the genes found on the inactivated X chromosomes do not operate in that cell, but the genes in the other X chromosome act in a fully normal manner. In abnormal conditions, when three X's are present due to nondisjunction, two of them are inactivated and there are two dark spots of sex chromatin. The test for this sex chromatin makes it possible to determine the sex of the embryo by examining embryonic cells before birth. It also makes it possible to detect some cases of sex chromosome abnormalities. The fact that all but one X chromosome are normally inactivated means that trisomies of the X chromosome are not much different from the normal cell with two X chromosomes; only one X chromosome is active in either case.

Patchwork females

According to the **Lyon hypothesis** now widely accepted, the inactivation of one of the X chromosomes occurs in the cells of the embryo after several divisions have occurred, when there are a number of embryonic cells present. In each cell, it is a random matter whether the paternal or maternal X is the one inactivated. The cells later derived from that cell retain the same inactivated X chromosome. Dr. Mary Lyon found that the adult female is a mosaic of patches of tissue expressing the genes of one or the other sex chromosome, a kind of patchwork quilt. The most dramatic expression of this is in cats, among which only females can be spotted black and yellow (tabby, tortoise shell, or calico), while normal males are either black or yellow. This is because the genes for black or yellow fur are alleles at a locus on the X chromosome. A male will have only one X chromosome, and can only be either black or yellow (unless other genes at other loci make him an entirely different color). But the female, having two X chromosomes, can be heterozygous for the black and yellow alleles, and will be spotted black and yellow, because the skin has patches of skin with one or the other X chromosome inactivated and thus with just one of the color alleles operating. The same phenomenon exists in individual human female cells grown in the laboratory and tested for something called G6PD enzyme, involved in a certain hereditary anemia. If the woman is heterozygous for the genes for the enzyme's presence and for its absence, some cells or clumps of cells will be positive for the enzyme, and some negative for it. All the cells of a man either have the enzyme or don't have it, however.

There is an important clinical consequence of this patchwork nature of women. If a woman is heterozygous for some harmful recessive gene on the X chromosome, it is not necessarily the case that exactly half her cells will express the normal gene, and half will express the harmful recessive one. By chance, most of her cells might have the same one X chromosome inactivated, and so she might be almost as greatly affected as if she were homozygous recessive, or conversely, hardly affected at all. If the gene in question has its effects mainly in one organ of the body, it is unlikely that that organ will be made up of evenly mixed cells expressing the two chromosomes; most will probably express a single specific chromosome. Thus women who are genetically heterozygous for a recessive X-linked gene have extremely variable phenotypes, from unaffected to seriously affected.

The inactivation of one X chromosome is obviously a device that was evolved along with the sex chromosomes so that the balance between X-chromosome genes and genes on the **autosomes** (all the other 22 chromosome types) will be much the same in the female as it is in the male, who has only one X chromosome. Any two alleles found on the two X chromosomes of the female will both be expressed somewhere in the whole body, because not all the cells will have the same X inactivated. Notice, though, that in any one cell, only one X-linked gene will be expressed. Since any extra X chromosome, beyond the normal two, will also be inactivated, such a condition will not usually be fatal to the early embryo, and the embryo will survive.

Abnormalties of the sex chromosome

The Y chromosome determines "maleness" in humans, and will have this effect even if (abnormally) two or more X chromosomes are present. If there are extra Y chromosomes found in cells because of nondisjunction during sperm production, there is no evidence that one of these is inactivated. But since the number of genes on the Y chromosome is very small, the presence of extra Y chromosomes is not fatal.

Because of all these properties, the sex chromosomes can show a great variety of abnormal combinations in living individuals, described in Table 2-1. (There are several more extreme combinations such as XXXY, but they are very rare.)

From the abnormal nature of the people with these arrangements of chromosomes, it is evident that the inactivation of the extra X chromosomes is not complete. If it were, XXY would be a normal male, and XO and XXX would be normal females. Some of the genes on each X must escape inactivation and so cause a slight imbalance—enough to cause some trouble. Many individuals with these sex chromosome abnormalities are normal in most respects and lead uneventful, useful lives. Some of these conditions are relatively common; men with Klinefelter's syndrome are said to constitute a significant fraction of those who seek help for absolute sterility (no remedy is available).

Table 2-1. Human sex chromosome abnormalities

Human medical genetic defect	Sex chromosome abnormality	Symptoms
Klinefelter's syndrome	XXY	Sterile, feminized male, often retarded, usually tall
Turner's syndrome	XO[a]	Sterile, sexually underdeveloped female, sometimes with "webbed" neck
XYY syndrome	XYY	Normal-to-low-normal intelligence, taller than average male, fertile, does not transmit condition
Triple-X female	XXX	Usually infertile female, occasional mental deficiency

[a] XO means one X and no Y.

Translocations

Most translocations result in attachment of only part of one chromosome to another. The result is a sort of trisomy or monosomy of only part of a chromosome. Since less than a whole chromosome is added to or subtracted from the normal form in these conditions, many of them are compatible with life. The seriousness of the condition is roughly correlated with the size of the chromosome segment translocated. Since any fraction of any chromosome can be translocated, the possible number of different partial translocations is very large, and each is individually very rare. To make the study of translocations still more complicated, it is technically more difficult to detect a chromosome slightly longer than normal than it is to identify an extra one. These and other defects involving parts of chromosomes probably account, taken all together, for about as many early abortions and about as many surviving malformed children as whole-chromosome trisomies and monosomies. They are, however, much less well studied because of the difficulties involved.

The gonads

Gametes are produced in special organs, the **gonads**, which contain some cells that were separated off fairly early in embryonic life and were never differentiated into ordinary working body cells. In the female the early embryonic gonad develops into an **ovary**, which remains in the interior of the body. In the male the corresponding embryonic organs become **testes**, and migrate to a position outside the main body cavity, where they are suspended in a fairly thin sac of skin and tissues, the **scrotum**. For some unknown reason, the sperm of man and of most other mammals cannot develop at normal body temperature, and normal production of sperm depends on the slightly lower temperature of the scrotum. (In contrast, the testes of birds are in the interior of the body, and sperm develop normally there despite the fact that the body temperature of birds is a degree or two higher than that of mammals.) If for some reason the testes of a man fail to descend into the scrotum, he will be sterile. Men of some primitive tribes are said to use frequent prolonged hot baths to avoid fathering offspring, since the high temperature interrupts the development of sperm.

Acquired characteristics are not passed on

In both sexes the cells that are to give rise to gametes seem fairly well protected from influences that may injure or wear down the person. The gametes pass on genetic information fairly independently of what happens to the individual in his lifetime. Hormonal abnormalities or diseases may interrupt the production of gametes, but if gametes are produced, they usually show no effects of accidents or other afflictions. There are, however, some influences that can extend into the gonad and bring about *permanent genetic change*. Such changes are called **mutations**. The most important of the influences that cause mutations are ionizing radiation (such as x-rays), a high temperature for the testes, and a few kinds of chemicals. The mutations are nearly random changes in the genes, with no obvious relation to the nature of the agent that caused them. Other structural changes that occur in our bodies are *not passed on* through our gametes to cause the same changes in our children.

Components of the gonads

The gonads of both sexes contain, besides cells that are capable of developing into gametes, a few kinds of cells that serve in some way to support the maintenance or growth of the gametes and of their precursors. Some of these secondary cells have still another essential function: They produce some of the hormones that regulate the production of gametes, the adoption of behavior that will lead to mating, and the preparation of the other tissues of the body for the support of the new embryo in the female.

The testis

The **testis**, inside its connective tissue outer wall, is made up of many coiled, thick-walled tubules that are connected through an elaborate multiple duct work to a final common duct. The walls of these tubules contain three principal kinds of cells (Fig. 2-13). The first are the cells that are to give rise to sperm. These include the **spermatogonia**, which are stem cells; when they divide, one of the new cells is just like the original and is capable of dividing on through endless cycles, while the other is somehow different enough so that the products of its divisions will become sperm, with a different fate. The cells in this developmental sequence, all found in the tubule wall, are called **spermatocytes, spermatids,** and mature sperm. The second group of cells making up the tubules are the support cells. They seem to be essential for the survival of the spermatids, which are becoming sperm, and of the mature sperm before they are released. The mature sperm are seen in the cavity of the tubule, lined up with their heads sticking into or against the support cells, and presumably receiving some food materials from them. The third type of cells distinctive to the testis is the **interstitial** cell. These are small cells, scattered (as their name suggests) among the other cells, and seemingly not important for their bulk or their structural function. However, these are the cells that produce the male sex hormone, testosterone. Without the many actions of this hormone, the sperm would never get to an ovum and would probably not even be produced.

The spermatogonia, in the testis, are dividing into more and more spermatogonia during most of a man's life. Before puberty the division is relatively infrequent, but when it occurs it apparently produces two stem cells in each division, since no sperm result. So the numbers of spermatogonia increase slowly up to puberty. After puberty, the frequency of division of spermatogonia is much greater, but half are transformed into spermatocytes, and spermatogonial cell numbers remain fairly constant. If some injury such as a

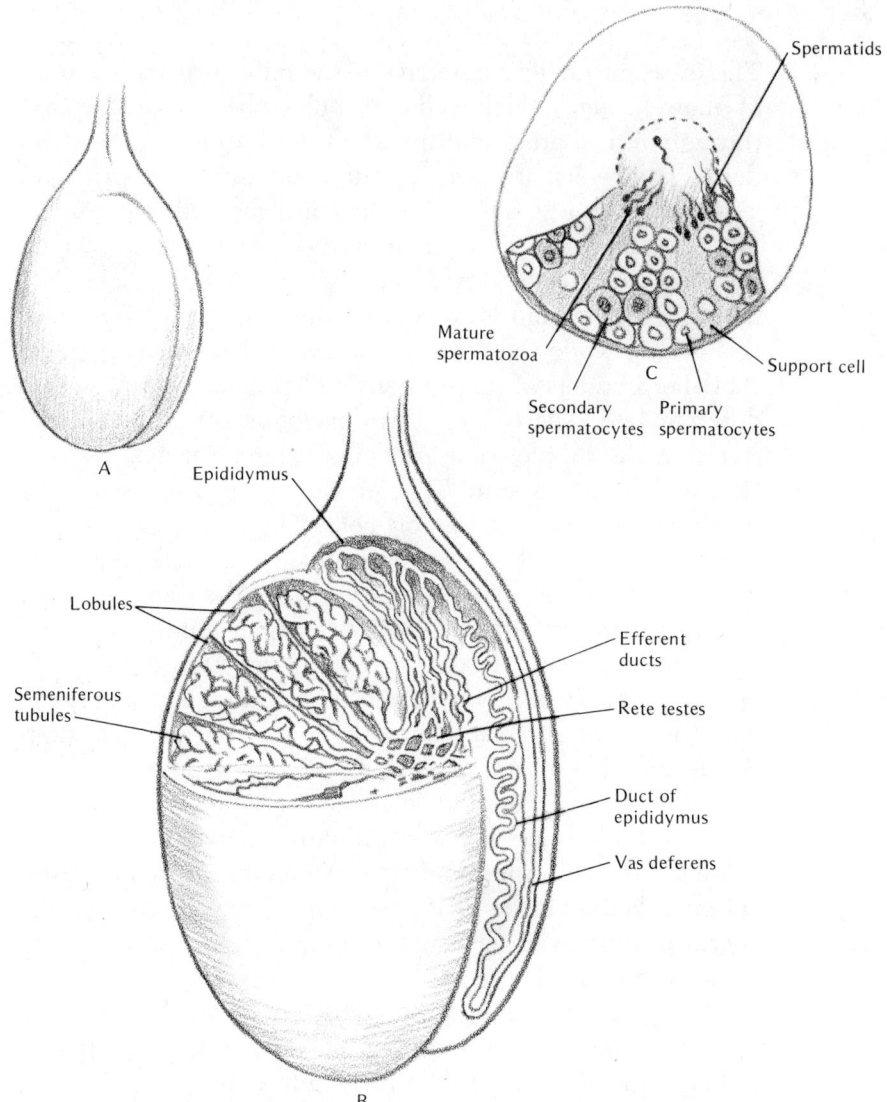

Fig. 2-13: Components of the testis. A. The whole gland. B. Cutaway diagram showing the parts of the testis. C. Cross-section of a tubule where sperm are produced.

strong dose of ionizing radiation destroys most of the spermatogonia in an adult, the few remaining can proliferate to restore the normal number.

The form of stem cell that divides by meiosis to produce sperm is called a **spermatocyte**. The cells formed by this meiosis are called **spermatids**. These immediately begin the transformation into mature sperm, which entails a drastic reorganization of structure, including a loss of most of the bulk. When this process is complete, the mature sperm is essentially a condensed bundle of chromosomes to which is attached a tail for swimming, and it has in the middle a compact form of the cellular machinery for supplying energy. Most of its essential cellular functions have been lost, so a sperm cannot survive for long even in an environment that would be perfectly healthy and supportive for almost any other isolated body cell. This is why it has to remain attached to the support cells of the tubules, and later must be nourished by special secretions in the sperm ducts of the male, where sperm are stored for a while. When finally released into uncontrolled environments, the sperm can live only a few hours, just long enough to fertilize an ovum—to fulfill its function of contributing a nucleus to the formation of a zygote and stimulating the zygote to divide.

The schedule of sperm production

The production of sperm is continuous from puberty on, although it may decline somewhat in later years. The whole process, from spermatogonium to mature sperm, requires from 74 to 90 days, depending on what is chosen as the starting point. Any one part of the testis produces a batch of sperm every 16 days and has a batch of cells in each of four or five different stages of maturity. Other parts of the tubules in the testis are at different points in the cycle, so some sperm are being produced every day, without any evidence of periodicity. Knowledge of the time taken to produce a sperm has one very practical application. Large doses of ionizing radiation, such as may be received from a fluoroscopic examination with x-rays, produce mutations of individual genes and even breakage and rearrangement of chromosomes. If this happens in the spermatocytes, spermatids, and sperm, where the nucleus is not being used very much for day-to-day needs, the defects will remain and may cause serious defects in the children ultimately conceived. Spermatogonia with damaged chromosomes are likely to die, to be replaced by uninjured spermatogonia, and the sperm from these spermatogonia are likely to be satisfactory. Therefore, a man who gets a large dose of x-ray (anything more than a chest shot for TB or a dental x-ray) should use contraceptives so as not to conceive a child for a period of 3 months after exposure.

The ovum

The ovum does not have to go through a process of preparation after meiosis has been completed, as the sperm does. The equivalent preparation has already taken place before meiosis is finished. The oocyte is contained in (attached to the side of) a fluid-filled space called the ovarian **follicle**. The follicle is lined with cells that do not themselves give rise to ova but do assist in the process. Some of the follicle cells are producing nourishment that helps the oocyte to accumulate yolk and to grow to the (relatively) enormous size it reaches. (Such follicle cells can be compared in function to the support cells of the testis, and they may even be derived from the same kind of embryonic cells.) Other follicle cells (perhaps corresponding to the interstitial cells of the testis) also act as an endocrine gland, producing the characteristic female hormones, of which the most plentiful is **estradiole**. These same cells, or their descendants, will change after an ovum is released. They change into a yellowish structure called the **corpus luteum** ("yellow body" in Latin) which produces the other principal female sex hormone, **progesterone** (Fig. 2-14).

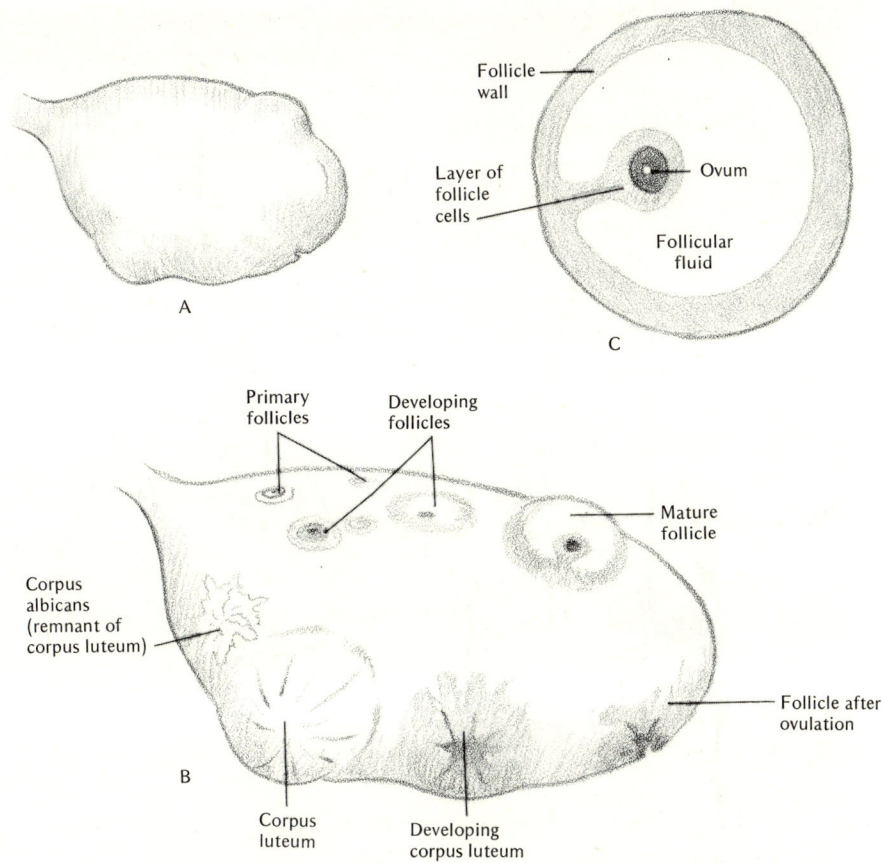

Fig. 2–14: Components of the ovary. A. The whole gland. B. Cutaway diagram showing the stages of egg development. C. Enlarged drawing of the ovum in its follicle.

Meiosis in females

At the time a baby girl is born, all her oogonia have been changed into primary oocytes. All the oocytes that she will ever have are already beginning to prepare for meiosis. They are poised for the first meiotic division, although it will be from 12 to 50 years before this division occurs. The first meiotic division finally occurs just before ovulation (release of the egg); the second meiotic division occurs just after fertilization, in the oviduct.

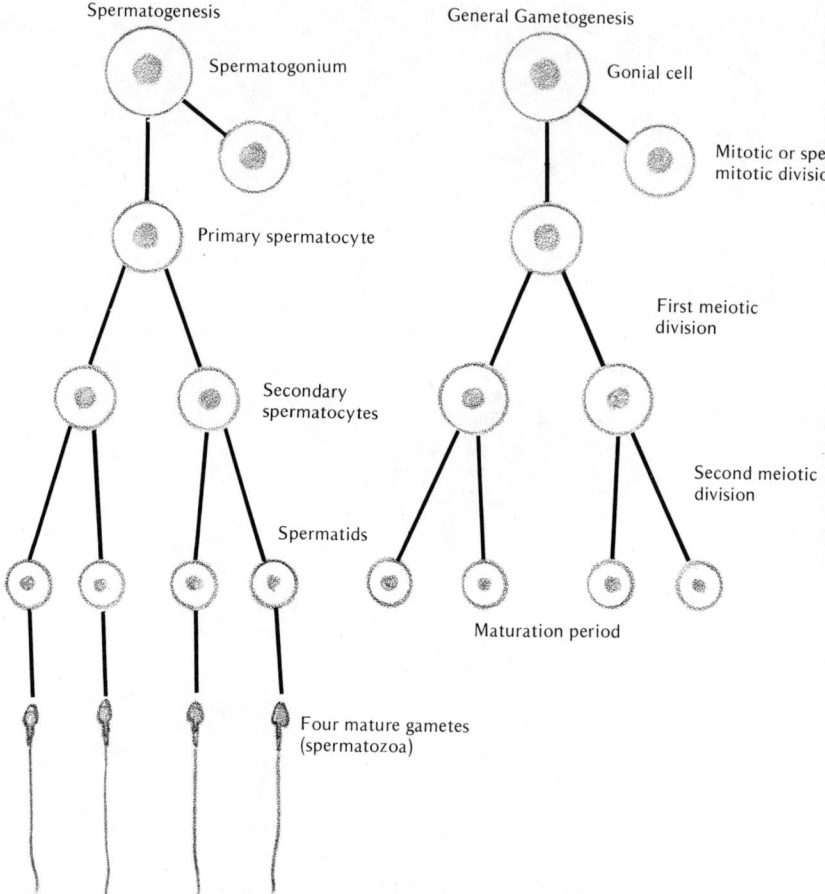

There is a striking and important difference in meiosis as it occurs in females and in males (Fig. 2-15). The **primary oocyte**, which undergoes first meiotic division, does not divide into two cells that are even in size. Instead, one of the resulting cells, called the **secondary oocyte**, has practically all the bulk of the cell, while the other product of this meiotic division, called the **first polar body**, is just a clump of chromosomes surrounded by a little cellular material. Similarly, in the second meiotic division, the division is again unequal, producing the ovum and the **second polar body**.

The significance of polar body formation is that, in this

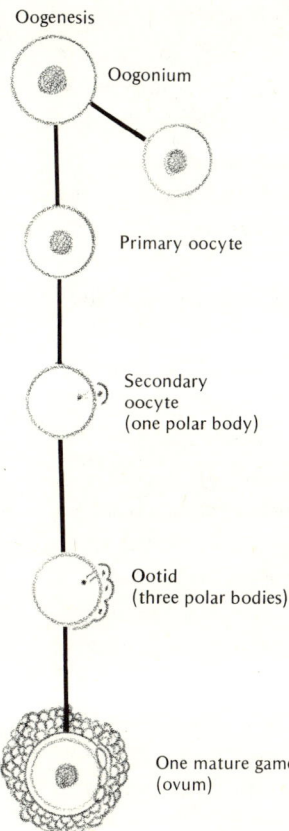

Fig. 2–15: Gametogenesis: A comparison of sperm production and egg production. Note that the mature ovum is shown with the layer of follicle cells that cling to it. The earlier stages are also, of course, within the follicle itself as shown in Fig. 2–14.

manner, the chromosome number is reduced to that appropriate for a gamete, but the ovum retains all the food material of the original oocyte. This food substance will serve to nourish the zygote if one should result.

The polar bodies usually disintegrate and are reabsorbed. (Some kinds of parthenogenesis may be due to recombination of a polar body with an ovum, as though it were a sperm.)

Growth of ova

The ovum is much larger than the original primary oocytes were at the woman's own birth. Some of the growth may have occured while the oocyte was quiescent for so many years, but most of the size increase occurs after the oocyte is part of a developing follicle moving toward ovulation. At this time the oocyte rapidly synthesizes the materials that will later nourish the embryo until it is well established. The oocyte even absorbs similar material manufactured in the surrounding follicle cells—a most unusual process. The mature ovum is then equipped to supply the initial nutrition (as well as one set of hereditary information) for the new individual.

The ovary of a baby girl at birth is more sensitive to ionizing radiation than is any other biological object we know. As little as 25 roentgens (about 1/20 the lowest fatal dose) is sufficient, in experimental animals, completely to sterilize the infant female. The full medical implications of this fact are not yet worked out, but for the present it would be a wise policy to avoid any radiation exposure of pregnant women. Ovaries of adult women are not especially sensitive to sterility from radiation, although mutations are always possible. The ovaries do seem to undergo a normal, gradual aging, in that the errors of division called nondisjunction (leading to trisomy of various chromosomes) are many times more likely to be found in offspring of mothers near the menopause. So it is advisable to have your family as early as you feel ready to be parents.

Despite all the things that could go wrong, the oocyte usually manages to divide properly and develop normally toward a mature ovum. While the oocyte is developing from a diameter of about 0.02 mm in the quiescent stage to about 0.14 mm in diameter just before release, the fluid-filled follicle around it has been growing from 0.04 mm to over 12 mm in diameter. The fluid pressure seems to rise within the follicle, and this may be one of the factors that cause it to burst and release the ovum to the outside: a process called **ovulation**. (While it was growing, the follicle had migrated to the surface of the ovary. Hence the bursting of the follicle releases the ovum to the outside of the ovary.) The ovum—strictly speaking, the oocyte, since the second meiotic division has not yet occurred—is released into what is really the abdominal cavity. However, the mouth of the oviduct is wrapped so closely around the ovary that the ovum is effectively released into the oviduct. The secretion of fluid by the walls of the oviduct and the beating of hair-like cilia on the surface of the oviduct cells cause a weak current that moves the ovum (really still an oocyte) slowly down the oviduct toward the uterus. If, during this journey, the oocyte is fertilized by a sperm, the two together become a zygote.

3 | Hormonal and nervous control of reproductive behavior

The production of gametes by the gonads would not result in their ultimate function, procreation, without the action of the sex hormones. The sex hormones, like the gametes, are produced by the testes in the male, the ovaries in the female. These hormones are responsible for the development and maintenance of nearly all the specific structures that aid in the process of getting the sperm and egg together and that distinguish members of one sex from those of the other. The hormones also act to stimulate the mating activity necessary for bringing the gametes together. Sex hormones belong to a larger group of hormones related chemically to a class of molecules called sterols (which include cholesterol). The hormones produced by the gonads are thus called steroid hormones, **sex hormones**, or steroid sex hormones.

Sex hormones in the female

The primary hormones produced by the ovary are named according to a behavioral effect that they have in many mammals—the production of sexual receptivity, called "heat" or "estrus"; thus the primary female sex hormones are called

estrogens (producers of estrus). There are actually several ovarian sex hormones that have this effect, but most of the estrogen concentration is of one called estradiole. As will be discussed in later chapters, estrogens are probably responsible for the construction of the oviducts, uterus, vagina, and external genitalia in the female embryo. Once formed, these structures then seem to grow in the young girl without requiring noticeable concentrations of estrogens. When the ovary begins to produce gametes and hormones in the girl's early teens, the dramatic structural changes of puberty are brought about by the estrogen that is released. The oviducts, uterus, and vagina all grow larger and acquire much thicker walls. These internal changes are the most important ones for reproductive function, but they are not as noticeable as the external ones. A dramatic change in bone structure occurs, so that the waist becomes narrower and the hips broader. Estrogen also affects the pattern of fat distribution, bringing about general increase in fatty deposits under the skin, and especially in the breasts. It also causes the hair follicles under the arms and surrounding the vaginal opening to become active. (The lower abdomen, where hair appears, is even called the **pubic region** from the Latin word for hair.) This growth of hair is different from the corresponding one in men, in that the top margin of the pubic hair of women tends to be horizontal, rather than tapering toward the chest as it does in men. Estrogenic hormones have a variety of other effects on the body, some of which may be only incidental and others important to reproductive function but not yet understood. For instance, the female sex hormones seem to be responsible for preventing a good many heart attacks in women, for men from 30 to 50 seem to have many more cardiac attacks than women in the same age group.

After ovulation, the ruptured follicle is changed into the **corpus luteum** (see p. 65), which then acts as an endocrine gland to produce a hormone called **progesterone**. Apparently, the cells of the follicle that had been manufacturing estrogens before ovulation change their synthetic machinery moderately so that after ovulation they produce progesterone.

This is chemically similar to the estrogens, but with some structural differences. Progesterone acts to maintain the structure of the thickened uterus, and it causes the development of secretory tubes (glands) in the uterine wall. Without this hormonal action, pregnancy would not continue even if an ovum were fertilized, because a decrease in the concentration of progesterone causes a breakdown and sort of peeling off of part of the uterine wall—the menstrual flow. The name progesterone means a chole<u>sterol</u>-like hormone that <u>pro</u>motes pregnancy, or <u>ge</u>station.

Both of the female sex hormones illustrate some of the standard attributes of hormones. They are produced by endocrine glands. ("Endocrine" denotes that the grandular cells discharge their product into the blood stream, by which it is carried to all parts of the body, in contrast, for example, to the salivary and sweat glands, which discharge their products through special ducts or tubes leading to the outside of the body or into some cavity within it.) Although the sex hormones are carried by the blood to all parts of the body, they do not seem to have very much effect on most of the cells, in contrast to their profound effect on certain "target" organs. The response of a certain kind of target cell to some particular hormone depends on the nature (heredity, age, past history) of the cell, as well as on the hormone. The effects of the female sex hormones on the organs of the body form the foundation for all the traits that we recognize as distinctively feminine, although these effects show a great deal of individual variation. The nature of the body's response to the sex hormones is not thoroughly understood. If the external structure of some particular woman is less obviously "feminine" than the average, she might have a reduced level of hormones, or else perhaps the genetic instructions in her cells caused some of her cells to be less responsive to sex hormones than the average.

Some of the effects produced by hormones seem completely accidental and are probably quite unrelated to the principal function of the hormones. For instance, many women experience an accumulation of water in the tissues (edema) during the time when considerable progesterone is being produced. Progesterone happens to be structurally similar to certain of the steroid hormones produced by the cortex of the adrenal gland, which promote retention of salt by the kidneys. The kidneys seem to "misread" the message of the progesterone as though it were a signal for retaining salt. With the retained salt an additional quantity of water is automatically retained by the body, causing the edema. When the progesterone level drops, the edema disappears. Hormones have many such useless or even harmful side effects.

Sex hormones in the male

In a parallel but less complicated manner, the hormones produced by the testes are responsible for male attributes. Such hormones are called by the general name **androgen**, meaning that they produce maleness, but most of the human androgenic effect is produced by just one kind of molecule, called **testosterone**. It is actually very similar structurally to progesterone, and somewhat similar to estrogens, yet it produces what we think of as opposite effects.

The name testosterone indicates a steroid produced in the testis. It is produced by the small interstitial cells lying among the tubules that produce the sperm. Like any hormone, testosterone is distributed to all parts of the body by the blood stream, but its most important effect occurs right in the testis. Both testosterone and a hormone from the pituitary gland have to act on the gamete-producing cells of the testis for the production of sperm to occur. In other parts of the body, testosterone is responsible for all the specifically male characteristics. It is responsible during embryonic life for the development of such male sex characteristics as a penis and the almost-external gonads, and during adolescence for the growth of a beard, the lower voice, and increased muscular development. Testosterone has an essentially continuous effect on the maintenance of the ducts that will carry the sperm to the outside, and it stimulates the glands associated with these ducts (principally the seminal vesicles and prostate gland) to secrete the seminal fluids. Less thoroughly understood is the effect on behavior patterns.

Effects of sex hormones on behavior

Hormone action before birth causes a permanent personality change in male infants. Baby boys are, *on the average,* more active and aggressive than girls, and this difference continues throughout life.

The behavioral differences between the sexes, present at birth, are reinforced by social expectations. It is impossible to say exactly how much of the contrast in behavior between men and women is due to the sex hormones and how much to the influence of society, since in most cases hormones and society cause similar actions, and every person is acted upon by both. Increased concentrations of sex hormones at adolescence and thereafter further enhance the behavioral differences between the sexes and help to initiate mating or reproductive behavior. Here again, it is not yet possible to sort out the effects of hormones from those of social expectations. Among all other mammals, the sex hormones are responsible for calling forth fairly complicated patterns of courtship and copulation. In humans, innate causes are certainly much reduced compared to those in other mammals, and our behavior is therefore much more plastic, more easily influenced by training. Most biologists who have studied the subject are convinced, however, that some of the behavioral differences between the sexes, and some of the common patterns of courtship and sexual intercourse that are found in all cultures, are built into the developing embryo under the influence of the sex hormones.

As will be discussed later with respect to general behavior, it is impossible to distinguish between the effects of nature (biological inheritance) and nurture (cultural imprint) in determining sex roles and behavior. The two factors act and interact creatively on and with each other. We do know, of course, that some aspects of sexual behavior—such as movements during intercourse and, in part, what people find sexually stimulating—are nearly universal in human cultures. These, at least, may be supposed to be innate and hormonally influenced.

Table 3-1. The reproductive hormones

Hormone	Endocrine gland where produced	Sex	Major effects
Estrogens	Follicle of the ovary	Female	Uterine growth, female characteristics, inhibits FSH secretion, stimulates LH secretion
Progesterone	Corpus luteum of the ovary	Female	Uterine maintenance, secretion; inhibits LH secretion
Testosterone	Testis	Male	Male characteristics, inhibits LH secretion
Follicle Stimulating Hormone (FSH)	Anterior lobe of pituitary	Female	Follicle growth and estrogen production
		Male	Sperm production
Luteinizing Hormone (LH)	Anterior lobe of pituitary	Female	Corpus luteum formation and secretion
		Male	Secretion of testosterone
Several "Releasing Factors"	Hypothalamus	Female	Pituitary secretion
		Male	Pituitary secretion

Control hormones

The steroid sex hormones (Table 3-1) which act to produce the structures and behavior necessary for getting the gametes together (via sexual intercourse) must be regulated so that they are supplied in proper concentrations and at times that coincide properly with the production of gametes. This regulation is carried out by the **gonadotropic** (acting on the gonads) **hormones** secreted by the **anterior** (frontal) **lobe** of the **pituitary gland**. The pituitary gland is a small but very important structure just under the brain, in the bone forming the roof of the mouth. It is really two different glands—called the anterior and posterior lobes of the pituitary—and it secretes a number of different hormones. All the pituitary hormones are small proteins—that is, small as proteins go, although large compared with other kinds of hormones (Fig. 3-1).

There are two important gonadotropic hormones in the human. One is called the **follicle stimulating hormone**, abbreviated **FSH**. As its name indicates, it causes the oocyte

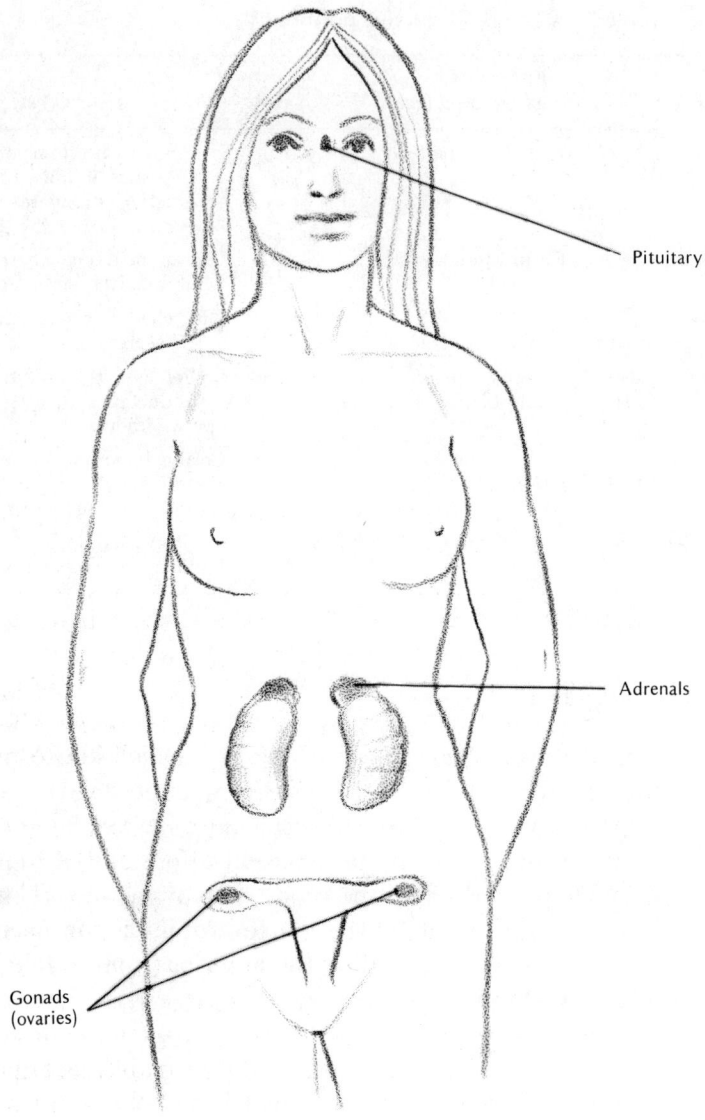

Fig. 3–1: The endocrine glands related to reproduction. The location of these glands in the male is the same except for the gonads.

and follicle to develop and enlarge in the female. In the male, the same hormone cooperates with testosterone to cause production of sperm. The other gonadotropin is called **luteinizing hormone**, abbreviated **LH**. It is responsible for the bursting of the follicle in ovulation, and it then causes the cells of the open follicle to be transformed into a corpus luteum that secretes progesterone. In the male, the LH causes the production of testosterone by the interstitial cells.

Unfortunately for those who want to understand it, the action of the gonadotropins on the gonads is not a simple one-way affair. Instead, there is a reciprocal reaction more complicated than anything else in the study of endocrine glands and their hormones. Since the action of the most popular contraceptive depends on this complicated control system, it is especially important to learn about it. The hormones are so interrelated (as well as so difficult to measure accurately) that their precise actions and reactions cannot be studied by simple observation—experiments are necessary. Drastic experiments on humans are generally not acceptable, so research workers must depend on data from experimental animals for most of their understanding. Results are checked wherever possible by observations on humans. Most of the time this works very well, since the same reproductive hormones (both gonadal and pituitary) are found in most vertebrates and have parallel actions in animals and humans. Of course, however, there are always some difficulties in applying animal data to human problems.

As just indicated, the general pattern of gonadal and pituitary control hormones and their interrelations are very similar in male and female. The one great difference is associated with the fact that men are continuously fertile. They produce sperm and can contribute them at any time of the day, the month, or the year. A woman, although she can be sexually active at any time, produces only one gamete a month and so can conceive at that time only. Thus the male hormones for the production and delivery of sperm are secreted steadily, while the female sex hormones go through a complicated and drastically varying monthly cycle, timed in relation to release of the ovum.

Hormone regulation in the male

The control of testosterone secretion follows a pattern familiar to biologists. This pattern is known as **homeostasis**—the maintenance of a steady state. Luteinizing hormone, produced by the pituitary, stimulates the interstitial cells of the testis to secrete testosterone, as previously described. Testosterone has, in addition to its other actions, the effect on the pituitary of decreasing its production of LH. Reduced LH means reduced testosterone production; reduced testosterone means increased LH production. But there is not the seesaw effect this would seem to imply; rather, things settle down to a steady level of secretion of both. If anything happens to disturb this steady state, its impact is minimized by the system. If testosterone is injected into a man, this has only a brief effect on the level of hormone in the body. The suddenly higher level of testosterone in the blood immediately reduces the production of LH, and as a result the production of testosterone by the testis is quickly cut down. Thus the testosterone level in the blood soon returns to very little more than it was before the injection. (Estrogens, if administered to a man by mistake or in life-saving cancer therapy, have a feminizing effect for two reasons. Firstly, the estrogens have a direct effect on the tissues of the body, and secondly, estrogen reduces LH and therefore cuts down on testosterone production.)

The level of FSH in the blood of a man is also quite steady, and this is reflected in the steady production of sperm. If some of the sperm tubules are destroyed by radiation, however, it is thought that the FSH concentrations increase, helping to restore normal production of sperm. It is not known just how this FSH level is regulated. Possibly there is some undiscovered hormone produced by the spermatocytes that serves this function.

How hormones regulate the female cycle

Hormone interaction in the female, necessary for cyclic action, is more complicated (Fig. 3-2). The first step in the

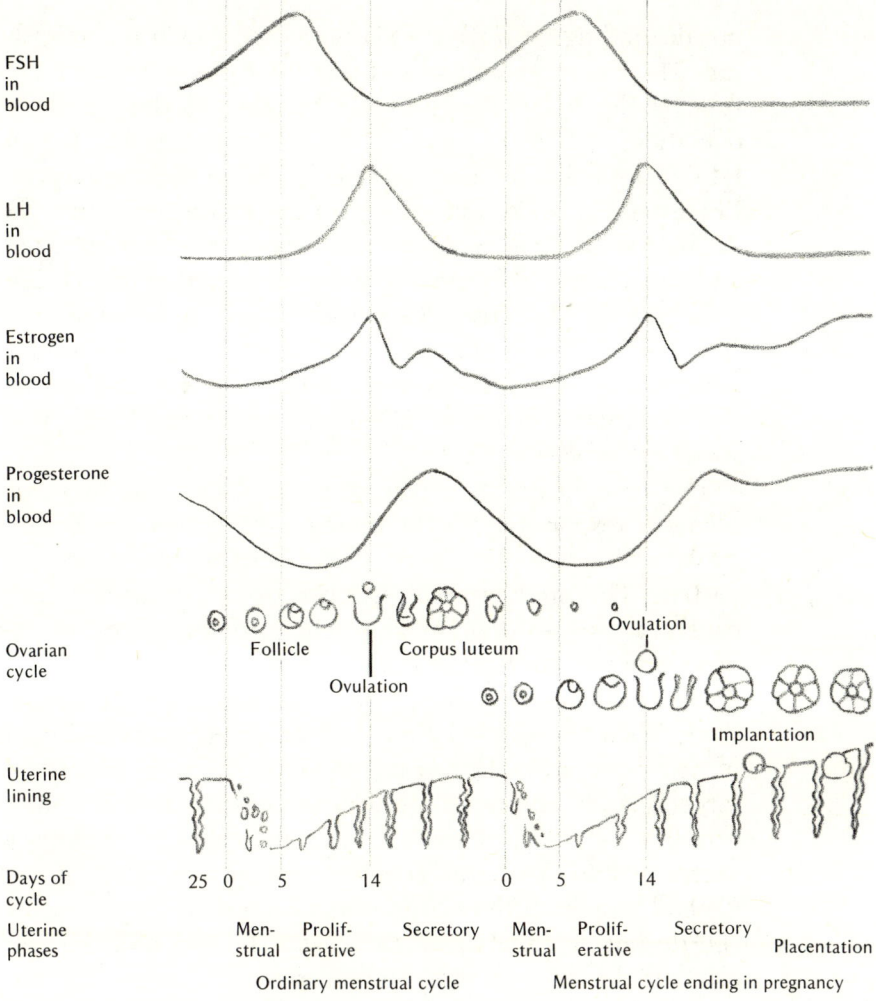

Fig. 3–2: The human menstrual cycle.

sequence is the production in the pituitary gland of FSH. The FSH then causes a few of the quiescent follicles in the ovary to begin to enlarge rapidly and to secrete estrogen. The estrogen, besides its effects on the uterus, vagina, and other tissues, has a feedback effect on the pituitary: It decreases the secretion of FSH but stimulates the secretion of LH. Once started, growth of the follicle seems to continue without depending on further FSH secretion, which is diminishing. The rising amount of LH first increases estrogen secretion by the follicle, triggers the bursting of the follicle at ovulation, and then causes the transformation of the follicle into a corpus luteum and starts its secretion of progesterone. Progesterone, besides its effects on the uterus, continues the depression of FSH secretion that was started by estrogen, and also slows and eventually stops the secretion of LH. The reduction in LH causes the corpus luteum to become inactive, and the cycle starts all over again. (If pregnancy has occurred, the membranes of the embryo secrete a hormone that stimulates the corpus luteum to persist and to continue secreting progesterone.)

This complicated interplay of gonadal and pituitary hormones is responsible for the timing of ovulation and for the cyclic changes in the uterus that occur in the same time pattern. The production of estrogen by the growing follicle causes a proliferation of the cells and structures in the uterine lining, so that the lining becomes much thicker. This is the "proliferative phase," which lasts, on the average, 11 days. Then comes ovulation; and the subsequent secretion of progesterone by the corpus luteum causes the newly enlarged glands in the uterine wall to begin secreting fluids—a 12-day "secretory phase." (If conception has occurred, these secreted fluids probably function for the nourishment and general support of the early embryo.) If no conception occurs, the decline in function of the corpus luteum and the consequent decrease in progesterone remove a necessary supportive influence from the uterine lining. The lining then disintegrates and is discharged in a 5-day "menstrual phase." Menstrual discharge represents the disposal of the cells that were built up in preparation for the pregnancy that did not occur. If conception occurs, the tissues of the embryo secrete a hormone that maintains the activity of the corpus luteum, so that the uterine wall does not regress.

Explanation of the cyclic pattern

The fact that secretion of sex hormones in the female occurs in such a dramatic cycle (in contrast to the steady state observed in males, who have several similar hormones) is due to at least five factors:

1. Estrogen secretion is by a growing tissue, the follicle, which, once it is started, continues to grow and to secrete more and more without further stimulation.

2. The effect of estrogen on LH production is stimulatory, and LH in turn stimulates estrogen production. This kind of positive interaction leads to higher and higher levels of both hormones.

3. The runaway increase in both estrogen and LH is brought to a halt by a sudden event, the bursting of the follicle. The burst follicle stops secreting estrogen and starts producing progesterone.

4. Under the new situation (a corpus luteum rather than a follicle), the hormone now being produced (progesterone) has a more ordinary inhibitory effect on LH production, and the high level of progesterone tends to stop pituitary secretion of LH. This allows the entire process to start all over again.

5. The cycle caused by the complicated interaction of hormones described seems to be reinforced (and further complicated) by a cyclic process in the pituitary itself. The early embryonic pituitary is probably potentially capable of either cyclic or steady production of FSH and LH. Early production of testosterone in the male embryo abolishes the cyclic tendencies of the pituitary, but in the female embryo, estrogen enhances these tendencies. The contrast between pituitary action in the two sexes may be increased by the influence of increased hormone levels at puberty. Probably either of the mechanisms for producing cyclic secretion of hormones in females could, by itself, maintain the cycles. This kind of "fail safe" mechanism, with two different control centers responsible for the regulation of a single process, is common in physiology.

Why a "lunar" month?

The specific duration of the menstrual cycle, about 28 days, is often described as a "lunar month." The resemblance to the cyclic changes of the moon is probably accidental. The average of 28 days for a woman's cycle is due to the growth rates and response rates of the various glands involved; there is no evidence that it is controlled by the moon at all. Other species of mammals have sexual cycles that may be longer or shorter than that of humans; there is no obvious general pattern. In women, the cycle does not always start on any certain phase of the moon. After a pregnancy, the time of the first menstrual period cannot be predicted at all by the moon's phase. The accidental similarity in timing of the menstrual cycle to the lunar cycle may possibly be responsible for the notion that certain phases of the moon are correlated with moods. Many hormones affect moods, and there is no doubt that many women experience a somewhat regular cycling of moods parallel with the reproductive hormone changes. Hormonally affected moods are presumably responsible (allowing for considerable male exaggeration) for the common belief in the changeability of women. In addition to a probable direct effect of hormones on the emotional centers of the brain, the side effects often cause physiological discomfort or other sensations. Progesterone and estrogen both can cause breast enlargement, and progesterone often causes uncomfortable puffiness or edema.

It is the very fact that the normal state of the female sexual cycle includes abrupt increases and withdrawals of hormones that make the widespread hormone administration in birth control pills sensible. Administration of a hormone that was normally in a steady state would cause undesirable fluctuations, some of which would be in a direction opposite to the effects desired. But administration of sex hormones in birth control pills is very close to "normal." Most of the features of the normal female cycle can be preserved in the administration of the pill. The side effects of the pill are simply the same as those of pregnancy, but usually less extreme.

Nerves and hormones

The two control systems of the body, the endocrine system and the nervous system, interact in many important but often subtle ways. These interactions seem to be especially important in reproduction. It has already been indicated that the sex hormones influence the nervous system to cause either male or female behavior, and that the fluctuating female sex hormones are responsible for some fluctuations in mood. The influences go both ways, and nervous influences on hormone production may be even more important than are hormone influences on the nervous system.

The nervous system affects the endocrine system principally through the pituitary gland. Although it is common to speak of feedback control between the pituitary gland and the glands under its control, the feedback actually (in the case of gonads) includes a part of the nervous system. The part of the brain known as the hypothalamus secretes specific small peptide hormones that are released near the pituitary gland and stimulate it (Fig. 3-3). There are specific hormones (called releasing factors) for each of the hormones produced by the anterior pituitary. When testosterone or progesterone is artificially introduced into the blood stream, what actually occurs, to restore a normal level, is inhibition by the steroid hormones of the hypothalamic neurons secreting the **luteinizing hormone releasing factor.**

The **hypothalamus** is not just another endocrine gland that adds to the complications. It is connected to the rest of the nervous system, containing nerve centers that send out impulses to the system, and to the internal organs, for responses associated with all emotional states. It is well documented that an emotional crisis may sometimes interfere with the normal timing of the menstrual cycle, presumably by altering the state of the hypothalamus.

In animals native to the temperate zones, ovulation and sexual activity occur only in limited parts of the year, commonly spring and summer. This timing is controlled by the length of day, operating through the eye and the nervous system, and causing the release of the gonadotropin releasing factors. It is possible that some of this pattern exists in man, despite his tropical origin, and is recognized as the "romance" of spring.

Another possible nervous effect is suggested by the fact that several mammals, including the cat and rabbit, ovulate

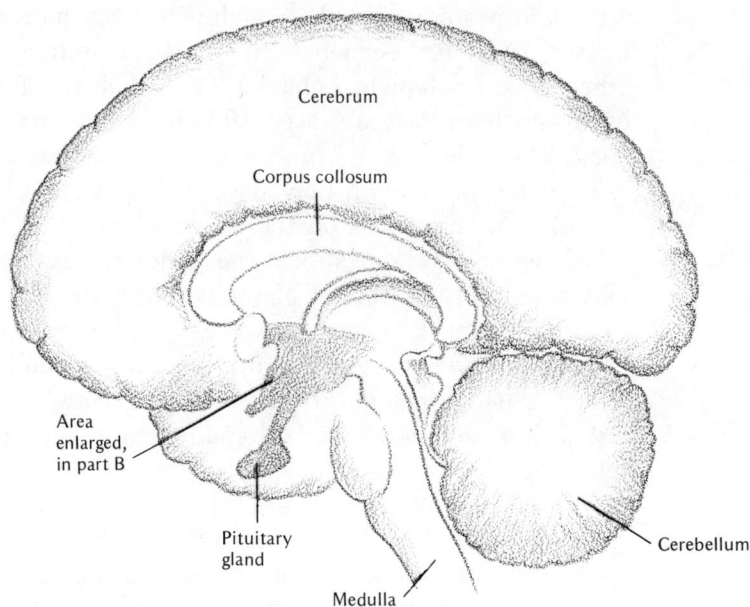

Fig. 3–3: The pituitary gland and the hypothalamus. A. Cutaway sketch showing the relation of the pituitary gland and the hypothalamus to the other structures of the brain.

just after copulation. The nervous stimulation of copulation causes quick release of LH, and that causes release of ova. A few investigators suggest that this may sometimes happen in women. If so, it would account for the surprising number of pregnancies that result from a single, isolated act of intercourse, since this would have a stronger emotional impact than usual. (It is still unlikely that rape and seduction result in pregnancy as often as fiction would suggest.) Still another possible effect (perhaps only folklore) is the common report that childless couples who adopt a baby after years of infertility bear children of their own soon after. The release of tension may affect gonadotropins as well as sexual activity.

The posterior lobe of the pituitary represents an even more direct link of hormones with the nervous system. The two hormones released there, **antidiuretic hormone** (acting on the kidney) and **oxytocin**, are manufactured in neurons based in the hypothalamus but extending into the pituitary. Appro-

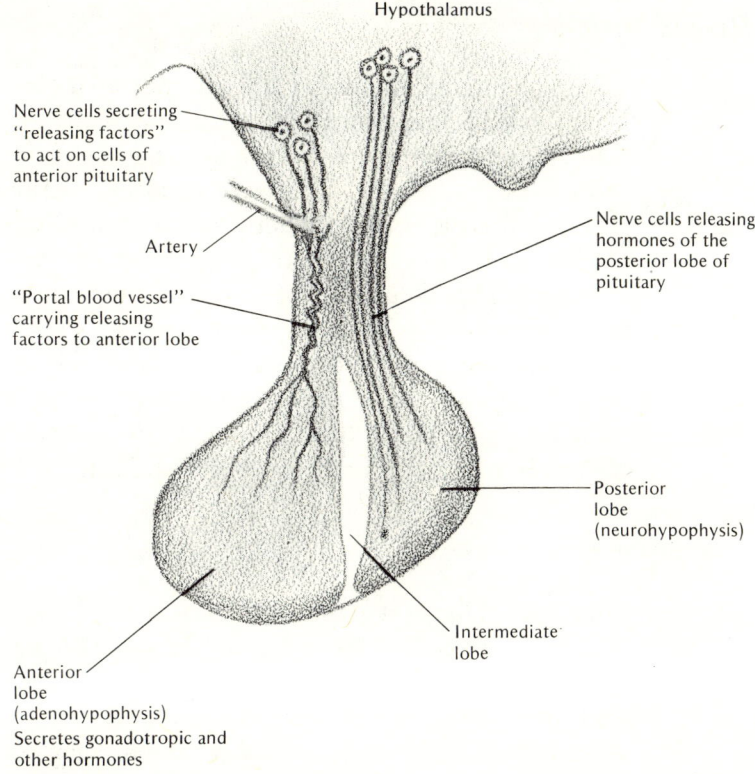

Fig. 3–3: B. Enlarged sketch showing the nervous and circulatory connections between the hypothalamus and the pituitary.

priate nervous stimulation can cause immediate release into the blood stream. Thus when a baby starts to suck the breast, the stimulation of nerve endings there causes impulses that bring about release of oxytocin into the bloodstream, causing automatic muscular transfer of milk into the nipple region. Immediately after birth, this hormone also increases uterine contractions, aiding expulsion of the afterbirth as well as the return of the uterus to its resting size; this action is stimulated by breast feeding. Conversely, worry and some other emotional states may prevent the release of oxytocin and prevent milk flow. Release of oxytocin, brought about by these same nervous mechanisms, probably contributes to the vaginal contractions of the female during sexual climax.

Human sexuality

The most intense emotional and interpersonal experiences of a person's life can occur in connection with sexual intercourse. For some people the whole range of their sexual activity is nearly always a source of pleasure, satisfaction, and renewal. For others, sexual activity may result in frustration, conflict, or guilt.

Human sexual intercourse is not a simple process. As a social activity, it is bound up with a person's social role and his self-image, and it is profoundly influenced by his training and experience. As a physiological experience, it involves both long- and short-term hormone effects and activity of the nervous system, both conscious and unconscious.

General sexual awareness

Most adults have a continual interest in the opposite sex and an awareness of its members. In any culture, men and women have a different manner of speaking and relating to each other than to members of their own sex. Every social relationship between a man and woman has some sexual overtones. An acquaintanceship between the two tends either to fade or to develop into a sexual attraction unless stabilized by careful social taboos.

Specific sexual arousal

Many different influences can cause a person's general sexual awareness to turn into much more specific arousal. Especially for a man (at least in our culture) such arousal may be initiated by even very minor influences—watching a pretty girl, seeing a suggestive picture, hearing a sexy joke, or even just a long delay between periods of sexual activity. Sexual arousal of both sexes comes about by flirting and caressing. Hand holding, kissing, and endearments lead to touching and caressing of various parts of the body, finally including the breasts and genitals.

The state of sexual arousal includes a powerful internal urge, driving toward intercourse. It may overcome people's inhibitions and even their good sense and real intentions. In marriage, when there is no obstacle to consummation of intercourse, the arousal itself is pleasurable. When intercourse is not permitted, either for social reasons or because of the disinterest of one partner, however, a strong arousal can lead either to regrettable actions or painful frustration.

The sexual act

When sexual intercourse is possible, the intimacy that leads to arousal continues in sexual foreplay—a great variety of tactile stimulation of each partner by the other—which may include kissing, manipulation of breasts and genitals, hugging and clasping.

Mutual stimulation by the partners causes a reciprocal increase of arousal, until the partners are ready for entry. For the woman, this means that the vaginal opening is relaxed and lubricated. For the man, it means that the penis is stiff (erect). **Erection** begins early in arousal but becomes more complete as arousal proceeds; erection of the penis is caused by the engorgement of spaces within it with blood, just as a flaccid balloon becomes stiff when filled with air (Fig. 3-4).

When the excitement of their arousal has become very intense, the man inserts his penis into the woman's vagina, and begins a series of pelvic thrusts, so that the movement of the penis within the vagina causes pleasurable sensations for both partners. The most common position for sexual intercourse is with the woman on her back and the man over her, but a variety of other positions are also used. The thrusting activity may alternate with periods of quieter sex play, but eventually the couple's excitement is so great that each may experience a sexual climax. For both partners this is an intense physical and emotional sensation lasting only a few moments but blotting out everything else and producing great satisfaction. The man's climax includes the **ejaculation**, or spurting of the seminal fluid into the vagina; this is neces-

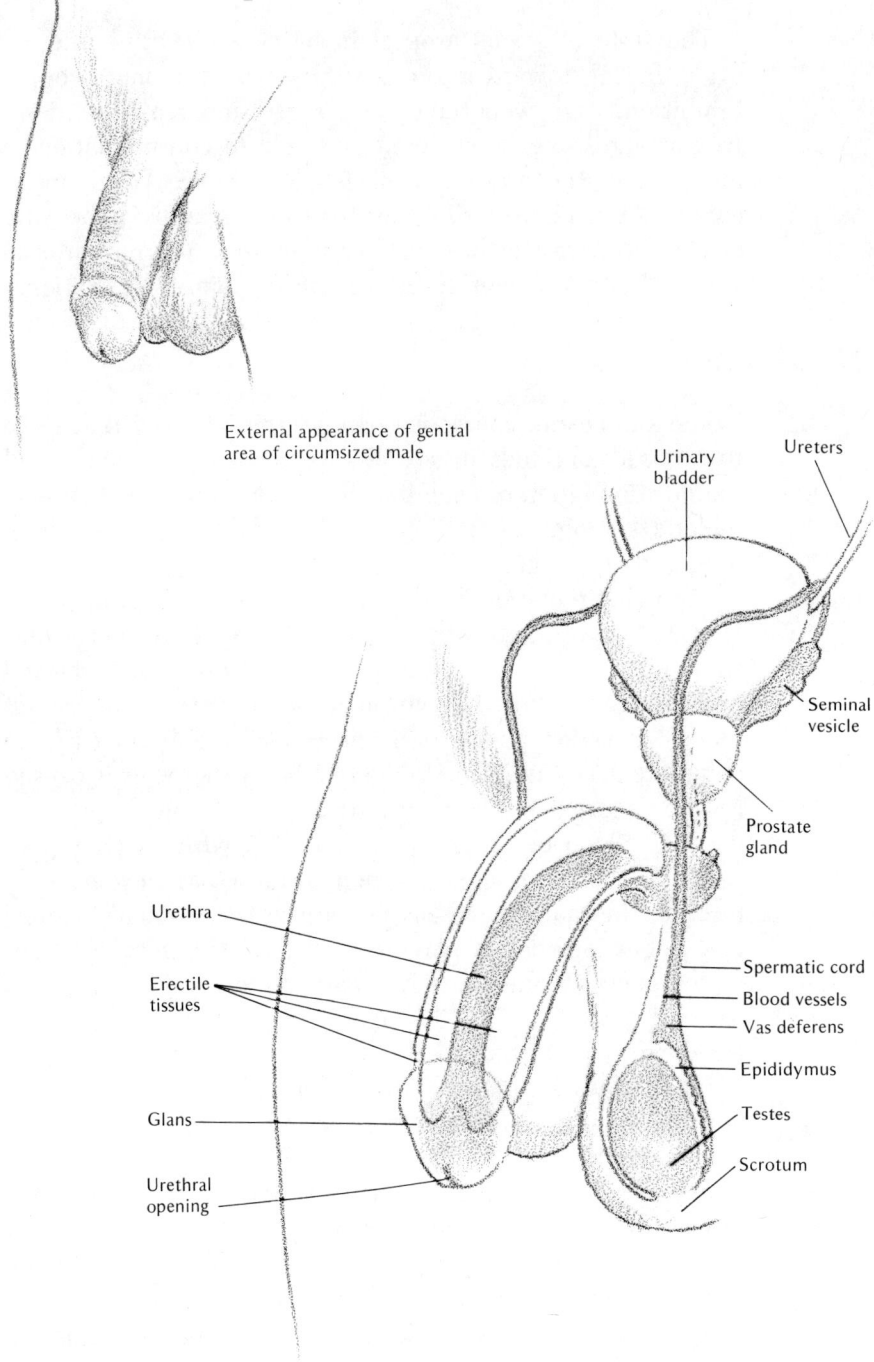

Fig. 3–4: A. Reproductive organs of the human male.

External appearance of
genital area of female

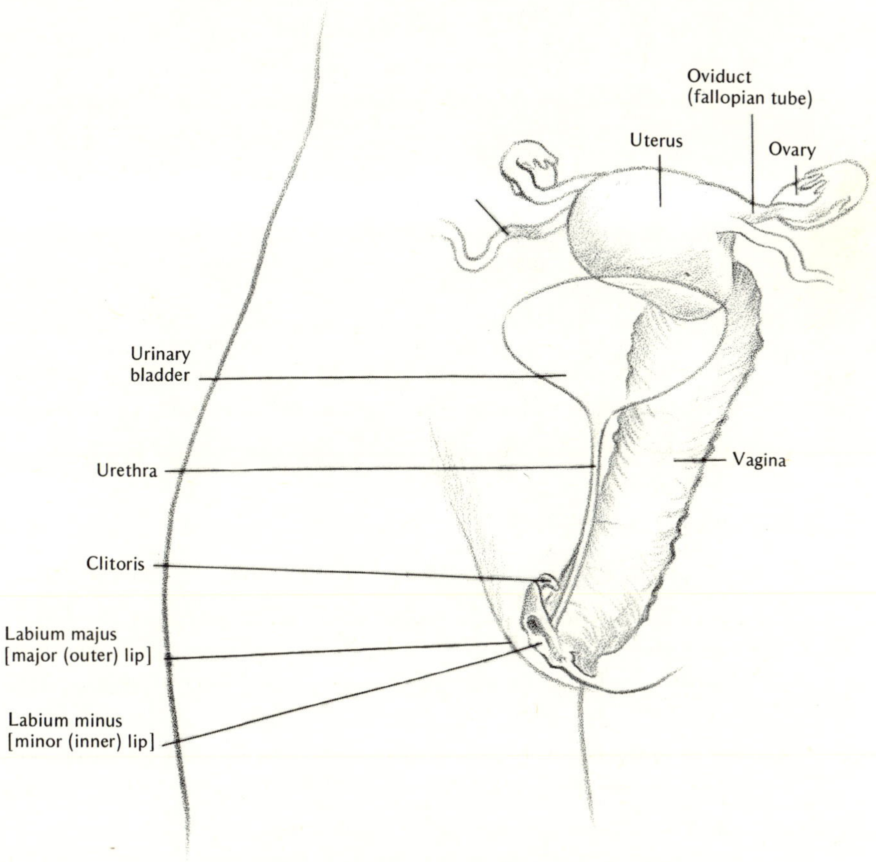

Fig. 3–4: B. Reproductive organs of the human female.

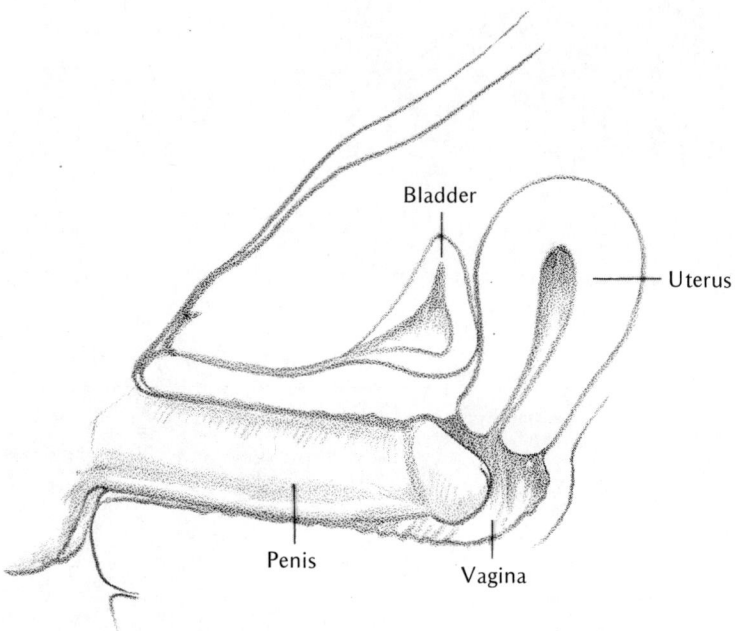

Fig. 3-5: Sexual organs during climax. The position of the erect penis in the vagina during intercourse is shown here. At ejaculation the semen is released into a "pooling" area just beneath the cervical opening into the uterus.

sary for conception to be possible as a result of intercourse (Fig. 3-5). After climax, the arousal of both partners subsides rapidly, and both are likely to be satisfied and sleepy.

A variety of patterns

There are so many possible variations of timing, position, and details of activity that sexual intercourse is extremely variable from couple to couple and from time to time. Couples who have frequent intercourse learn what is mutually pleasurable, but even they usually vary their actions from time to time. There are many good marriage manuals that can be consulted for fairly detailed descriptions of the different positions and other common variations. There are also now many more accounts of intercourse that are written just so that they will be bought for their ability to stimulate. These often give quite inaccurate information.

What can go wrong

As with any complicated process, sexual intercourse does not always go forward as desired. Sometimes the man's penis does not become and remain stiff enough to penetrate the vagina; in this case he is said to be "impotent." Even without this problem, a man sometimes fails to reach a climax, especially if he is tired or worried. A woman need not be sexually aroused to participate in intercourse, but she must experience an increasing arousal for intercourse to cause a climax. A woman who never has a climax during intercourse is said to be "frigid"; usually this is due to the fact that her sexual impulses have been repressed for all of her life, and she has not readjusted her attitudes enough to be ready for satisfying sexual intercourse.

[handwritten annotation: pre-orgasmic]

The sexual control center

Human sexual activity is influenced by a confusing array of different agents whose interaction can best be understood by considering the parts of the nervous system involved and how they work. Central to the process is a specific nerve center (a cluster of nerve cells or neurons) in the brain that is especially concerned with sexual behavior. This sexual response center is located in the hypothalamus, which is in the basal part of the brain. The hypothalamus, although small, also contains centers for control of respiration, heart rate, eating, water balance, and pituitary hormone secretion, in addition to being the sexual response center (Fig. 3-6).

Any nerve center in the central nervous system sends out impulses to other nerve centers and sometimes also to glands and muscles, and it receives impulses from sense organs and from other nerve centers. Impulses coming into a nerve center may either inhibit the center (make it less likely to send out impulses) or stimulate it, depending on which nerve fibers have carried the incoming impulses. The balance of

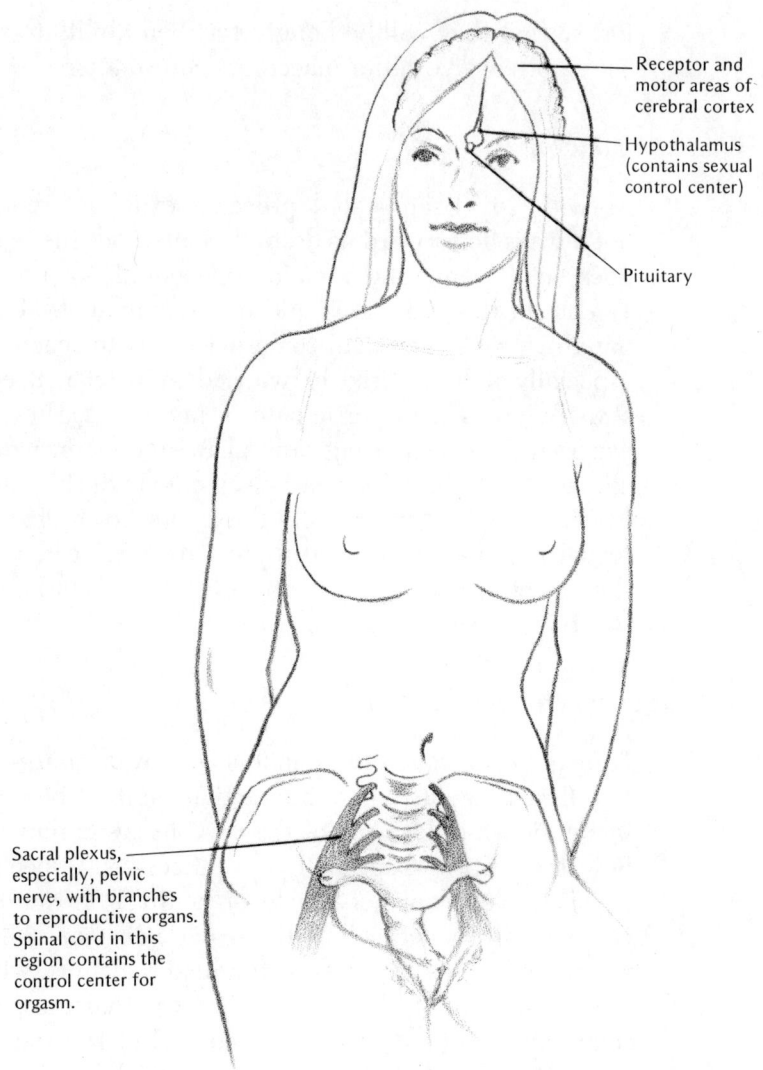

Fig. 3-6: The centers in the nervous system that are concerned with sexual response.

stimulatory and inhibitory influences determines the frequency of impulses sent out by the center.

The sexual center in the hypothalamus probably has a standard level of activity, a basal rate of sending out impulses, that is established by the presence of one or the other sex hormone. (After sexual habits are well established in an adult human, the hormone level seems to be less important and not always necessary. In mammals other than man, the sex hormones appear to be absolutely essential.)

Effects caused by the sexual control center

Impulses from the sexual center go to the cerebral cortex (the seat of consciousness), to the internal organs, and to the genitalia—the penis, ducts, and testes of the man and the vagina, clitoris, and uterus of the woman. These impulses to the cerebral cortex cause the person to dwell "in his mind" on sexual themes, to notice members of the opposite sex, to put a sexual interpretation on events and actions, and to consider sexually related courses of action. The cortical cells become more sensitive to impulses from the genitalia and from certain other parts of the body (earlobes, nipples, face) and somewhat less sensitive to pain signals. Impulses going to the respiratory and circulatory centers (there, in the base of the brain) cause an increase in the rapidity and depth of breathing and in heart rate and blood pressure. Impulses going to the genitalia cause an expansion of the arteries carrying blood into the sexual organs, with only a slight expansion of veins carrying blood out. This brings about a local increase in skin temperature and a turgidity and swelling (usually called erection) of several parts, notably the penis and the vaginal wall.

Influences acting on the sexual control center

The sexual center receives impulses from the cerebral cortex and sensory impulses from general sense receptors such as the eyes and ears, the genitalia, and the other sensitive areas, such as earlobes. (Probably all these sensory impulses come directly to the sexual center as well as to the cerebral cortex.) Sexual thoughts and memories cause the cortex to increase its stimulation of the sexual center; and impulses from the genitalia, as well as stimulatory sights, sounds and touching, increase the center's activity, too. When the activity of the sexual center is high for any of these reasons, the center sends out more impulses to increase the changes described in the preceding section, which in turn cause more stimulatory impulses to be sent to the control center.

On the other hand, if the cerebral cortex is occupied with a sexually unrelated task, is involved with feelings of fear or rage, or is receiving many impulses from pain receptors, it sends fewer stimulatory impulses, or more inhibitory ones, to the sexual center. The cerebral cortex also sends inhibitory impulses to the sexual control center if the cortex contains negative memories, associations, or training about sex. Strong pain impulses, especially from the genitalia, probably go directly to the sexual center and inhibit it, as well as acting through the cerebral cortex. Any of these inhibitory influences so depress the activity of the center that it sends out fewer impulses, and the genitalia, for instance, become less sensitive. Thus fewer stimulatory impulses come back to the control center, and its activity level sinks still lower.

Why sexual response is unpredictable and changes rapidly

Two things should be clear from this description of the nervous control of sexual response. First, the influences are extremely numerous, diverse, and interrelated. There can be no simple prediction of what will happen with regard to sexual desires and behavior. This, coupled with the intrinsic interest of the subject, is the reason for the vast amount of folklore that surrounds the subject of sexual intercourse; people like to think that they understand anything important to them, even if they really don't. The second obvious feature of the sexual control system is that it works to amplify any change: A small stimulus that increases a person's sexual excitement can reverberate so that the effect is far out of proportion to the original stimulus, and inhibitions are swept away. Conversely, a small negative influence may so damp down the back-and-forth stimulation pattern that desire suddenly evaporates, and even strong sexual stimuli are ineffective thereafter.

Is sex instinctive?

The extent to which instinct rules our sexual lives is still unknown—we may or may not have complicated inherited behavior patterns of reproductive activity. It is clear at least that we are physically equipped to pay attention to certain kinds of stimuli—finger touch, sounds, sights—and to respond to them. It is reasonable to suppose that certain touches, sounds, odors, and sights are conducted almost directly to the sexual response center, as well as to the cerebral cortex, because the necessary pathways are present at birth. But it is still being argued just how much instinct we have beyond a primitive kind of intrinsic sensitivity to warmth, soft pressure, stroking, and other such generalized stimuli. Certainly it is the structural connection to the sexual center from nipples, earlobes, and lips that makes these areas (as well as the genitalia) qualify as "erogenous zones," capable of causing sexual arousal.

Physical readiness for sex

The sensory nervous connections from the genital organs themselves certainly belong in the category of sense organs directly connected to the sexual center, but they are important enough to be considered by themselves. First, the general condition of the genitalia and ducts is transmitted continually to the brain. A man who has not had a discharge of sperm for some days or weeks is likely to be easily aroused to sexual excitement. At least part of this increased sexual desire can be explained by the gradual accumulation of secretory products in the seminal vesicles and prostate gland: The pressure in the swollen glands is transmitted as sexual readiness. Strangely enough, this effect may sometimes be mimicked by mild irritation elsewhere in the genitals, ducts, or urinary system. The aphrodisiac called "Spanish fly," which is the irritating secretion of the blister beetle, can actually stimulate sexual readiness on some occasions, just as the folklore says. When this chemical is taken by mouth, some is eliminated unchanged through the bladder. There it causes a local irritation that is interpreted by the sexual response center as though it were "swollen seminal and prostate glands." A more commonplace example is the fact that men often have an erection when they awake in the morning, because of the need to urinate.

The final sexual climax is even more strongly influenced "locally" than is general sexual excitement. The circulatory congestion and sensory sensitivity of the sexual organs increase. The sensations from the penis and clitoris and vagina stimulate the sexual control center in the brain to increase rapidly in activity so that a center in the **spinal cord** is stimulated to send impulses to the muscles lining the reproductive passages. In the woman's orgasm there are waves of contraction passing along the vagina, and probably up the uterus, too. In the man, the climax occurs in two steps. First, the muscles of the testes contract, forcing sperm into the ducts, and similarly the walls of the seminal vesicle and the prostate gland contract, discharging their contents into the ducts. (This process is perceived by the man as a feeling of "inevitability"—the climax is surely coming.) Then waves of contraction pass over the **vas deferens** and the **ejaculatory duct**, causing the ejaculation of the mixed seminal fluids. These climaxes or orgasms give rise to intensely pleasurable feelings and emotions, which, besides pleasing the person, seem to satisfy temporarily the sexual urges. The process of sexual excitation may occasionally be rapidly repeated in the woman but requires some delay in the man.

(The local nature of the final sexual climax is emphasized by reports that some paraplegic men, whose spinal cords were severed and who hence have no connections between the brain and the lower parts of their bodies, have erections and even ejaculations, and have, rarely, fathered children. They, of course, receive no direct sensory pleasure from the genitalia.)

Influences that help or hinder sexual fulfillment

Any of several kinds of influences are sufficient, in themselves, either to cause sexual excitement and even climax, or to prevent it.

First, the condition of the sexual organs, and their stimulation, can cause a buildup of stimulation both in the sexual center and in the response center in the spinal cord so that all the phases of sexual response are accomplished. Conversely, ineffective stimulation of the genitalia, and especially painful stimulation, can inhibit and prevent stimulation of sexual response by other influences.

Second, events and sensations such as sight, sound, and touch of another person, and that person's response to certain signals with other appropriate signals, may build up the sexual excitement so that little or no genital stimulation is required. But a miscuing of signals with the sexual partner, or the perception of distracting or worrying sights and sounds may prevent an otherwise successful act of lovemaking.

Third, it is possible for some people to fantasize, to imagine sexual stimulation on the basis of their memories and their training, so that sexual climax may occur without another person. Under normal circumstances, good feelings toward the partner, pleasant memories, expectation—all contribute to sexual fulfillment. A learned feeling that sex is dirty, a dislike of the husband or wife, or a distracting memory or thought may spoil everything.

Usually a happily married couple has everything going for them, but it is evident that many things can go wrong in sex, and the problem may be hard to pin down. Luckily, a knowledge of all this detail is necessary only when problems arise. Ordinarily, it would be a nuisance if kept in mind during sexual activity!

Factors influencing human sexual activity

Although human sexual intercourse can be accurately described as a sequence of muscular, circulatory, and sensory changes, it has influences on the rest of human life and is in turn influenced by the other aspects of life to an extent that is difficult to exaggerate. The very vocabulary is not neutral, since words are related to the philosophical beliefs and attitudes of those using the words. Talking of "human sexual activity or behavior" is relatively neutral, but might seem to imply a detachment, a cold attitude inappropriate to the subject. Speaking of "sexual response" acknowledges that it takes two, but still focuses on one person at a time, while talking about "interaction" may be closer to the actuality. The term "sexual performance" indicates that two are involved, but it focuses so heavily on what one person does that it rather ignores the other except as spectator or judge; the very use of this phrase may interfere with satisfactory interaction. The various obscene words for sexual intercourse are not merely out of style or considered "lower class" (as is the case with the Anglo-Saxon words for urination and defecation) but are transitive verbs, indicating something that one person does *to* (not *with*) another. Thus they are akin to "rape," but with the implication that rape is the normal mode. The use of these words as terms of abuse demonstrates these implications.

One way of realizing the complexity of sexual interaction is to see it as a collection of opposites or contrasts. It includes voluntary actions and involuntary responses. Some of the involuntary responses are inborn, some are learned. Sexual activity is privately done but socially influenced. It is a highly personal activity but normally involves intense interpersonal relationships.

An understanding of sexual behavior is handicapped by a wide range of contrasting assumptions about the origins of what we do. At one end of this spectrum of ideas is the view of popular sociology and psychology that all behavior (including sexual) is socially determined—that the newborn is a "blank slate" on which experience writes, and that all personality is cultural imprint. One form of this theory holds further that the cultural imprint, once made, is indelible, or nearly so. Thus we are pawns of our culture. In this view, expectations or prohibitions by society that influence sexual behavior are only repressive, and there should, ideally, be no guidelines.

Diametrically opposed to this theory is the idea that the range of proper sexual activity is extremely narrow, that what is right has been determined permanently and inflexibly, perhaps even by God, and often (but not always) that sex is a necessary evil to be avoided as much as possible.

Still another view, popularized by Desmond Morris, is that human behavior (including sexual) is heavily influenced by evolution and is largely inborn. Although he acknowledges the plasticity of human action and the possibility of learning, he chooses in his books to emphasize the parallels between animal and human actions, and the variety of inborn influences.

As with most arguments, all extreme positions are distortions of reality. I feel the conclusion is inescapable that we have a large heritage of behavioral tendencies. These behavioral tendencies have been derived by evolution but are developed and greatly modified by our experience and by the social forces acting on us. All of our collection of behaviors at any one time is subject to change by learning, some easily, some with great difficulty. Even if this middle position is accepted, however, it just sets the stage for fruitful argument and research on how these diverse elements in our personality act and interact.

Table 3-2 summarizes data published by William Masters and Virginia Johnson in their book *Human Sexual Response*.* Essentially all the accurate information on physiological events during sexual intercourse comes from the work of Masters and Johnson, published in *Human Sexual Response* and several other books and papers. They have also pioneered in using a therapy for sexual problems that is based on their findings. A very different approach to the subject is provided by Desmond Morris, who deals with the evolutionary

* Little, Brown, Boston, 1966.

Table 3-2. Sequence of events in human sexual intercourse

Phases	Changes in the female	Changes in the male
Excitement (foreplay) Most of these states increase throughout the phase	M—vaginal elongation and relaxation, increased breathing rate; increased general muscular tension C—vaginal swelling and lubrication, swelling of labia and breasts, flushing of skin, increased heart rate. S—increased sensitivity of clitoris; many kinds of stimulation by touch become more pleasurable	M—muscular tension and breathing rate as in female C—erection of penis; flushing and heart rate changes as in the female S—penis becomes more sensitive to touch, less sensitive to pain. Same increased sensitivity to touch as in the female
Plateau (entry and coital movements) Level of excitement increases more slowly	M—retraction of clitoris (painful sensitivity), even more muscle tone; pelvic movements become involuntary C—erectile narrowing of vaginal opening, continued high heart rate S—little additional change	M—increased muscle tone, pelvic movements become involuntary C—heart rate, flushing, maintained or increased S—little additional change
Orgasmic (climax) Brief duration	M—3 to 15 vaginal contractions; involuntary movements, spasm (involuntary contraction) of many or most muscles; breathing three times normal rate C—heart rate 3 times normal, extreme flushing S—flood of sensations seeming to come from all parts of the body	M—muscular discharge of testis, epididymus, seminal vesicle, prostate into ducts, producing a feeling of inevitability; peristaltic discharge of seminal fluid from ducts: ejaculation; body movements as in female C—swelling of testes; flushing and heart rate as in female S—flood of sensation
Resolution (relaxation)	M—great relaxation, sleepiness C—rapid return to normal S—sensitivity immediately reduced, but does not reach normal at once; multiple orgasms possible	M—similar to female; partial erection may persist for a while C—return to normal S—refractory period of minutes to hours, depending on age, individual factors

Note: M indicates muscular changes, C circulatory changes, and S changes in sensation.

origins of human behavior in a number of popular books, including *The Naked Ape** and *The Human Zoo*.†

Normal sexual activity

The concept of normal versus abnormal sexual activity is meaningless to anyone who thinks that human behavior, specifically sexual behavior, is totally a cultural artifact. To one who believes in divinely inspired rules, "normal behavior" is a rigid concept not subject to discussion. But on the basis of the analysis of the origin of sexual behavior presented earlier, it is possible to talk usefully about normal sexual behavior, although the idea is a little vague.

What is normal or natural in sex? What I call "natural" is what is generally found in most contented people. It usually has parallels in animals, or has an apparent evolutionary advantage. It will be found common to most cultures. Such normal behavior has been caused by evolution both biological and social, with selection of the kind of behavior that enhanced the survival of the group by increasing reproduction, improving child care, and aiding general social cooperation. Desmond Morris suggests that the sexual aberrations in our society (such as homosexuality and masturbation) are similar to those observed in animals penned in a zoo, but are not commonly found in nature. He assumes reasonably that our instinctive responses were evolved while our ancestors were still widely scattered hunters and gatherers (the human condition for most of our history). He then proposes that under conditions of greater population density and social organization—including agricultural and all more complicated arrangements—the instinctive sexual responses of the human, which were evolved for connecting scattered people, are continually overwhelmed and overstimulated, and that therefore many of our responses are inappropriate. This is the reason that social controls over sexual activity, however unfortunate their effects in individual cases, have been necessary. He suggests that even these controls were developed for

* McGraw-Hill, New York, 1967.
† McGraw-Hill, New York, 1969.

a less stressful, agricultural society, and that we will have to work out some new ones for our new situations, present and future.

Some of the behavioral patterns that I here include as natural or normal are not necessarily those which I would approve of or advocate for myself or others. Within the range of normal activity there may be (and I believe there are) certain courses of action that lead to long-term happiness (and are therefore moral) and others that lead to eventual unhappiness (and are therefore immoral). Some of these immoral actions lead to injury of others and to social disorder, and so they should be condemned by society. Those which are injurious only to the person performing the act are his own affair, although we may want to advise or admonish him against them as our personal, not social, responsibility. Normal human sexual activity can be described under the following headings.

Relationship. Stable, long-term affection between a man and a woman, each getting pleasure from the other's enjoyment (love). The standard that is approached in most societies, and that is explicitly required by the Jewish and Christian traditions, is a lifetime, one-to-one relationship, with complete fidelity. Variations of this pattern are quite common and have to be included within the normal range: polygamy, either serial or simultaneous, occasional infidelities, or a series of moderately long liaisons. These are all normal if they include lasting affectional relationships.

Frequency. As desired by the couple. There is no reason for any specific frequency; intercourse may vary from several times a day to only a few times a year. Twice a week is often mentioned as an average among happy couples and might be a good guideline when desires of the partners differ. It happens to be about optimum for fertilization, too. There may be a reason to suspect emotional difficulties when the frequency of intercourse is less than a couple of times a month (getting pretty cold), or more than once a day (trying to prove something). There is no reason to criticize even an unusual pattern if both partners are satisfied.

Normal events in intercourse. Some minutes of foreplay, with mutual caressing designed to give pleasure, followed by vaginal intercourse. In most individual instances of intercourse (but not all) a well-adjusted couple will each reach climax. Simultaneous climax is probably appreciated by most couples when it occurs, but it is not very common or really important. If the husband is the one to reach climax first, it may be difficult for his wife to reach hers. So trying for simultaneousness may cause more trouble than it is worth.

Variations. Many positions are used, with no reason for preference except individual experience. A preference by a person or a couple for certain rituals or types of contact may be the result of unusual experiences or attitudes, but they may still be successful in giving dependable mutual pleasure and are therefore good. A wide range of literature is currently available on the subject.

Treatment of sexual inadequacy

Most of the sexual problems treated by Masters and Johnson were found to stem from difficulties in the relationship between the husband and wife, to the couple's having learned an inappropriate pattern of sexual behavior, or to an interaction of these factors. An exception is painful intercourse, which is often due to irritation of the genitalia from infection or allergy or to malformations or injuries. Problems (or symptoms of problems) treated by Masters and Johnson included the following: premature ejaculation, inability to ejaculate, impotence (insufficient erection for vaginal penetration), lack of orgasm in the woman, and vaginismus (a muscular spasm of the vagina that prevents intercourse).

In treatment a few direct physical treatments are used. For premature ejaculation the wife is instructed in a kind of manipulation that will delay her husband's climax, and for vaginismus a graduated series of mechanical dilators is employed. But for these and all other problems the major treatment is a combination of conservative psychotherapy with a program for relearning how to go about intercourse.

In the counseling sessions, the couple is encouraged to

talk about their problem, their backgrounds, their feelings. They talk individually to both the male and female member of the therapy team, and with both in group sessions. They are encouraged to talk out and work out their own relational problems, and the suggestions given by the therapists are designed to fit the value systems and the deep feelings of the particular patients.

During the two weeks of therapy, while the interviews are going on, the couple is given instructions for therapeutic activities to be carried on privately in their hotel room. The program is really an exaggerated, prolonged model of sexual intercourse, designed to implant new and more satisfying habits and responses. It starts after two days of counseling, and on its first day the partners each learn how to give the other pleasure by caressing. They are instructed to take turns "pleasuring" each other by touch, but avoiding specific stimulation of genitals or breasts. In this exercise each person learns to focus on the feelings of the other (rather than paying attention to their own feelings), and to give the other pleasure without pressure to "perform" sexually. They are each helped to learn to get enjoyment for themselves from the pleasure of the other. The next day each partner, when being "pleasured," is instructed to direct the caresses by touch signal, so that the couple will learn to communicate skillfully and immediately without the need to verbalize. At these sessions the sensuous touching is extended to include the breasts and genitals, but with no advance to actual vaginal intercourse. (On each day, the counseling sessions have been continuing, with the patients reporting their feelings and progress, and discussing with the therapists how problems might be solved.) The instructions on succeeding days vary with the particular problem but involve further stimulation for a day or two, and usually vaginal penetration without any attempt to achieve climax. By this time sexual tensions have usually built up to the stage where complete intercourse with orgasm is inevitable and satisfactory to both. So for the next several days the couple is given advice about experimentation with various kinds of mutual stimulation, with different positions, and with different sequences and tempos, all in order to learn the optimum amount and sequence of stimulation for maximum satisfaction. Most important, each partner will have learned, by doing, to make involuntary responses to the actions of the other in such a way as to satisfy both of them.

Masters and Johnson report much higher success rates with their form of therapy than users of any other treatment techniques report, but they are by no means always successful. About one-quarter of the cases were classed as failures, although some progress was made in a portion of them. It is the assumption of Masters and Johnson that all such problems are treatable, that people are capable of learning more appropriate actions and responses. So obviously the program of treatment was not sufficient in the unsuccessful cases. It was possible to identify some types of patients for whom the program was most likely to be ineffective: (1) Some patients were suspected of having been unwilling to face and discuss traumatic or shameful experiences in their past. (2) Some for whom the treatment failed had backgrounds of severe religious condemnation of sex. (3) Several of the failures had a history of homosexual as well as heterosexual activity, and had become, and remained, unable to function heterosexually. (4) Although couples were accepted for therapy only if both seemed highly motivated, some later seemed to lack sufficient motivation. (5) Some errors of judgment by the therapists could be identified. These five explanations did not really clear up the reasons for therapeutic failure, since other couples who seemed to have equally severe problems in the same areas were treated successfully.

Some elements of this therapeutic program should be useful to more fortunate couples, helping them, on their own, to solve small problems or to prevent them. The partners could discuss their backgrounds and feelings with each other, learn to give sensual pleasure to each other without feeling pressure for "personal performance," and they could experiment with variations that proved to give mutual pleasure.

Sexual aberrations

Aberrations have the same two causes as sexual inadequacy: failure to form a stable relationship with a sexual partner, and having learned an inappropriate pattern of sexual behavior. From evolutionary considerations, most people would be expected to have a normal tendency to form what Morris calls a "pair bond," and would respond normally to sexual stimulation in an appropriate manner. So aberrations can be presumed to be due to emotional damage or to overstimulation by inappropriate stimuli.

Masturbation. **Masturbation** is sexual self-stimulation using the hand or some other object as a substitute for the penis or vagina of a partner. It is common among men, especially before marriage, but considerably less common among women. Traditionally it has been strongly condemned, and it has been claimed to cause insanity and a host of lesser ills. There is absolutely no evidence, however, that masturbation is physically harmful. The phases of stimulation and the internal events during stimulation and orgasm are normal; in fact, the intensity of orgasm may be greater than during intercourse. As a means of releasing tension for the unmarried, and for the married during a period when sexual activity is impossible, it is probably helpful, and preferable to promiscuity. But as a dominant form of sexual activity it is an aberration; it represents the failure to establish a relationship with a partner, and all the attention is focused on the self. Thus it doesn't give other, nonphysical satisfactions and is inherently dull.

Promiscuity. Promiscuity means indiscriminate or nonselective activity of any kind, and it has come to refer especially to the practice of sexual intercourse with many partners, indiscriminately. Despite its glorification in pornographic literature and in adolescent folklore, this is an aberration that is really very similar to masturbation. Here the other person is just a device, a masturbation aid. No stable relationship is formed with the "playmate." The aspects of promiscuity that go beyond masturbation are not much more attractive. There is a strong element of agression evident, as illustrated by talk of a new "conquest," or the other person is used as a means of private or public role playing—a way of bragging. Even if two promiscuous persons get together, there is the danger that one will become emotionally attached and will be injured when the other leaves.

Sadism, masochism, fetishism. These all can be heterosexual activities, even within marriage. They represent failures of relationship, since the partner is not valued as a personality but only as a device to give or receive pain, or to receive the behavior generated by the fetish. They also represent abnormal learning, in which pain or some unrelated object becomes fixed in the mind as associated with sexual stimulation. Hopefully, these will become more treatable in the future.

Mouth–genital contacts and anal intercourse. These are illegal in most states, although rarely prosecuted. There is no reason why they should be any concern of the state, however distasteful they may be to individuals or groups. As playful variations of pattern they are mutually attractive to many but not all couples. As the sole or major kind of sexual stimulation they must be regarded as abnormal—the result of mislearning.

Homosexuality. Homosexuals feel their sexual urges toward members of the same sex and carry on their sexual activities with them. Since actual sexual intercourse is anatomically impossible, arousal to climax is carried out by mutual masturbation of some sort; for men, oral or anal intercourse may be used. Despite the propaganda of the Gay Liberation movement, I still do not see homosexuality as normal sexual activity. This does not mean that there is any excuse for the harsh laws against homosexuals still on the books in most states. When the threat to society was underpopulation, such laws may have made some sense. With the population explosion, such nonreproductive activity is hardly a threat any longer. It is suspected that homosexual activity by adolescents may sometimes act as an emotion-driven learning experience that interferes with normal heterosexual activity later. The relative infrequency of such problems in spite of the high frequency of early homosexual experience suggests that our instinctive behavioral tendencies must be pretty strong. Homosexuality is certainly not inherited in the usual sense of the word, since it has such a strong biological selection against it. The best guess is that it comes from mislearning of two kinds: a mixup of role-identification because of sick relationships in the family, and an overstimulation of the residual behavioral patterns of the opposite sex that are normally not activated. This overstimulation is strongly favored by extreme sexual segregation in our society, such as exists in English residential boy's schools, the army, and prisons.

Homosexuality is biologically absurd and anatomically ridiculous, but it still may represent a real human relationship. One graduate of an English boy's school, who had not himself become sexually involved there, nevertheless reported that the homosexual relationships afforded about the only occasions for affection and mutual caring in a cold environment. It may be that this is the best situation for some people as they are now constituted. I do feel, however, that humans are capable of learning a variety of important things, that it will some day be possible for homosexuals to learn to function normally, and that this would be desirable.

The Oneida experiment

The problems of a famous social experiment support the idea that formation of a pair bond is a natural, instinctive human tendency. The Oneida Community that existed for about a generation in the last century was like several other communities in practicing Christian communal economics. But the Oneida community expressed the remarkable theory that the Biblical admonitions for Christians to love all their fellows included complete sexual sharing. So in this community, sexual intercourse was to be entered into with many other people; it was considered selfish and sinful to concentrate all one's favor on just one other person. Elaborate indoctrination, training, rules, and taboos were used to enforce this ideal. Yet one of the forces that led to the dissolution of the community was the fact that young couples fell in love and wanted to love each other exclusively, against all social and moral pressure. Even in a group where "free love" was not only socially acceptable, but morally desirable, there seemed to be a real desire to form stable sexual relationships. Perhaps the institution of marriage is supported, at least to some degree, by the nature of man as well as by the pressures of society.

4 Conception and Contraception

In all mammals except man, copulation is timed to occur only when conception is likely. For many birds and mammals in the temperate zones, both mating activity and the production of gametes occur during only part of the year—usually during the spring and summer months, when offspring are most likely to survive. Among most tropical animals and domesticated animals, reproductive activity occurs at any time of the year. But whether reproduction is seasonal or year-round, the transfer of sperm from the male normally is synchronized with ovulation in the female in one of two manners. In the majority of mammals, the female will not accept the mating approach of the male unless she is in a special state of sexual readiness, called "heat" or "estrus." This state of estrus is brought about by the estrogen hormones that are released at the time of ovulation. In several mammals, however, including the cat, the rabbit, and the ferret, the ovarian follicles develop to a stage of readiness for ovulation but are not released. The female is thereafter in a state of continuous estrus—she will accept the male at any time. The activity of the nervous system during copulation then stimulates the hypothalamus to produce the "LH releasing factor," which causes the pituitary gland to secrete LH (luteinizing hormone); this in turn causes the ovarian follicle to burst and release the ovum. All this is accomplished within an hour or so after copulation. In either of the standard mammalian conditions, whether with estrus after ovulation or with ovulation induced by mating, the transfer of sperm and the release of the ovum take place at practically the same time. Fertilization of the egg by one of the sperm is therefore much more likely than it would be if sperm and egg release were uncoordinated.

The human condition seems to be unique among higher mammals, and probably among all animals—humans are physiologically ready for sexual intercourse at any time. (Many mammals, especially primates, use some of their copulatory movements outside of estrous periods to signal social dominance or submission.) Sexual activity is indeed regulated by all human societies, but this regulation is not timed for fertilization—it serves other purposes of the society instead. Sexual intercourse and the transfer of sperm into the woman may occur at any time in the female cycle (usually with the exception of the time of menstrual flow). Conception occurs only if the arrival of the sperm happens, by chance, to coincide with ovulation.

The time of human sexual intercourse is not related to the timing of ovulation. This is a dramatic indication of how thoroughly human sexual behavior is dominated by its societal functions. Sex is still necessary for human reproduction (although technology is changing even that fact), but the reproductive function is sufficiently accomplished almost as a side effect of sexual activity. Continuous and pervasive sexuality was probably a human evolutionary adaptation to bind the family together so that the father would help in the support of the immature child. Human infants are helpless for a longer period of time than the young of any other animal; this makes it unlikely that a woman alone could provide for both herself and her child. Such independence might be possible in an unusually lush subtropical area with lots of fruit available for the picking, but it would not be possible in the harsher environments where, for millions of years, most of our ancestors eked out a living by hunting and gathering. Only in the most affluent of modern societies are single-parent families practical, even now. Desmond Morris suggests that development of continuous sexuality may have occurred when our ancestors moved into areas where they could live only by hunting—something a female carrying a baby could hardly do by herself. Whereas most groups of carnivores have an instinctive pair bond (mated pairs staying together and cooperating both in and out of mating season), our fruit-eating primate ancestors had no such instinct. Morris thinks that a kind of hypersexuality evolved in these early people to encourage the development of pair bonds. He says that we are unusually well supplied with sexual symbols and anatomical specializations for making sex more enjoyable.

Whatever the exact sequence of events in the evolution of human sexuality, human sex is now (and may have been for as long as humans have existed) devoted primarily to attraction between the partners and enjoyment of the relationship. In all societies sex has always been an important part of status, culture, and customs, just as in our society it is the primary sales gimmick of the television commercial.

It is a sign of ignorance of the human condition to believe that the overriding function of sexual intercourse is reproduction, and that all other functions are secondary or even shameful. The sexual urge is such a powerful one that it must be regulated by the individual for his own well-being, and by society for its stability and health. But sex for mutual pleasure rather than for producing children is not a perversion—it is part of human nature.

The frequency of human sexual relations and the rate of production of sperm are such that humans are fertile enough to survive, despite the fact that intercourse is not concentrated at the time of ovulation. An otherwise fertile couple rarely goes for more than a few months of intercourse without contraceptives before pregnancy ensues. In any month, however, the factors deciding whether or not conception occurs are the timing of intercourse in relation to ovulation and the longevity of the gametes.

How long can the gametes function?

Sperm storage. The mature sperm produced by the testes are normally stored for a period of some weeks in the **epididymis**, a coiled tube lying next to each testis within the scrotum. The length of time spent in the epididymis probably depends somewhat on the frequency of ejaculation; if the sperm are retained for a long time some of them may be absorbed and destroyed by the cells lining the epididymal tube. During storage the sperm do not move; the secretions of the surrounding cells prevent activity and supply nutrients. A stay of moderate length here seems to be necessary for the sperm to function in fertilization—there is some kind of maturation during the storage period. Sperm that have been retained for the maximum 3 or 4 weeks seem less likely to be capable of fertilization.

The male accessory glands. The bulk of the semen that is expelled from the penis comes from the accessory glands; the sperm represent only a tiny part of the bulk of the fluid. The most important of these glands are the **seminal vesicles** (one gland each connected to the ducts coming from each testis), and the single **prostate gland**, located where the two sperm-bearing ducts join the urethra. The secretions of these glands and of two or three other smaller ones are thought to be important in stimulating the sperm so that they begin to swim, and in supplying nutrients for them. The secretions are probably also important in neutralizing the rather acid secretions of the vagina. The secretion of the prostate gland was the place where the **prostaglandins** were first found (hence their name), although they are now known to occur in many other tissues as well. These hormone-like chemicals are only now being studied and found to have a variety of important properties. The prostaglandins in the semen may cause contraction of the vaginal and uterine walls, and promote the mechanical transport of sperm up into the oviducts. Many other components of the glandular secretions are unknown or poorly understood; no one knows what function most of them play. A principal reason for this uncertainty is the fact that various species of laboratory and domestic mammals differ so greatly in their accessory glands. The seminal vesicle may secrete most of the seminal fluid in one species but hardly any in another species that has a large prostate. In still another species the major secretion may be produced by a gland that is so small in the human that it is hard to find. The functions of the secretions may be vitally important in each species, but the conflicting data from different animals are little help in determining the functions of these glands in humans. Further research will undoubtedly answer some of the questions and will probably demonstrate how a deficiency in some component of a glandular secretion can be responsible for certain cases of sterility.

Within the vagina

Once the sperm are deposited in the vagina, they are subjected to still another variable influence. Not only does the vaginal fluid usually cause the sperm to swim much faster, but it seems to make them capable of carrying out the fertilization of the egg, which they could not do without this influence. This effect, called **capacitation**, is not well understood and may be unimportant in the human. If the sperm remain in the vagina more than a few hours, the effect on them is quite the opposite; they are no longer capable of fertilizing the egg even if artificially transported to it. But sperm that manage to get to the oviduct may retain their fertilizing ability for 1 or 2 days.

The sperm reach the oviducts, where an egg might be, by a combination of their own swimming and the pumping action of the vaginal, uterine, and oviduct walls. In large animals such as the cow, the muscular movements transport the sperm to the oviducts. Sperm are found there only a few minutes after copulation; they could have swum only a small fraction of the distance in that time. In the mouse, on the other hand, the sperm probably swim all the way to the oviducts. The human is rather intermediate in size, and the best guess is that both muscular transportation and the swimming of the sperm are significant in the journey to the oviducts.

When the sperm arrive, by whatever means, in the vicinity of the ovum, they need to be capable of swimming to reach the egg.

Use of the sperm

Of the 100 million to 1000 million sperm that might be deposited in the vagina in a typical ejaculation, fewer than 100 are present in the oviducts at the time of fertilization. Just one sperm fertilizes the egg, but many more may somehow participate in the process. The astronomical number of sperm transferred seems to be necessary for fertility, since men with sperm counts much below the average have low fertilities. A frequency of intercourse greater than once every couple of days seems to be less likely to lead to pregnancy, presumably because too few sperm are transferred each time. So the millions of sperm seem to be necessary, for some reason. This may seem like a very wasteful method, but the actual volume of cellular material used up is very small. The many cells discarded as sperm are only a small number when compared with the 100 billion intestinal cells worn off and replaced each day, or the billions of skin cells similarly discarded. There was a myth spread around, a generation or two ago, to the effect that loss of semen led to exhaustion and debilitation. This was published as propaganda against masturbation and was believed by many. There is no good evidence for such a belief. The bulk of the seminal fluid is from the accessory glands. The components of those fluids may include some unusual mixtures of compounds, but none is particularly demanding for the body to manufacture. The volume of ejaculated fluid (about a teaspoonful) is really rather small, considering the total body economy.

Migration of the ovum

When the follicle ruptures, the ovum (really the secondary oocyte) is released into the abdominal cavity and passes into the oviducts. Along with a small quantity of fluid, the ovum moves slowly toward the uterus, propelled by the beating of cilia on the cells of the oviduct lining and by the slow contractions of muscles in the oviduct walls. The time required for the ovum to reach the uterus is roughly similar to the time during which fertilization may take place. The passage to the uterus takes about 3 days. Fertilization generally must take place within a somewhat shorter time, perhaps within as little as 1 day after ovulation.

Fertilization

Just as tiny, blind, free-living one-celled animals are able to swim to a source of food or away from a harmful substance, the sperm in the oviduct probably are able to swim toward the ovum. This is presumably because the ovum releases some guiding chemical. Whatever the means of guidance, a number of the sperm manage to hit the target. The ovum as released from the follicle still has a covering of small follicular cells surrounding it. The sperm penetrate these cells and push in among them. There is good evidence that the sperm release a specific digestive enzyme that dissolves the adhesive substance holding the follicle cells together. The combined enzymes of many sperm are necessary to clear a passageway for a single sperm to reach the egg. This successful sperm penetrates the membranes covering the egg by its own actions, but then the surface of the ovum seems to expand outward to assist in engulfing the sperm. As part of the same process the outer membrane of the ovum becomes resistant to entry, so that no more sperm may enter. Stimulated by the sperm, the egg undergoes the second (final) meiotic division necessary to give it the half-of-normal number of chromosomes. The sperm disintegrates as it enters the egg, and the condensed spermhead expands again into a nucleus. The sperm nucleus and the egg nucleus then move together and fuse. There is now a single cell with a single nucleus containing 46 chromosomes. It is the zygote, the first cell of the embryo that can grow into a new individual (Fig. 4-1).

The ovum has been, for some hours previous to fertilization, in a state of blocked cell division. It is poised and ready

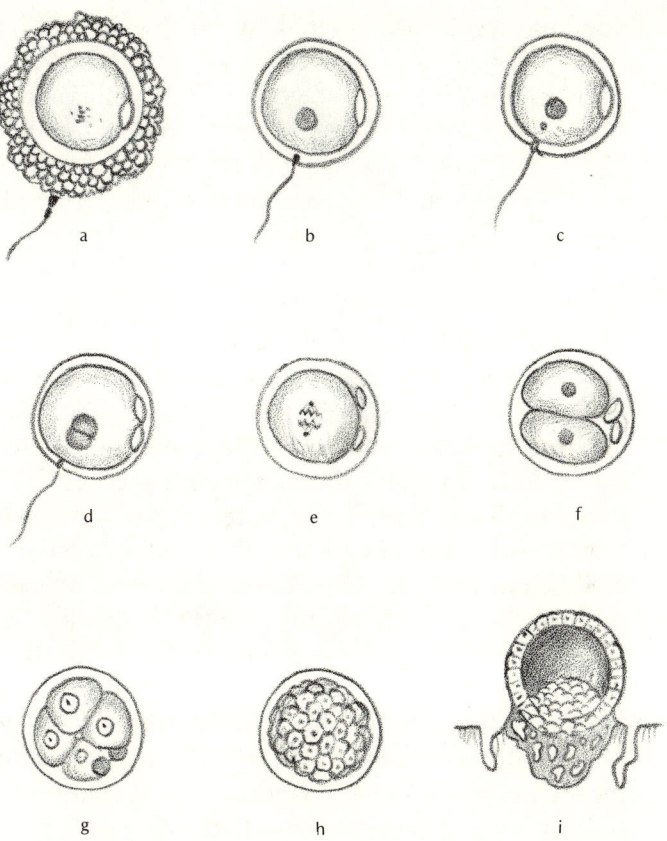

Fig. 4–1: From gametes to embryo: A. Sperm begins entry; second meiosis of oocyte. B. Sperm entry complete. C. Sperm absorption. D. Fusion of half-nuclei to complete zygote formation. E. Mitosis of zygote. F. Two-cell stage. G. Four-cell stage. H. Multicellular stage. I. Blastocyst, including trophoblast and inner cell mass, begins implantation.

for division and development, but if it is not fertilized it will normally not divide. The sperm breaks this division blockage and initiates the cell divisions of early embryonic division. (Under experimental conditions, unfertilized ova have been induced to divide and to develop into adult individuals by being artificially activated by extreme temperature or by a pinprick. This has been done with many frogs, with many turkeys, and with one litter of rabbits.) By the time the dividing zygote has entered the uterus, it may have something like 50 cells. By 5 or 6 days after fertilization, there are about 200 cells. At this stage, the early embryo, now usually called a **blastocyst**, becomes **implanted** in the uterine wall. Up to this time it has been nourished by a combination of its stored yolk and the secretions of the uterine glands.

Implantation

The hollow sphere of embryonic cells called the blastocyst must become imbedded (implanted) in the wall of the uterus if it is to survive. It begins this process by secreting protein-splitting enzymes that loosen the tissues and allow the blastocyst to burrow into the maternal tissue. As the blastocyst becomes nearly buried, the maternal cells respond by growing around it and completing the implantation. The success of this process depends on the fact that the cells of the uterine wall have been prepared by the hormone changes that have been taking place to respond properly to the actions of the blastocyst. When blastocysts are surgically transplanted from one uterus to another (this has been done in a variety of animals, but not yet in humans), the blastocyst and the uterus must be in corresponding stages of development for implantation to be successful. If they are as little as one day out of synchrony, the proper establishment of the embryo may not occur. When properly implanted, the blastocyst continues to receive nutrition from uterine secretions while it begins to send out tissues that will grow together with parts of the uterine wall into the **placenta**, an organ that will allow a massive exchange of materials between the uterus and the embryo.

Infertility

It is obvious that in such intricate processes as the formation and combination of gametes into a zygote and the successful establishment of the embryo in the uterus, there are many opportunities for accidents, defects, or mistakes in timing to prevent success. It should not be surprising, therefore, that almost 10 percent of married couples are troubled by infertility or low fertility; they are not able to have as many children as they want, when they want them. Before antibiotics were available to treat venereal disease, it was very common for the oviducts to be scarred shut as a result of infection with gonorrhea. Now gonorrhea is usually cured early enough to prevent such scarring, and there is no single condition that explains a large fraction of the cases of infertility. Practically any of the problems that can be imagined contribute to this difficulty in one case or another. Gonadotropins may not be secreted properly; the precursors of the gametes may not respond to hormonal stimulation; the secretions of the accessory glands may not do their job properly; there may be some chemical incompatability between the sperm and the vaginal secretions; the number of sperm reaching the oviducts may be insufficient for fertilization; the timing of intercourse and ovulation may always be just wrong for success; the lifespan of sperm and egg may both be at the low end of the normal range, so that proper timing is difficult to arrange; and so on. The source of the problem cannot be detected by the couple themselves, and even physicians skilled in such work fail to find and cure the problem in a large fraction of the cases. One new therapy that has caused considerable interest is used in the few cases where the woman fails to ovulate because the gonadotropic hormones are insufficient or improperly timed to cause the development and rupture of follicles. Gynecologists have recently been treating this condition with gonadotropins obtained from mares and have had success in causing ovulation with this therapy. Unfortunately, it is difficult to be sure exactly how much of the hormone to administer, and it often happens that too much is given, so that several follicles rupture at once. This can lead to the conception of multiple embryos, sometimes so many that they overcrowd the uterus and are unable to survive. There have been several recent cases of five or even six babies being born as a result of these fertility drugs; they usually do not survive.

For most of the causes of infertility or low fertility, there is little that anyone can do now, and nothing that the couple themselves can do. There is one thing, however, that a couple trying to have a baby could attempt before they resort to professional help. If they will calculate the probable time of ovulation (about fifteen days after the onset of the menstrual flow), they will increase the chances of conception by having intercourse frequently for several subsequent days. However, intercourse more than once a day might decrease the sperm concentrations and thus work against their purpose.

Prevention of pregnancy

Despite the many obstacles to conception, the concern of most couples is to prevent it, or to conceive only at the desired times. Most human cultures have attempted to accomplish this, but effective methods have been available for only about 50 years. New and better methods are being developed, and progress will probably continue, although not as fast as many desire.

Contraceptive methods can be divided into those which prevent the production of gametes, those which adjust the timing of intercourse so that fertilization is unlikely, those that interfere in some mechanical way with the union of sperm and egg, and those that prevent the implantation of the zygote.

The pill

The contraceptive administration of hormones, commonly called "the pill," is one of the newest of the common means of contraception, but it can be argued that it is the most "natural" one. During pregnancy the estrogen and progesterone produced by both the maternal tissues and the placenta have an inhibitory action on the production of FSH (follicle stimulating hormone), so that the growth and maturation of follicles does not occur. The administration of synthetic sex hormones as contraceptives similarly inhibits the production of FSH, and ovarian follicles do not mature. The synthetic hormones administered have much the same effects on the uterus and on the rest of the body as the normally produced estrogen and progesterone would have. The first such contraceptive pills were made with natural hormones obtained from animal sources. Then it was discovered that similar hormones could be chemically produced by treating extracts from some Mexican plants. Practically all the contraceptive hormones marketed have come from a company formed to produce these synthetic hormones from the plant material. The synthetic hormones now used were chosen to have a contraceptive effect at a low dosage, in order to reduce the side effects that have troubled some of the users. The pills are taken, one each day, for 20 or 21 days. Within a few days after the last pill is taken, the menstrual flow begins, much as without the pill, but perhaps lasting for a shorter time. A week after the last pill was taken, the first pill of the next series is used. Most of the drug companies marketing a contraceptive pill package it in such a way that each pill is labeled to tell the day it is to be taken, to minimize the likelihood of forgetting to take a pill each day. The most commonly used type of administration, called the "combined pill," contains a synthetic estrogen and a synthetic progestin (having the same kind of effect as progesterone). The other form of administration, the "sequential pill," attempts to mimic the natural condition a little more closely; for 7 days the pill contains only estrogen, and for the next 14 days it contains both hormones.

The great popular appeal of the contraceptive pill comes from its convenience. Taking the pill is quite separated from lovemaking, and so creates no interference. The previously standard methods, such as the diaphragm and condom, required foresight, interruption, and self-control during a spontaneous act of intercourse, but the pill can be taken at any time of day, as a part of the woman's regular routine. This noninterfering property of the pill contributes to its effectiveness: the user doesn't get "carried away" and decide not to bother with it. A woman who forgets to take a pill is advised to use some other means of contraception for the rest of the month, and she or her husband should follow that advice. But the probability of pregnancy in such a case is really not very high, because the hormones have probably made the stage of development of the uterine wall a little out of phase with the condition necessary for successful implantation. In fact, it is now recognized that the success of the pill in preventing pregnancy is partly due to this out-of-sequence effect on the uterine lining. Because of some unusual interaction, there may occasionally be an ovulation even when the pill is taken faithfully. But if the uterine wall is not in the right part of the secretory phase, the zygote will not implant. This is probably the explanation, too, for the slightly higher failure rate for the sequential pill; the uterine wall is likely to be developing in normal sequence, so the zygotes from occasional stray ovulations are able to implant.

Side effects and dangers of the pill

As popular as the pill is, there is some concern about side effects and dangers, based both on the legitimate, cautious attitude of some physicians and on some irresponsible scare articles in national magazines. There are, indeed, side effects of the pill. Since it is a little like a pseudo-pregnancy, and the pattern of hormone concentrations resembles the pattern during pregnancy, the side effects tend to be those of pregnancy, although much milder. Side effects can be studied properly only with a "double blind" experiment, in which half the subjects received the medication, and half received what is called a "placebo," an identical-appearing pill containing only sugar or other harmless, ineffective ingredients. The experiment must be "double blind" in the sense that only some clerical person knows which patient is receiving the pill being tested and which receives the placebo; both the subjects and the doctors who examine for effectiveness or for side effects are kept "in the dark" until all the data are gathered. Such an investigation can properly be made only when a medicine is new, and when probable benefits and possible dangers just balance each other. So the only good data of that kind on the contraceptive pill come from the 1950's, when the pill was undergoing its early mass trial in Puerto Rico. Both groups of women reported side effects (usually not serious enough for them to drop out of the program). The women actually receiving the hormones reported somewhat fewer of the undesirable side effects and more desirable ones than did those receiving placebos. (Remember that neither group knew which pill they were taking.) It is evident that most of the side effects reported were part of normal variation in the way people feel; the women paid attention to them because they were in a special testing program. On the whole, the pill actually came out ahead in the desirability of the effects actually produced by its use. To be sure, individual women both then and now have genuinely unpleasant reactions to contraceptive pills, as happens with any type of medication. The women who have such unpleasant reactions will want to try another brand, or a sequential instead of a combined pill, or finally choose some other method of contraception. The commonest of the unfavorable side effects reported are a gradual darkening of the skin and a greater tendency to migraine headaches. Allergies may be affected in either direction. Benefits commonly reported include elimination of acne and decrease of menstrual cramps. Weight control, water balance, and sexual desire are sometimes reported to improve, sometimes to worsen, so it is likely that they are usually not really affected.

The most alarming reports about the pill are those that allege that it may cause serious, even fatal, complications. Allegations that it causes cancer seem to be completely unfounded—it may, in fact, make several kinds of cancer less likely. Women who already have cancer, liver malfunctions, or diabetic tendencies are usually not given prescriptions for oral contraceptives.

Of all the serious problems sometimes blamed on the pill, the one danger that seems to be really increased is that of blood clots. Data gathered from thousands of women taking the pill over a number of years indicated that taking a contraceptive drug did increase the minute likelihood of spontaneous blood clots. This disorder is called thromboembolism, and it can be crippling or even fatal. The statistics on which the conclusion was based, however, were obtained from women who were taking the pill in one of its earlier forms. Oral contraceptives now in use contain only a fraction of the quantity of hormone included in the earlier ones and probably have correspondingly fewer side effects, including blood clots.

Among women not taking the pill, about 1 per 100,000 per year may be expected to die from thromboembolism. Among women taking the pill, the statistics indicated a mortality of about 3 per 100,000 per year. (Pregnant women have a similar or even greater risk of thromboembolism.) There is indeed a dramatic increase—threefold—in fatal thromboembolism among pill users. The overall risk of death is not nearly so much increased, however, as these statistics might imply. For women in the childbearing ages of 14 to 44, the overall mortality rate from all causes is about 100 per 100,000 per year. The increased mortality from thromboembolism from the pill represents about a 2 percent increase in risk.

Does this mean that the pill is dangerous to use or safe? How can a woman decide whether or not to use this form of contraception? Garrett Hardin suggests that the best way to look at this small but real risk of death is to compare it with the risk of some alternative action. One of the best alternative methods of birth control is the diaphragm, which has no dangerous direct side effects. The practical contraceptive effectiveness of the diaphragm, however, is much lower than that of the pill. A woman using the diaphragm instead of taking a contraceptive pill faces an added risk from the complications of pregnancy; this risk can be calculated for comparison.

The failure rate (expressed as the number of women getting pregnant per 100 women using the method for 1 year) is one for the pill and ten for the diaphragm. If a woman does become pregnant, whether because she did not use contraceptives or because the contraceptive failed to work, there is a mortality rate from complications of pregnancy and childbirth of 25 per 100,000. The risk of death from pregnancy while using a particular contraceptive is the yearly failure rate times the mortality from pregnancy. The comparative risks can be calculated as follows:

For the pill:
	From thromboembolism			3.0 deaths per 100,000 users per year
	From complications of pregnancy
	(25/100,000 × 1/100)			0.25 deaths per 100,000 users per year
		Total				3.25 deaths per 100,000 users per year

For the diaphragm:
	From thromboembolism			1.0 deaths per 100,000 users per year
	From complications of pregnancy
	(25/100,000 × 10/100)			2.5 deaths per 100,000 users per year
		Total				3.5 deaths per 100,000 users per year

For sexually active women using no contraceptives (assuming that 90% will get pregnant within a year):
	From complications of pregnancy including thromboembolism
	(25/100,000 × 90/100)			22.5 deaths per 100,000 women per year
		Total				22.5 deaths per 100,000 women per year

The calculations show that it is almost exactly as dangerous to use the diaphragm as to use the pill. On the basis of risk to female life, the "danger" of the pill is largely an illusion.

Other contraceptive preparations of hormones

Since they were first developed nearly 20 years ago, contraceptive pills have been modified by the researchers and drug manufacturers to make them even more convenient and to reduce side effects. A major modification still being evaluated is the so-called "minipill," which is a low dosage of progesterone or the equivalent taken every day, without stopping at the end of the month. This pill probably does not prevent ovulation; its effects on the secretion of pituitary hormones are erratic and not well explained. The contraceptive effectiveness of this progesterone pill probably comes from its effect on the uterine lining and on the muscles of the oviduct and uterus. The ovum is propelled down the tubes so rapidly that it is expelled before it has time to be fertilized; or, if fertilized, it does not become implanted. These minipills lack most of the side effects of the standard pill, both the good effects and the bad ones. But effectiveness is not very high; the failure rate is about that of the diaphragm. Other more convenient forms of contraception related to the minipill are being investigated; they are generally some form of once-a-month injection of a slowly absorbed form of progesterone. The effectiveness would probably not be very high, and users subjected to an injection would probably disagree about how convenient it was.

Prevention of spermatogenesis

There have been a number of attempts to find a drug that would suppress spermatogenesis as the pill suppresses ovulation. Several chemicals have shown limited success but have not been adopted because of various serious side effects, or because their effects were not reversible. The normal physiology of spermatogenesis, as it is understood now, does not suggest any normal constituent of the body that could be administered to prevent sperm formation. Attempts at producing a chemical contraceptive for men have therefore not proceeded from a good theoretical understanding of what should work. No one knows when an acceptable method will be developed.

Contraception by timing of intercourse: the rhythm method

Since conception in the human depends so totally on a chance coincidence of ovulation and intercourse it would seem reasonable that conception could be prevented simply by avoiding intercourse at the time of ovulation. This is the basis of the contraceptive method called **rhythm**, or safe period. However, its effectiveness is extremely low, and the rhythm method is disliked for other reasons, too. The only reasons that rhythm is widely practiced are its low cost (usually nothing), and the fact that it is the only contraceptive method allowed by the official statements of the Roman Catholic Church. The basic reason for the ineffectiveness of the rhythm method is that it is so difficult to know when ovulation is going to occur. Few women actually ovulate at the "average" time every month. Intercourse that is supposed to be in the "safe" period may come close enough to the time ovulation actually occurs to allow either sperm or egg (whichever is earlier) to survive until the other becomes available. (Thus the joke: "What do you call people who use the rhythm method?"—"Parents.") Much of the failure rate, too, is due to the considerable restriction that rhythm puts on marital sex. To be reasonably safe, even when a woman has a very regular monthly cycle, the couple must abstain from intercourse for about 8 days around the time of probable ovulation. (This is because ovulation always varies within a day or two, and the gametes may survive for 2 days or so.) If the wife's cycle is a little irregular (which is common), the period of abstinance necessarily increases. Counting the time of menstrual flow, when intercourse may be uncomfortable or unpleasant to many couples, there may be only about a week left during each month when intercourse will not be likely to result in pregnancy. Couples using the rhythm method also complain that, even when the number of days available for intercourse might not fit too badly with the desires of the couple, the idea of impulsive love being tied to the calendar is extremely frustrating.

The reason the Catholic Church accepts the rhythm method seems to be as follows: The purpose of sexual intercourse is producing offspring, and it is perverse and sinful to scheme to have the one without the other. However, it is permissible to have intercourse during early pregnancy and during those parts of the month that turn out to be infertile periods. Further, abstinence from sexual intercourse by mutual consent is not sinful (and, according to some writers, may be virtuous). Therefore, periodic abstinence—rhythm—

is acceptable. The Catholic Church has historically been strongly against the very idea of population control. That attitude seems now to be slowly changing. The specific objections raised against more effective contraceptive measures are that they are "unnatural," whereas rhythm utilizes a time of natural infertility. Many Catholic couples have decided that the sex-by-the-calendar required for "Vatican roulette" is even more unnatural. Young Catholic couples in the United States now use the same kinds of contraceptive methods as their neighbors, in almost the same proportions. It has been claimed that the small number of couples who have found the rhythm method effective for contraception are members of that 10 percent of the population that is infertile, or nearly so, anyway.

Mechanical obstruction to the union of gametes

Several different means may be used to impose a mechanical barrier that will prevent the sperm from reaching the ovum (Fig. 4-2). All work fairly well, none has any dangerous side effects, and all are somewhat inconvenient. Of these, only the diaphragm requires paying for the help of a physician. A diaphragm is a thin, shallow cup of rubber sheeting, supported by an elastic rim. It is inserted into the vagina near the opening of the uterus, and the springy rim holds it tight against the walls of the vagina, preventing the passage of sperm into the uterus. A doctor's help is needed in choosing the right size and shape of diaphragm for each woman. The diaphragm must be inserted before intercourse and left in for several hours afterwards.

The male counterpart of the diaphragm is the condom, a thin sheath of rubber that is placed over the erect penis in preparation for intercourse. The condom has a special advantage for someone with several sexual partners, in that it helps to decrease the spread of venereal disease. (This is sometimes an emotional disadvantage, since some people associate condoms with improper conduct.) Both diaphragms and condoms have a slight risk of being misplaced during vigorous intercourse. The other effective mechanical methods

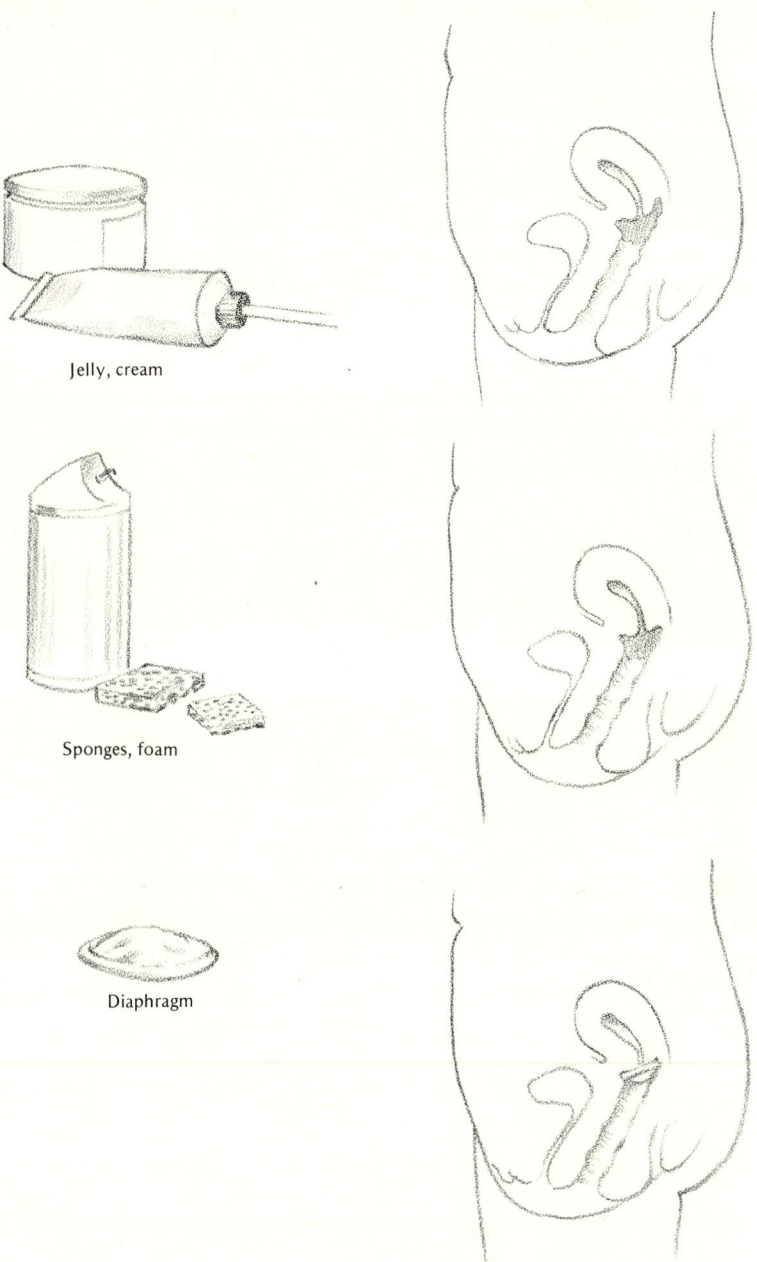

Fig. 4–2: A. Mechanical means used by women for preventing conception shown, on the left, as purchased, on the right, in place for use.

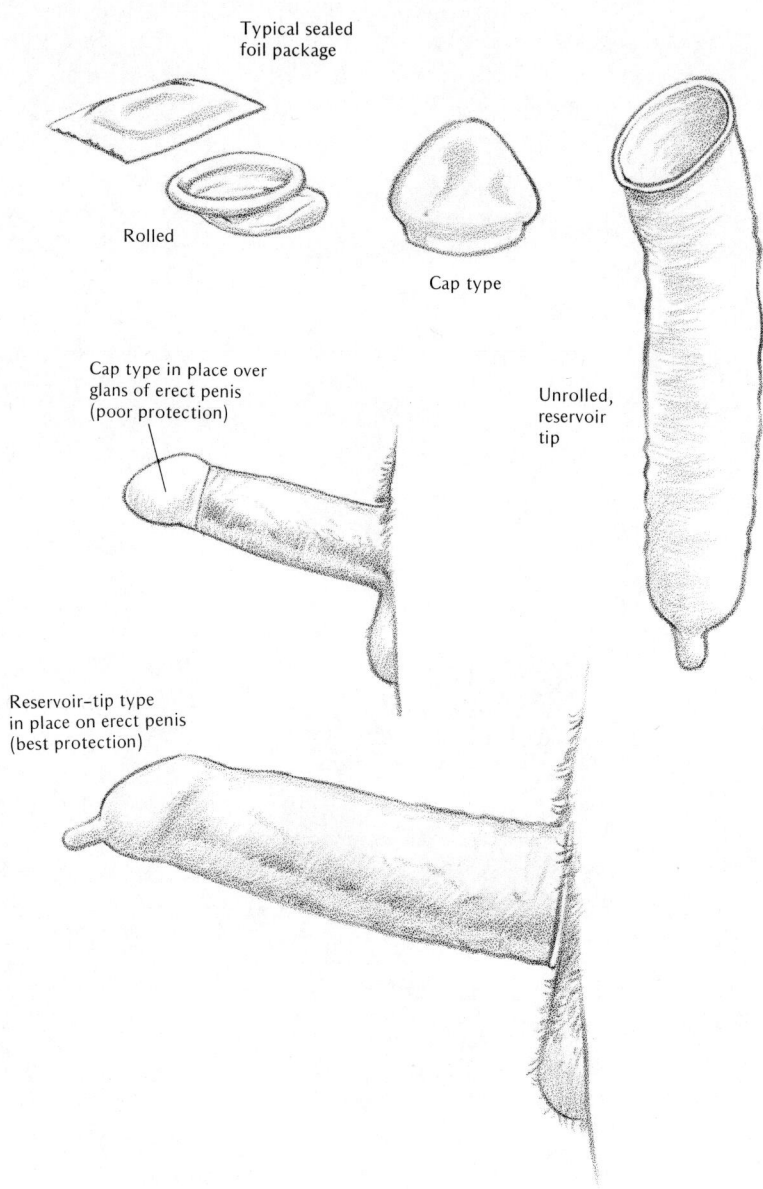

Fig. 4–2: B. The mechanical means of preventing conception used by men. The condom ("rubber") is also used to prevent transfer of venereal disease.

of contraception are jellies, foams, and suppositories that are inserted into the vagina before intercourse and which sperm are unable to penetrate. Some of these contain compounds that immobilize the sperm, as well. For increased effectiveness, one of these chemical barriers should be used along with the diaphragm or condom. All these effective mechanical methods require that the couple interrupt their mutual fascination and rising excitement to "remember" to use a contraceptive. It is easy to get carried away and decide to take a chance. This accounts for many of the failures of these methods to prevent conception. Condoms, in addition, are reported by some men to decrease the pleasurable sensations of lovemaking. All these methods require that some product be available and on hand every time a couple has intercourse, and this may not always be the case.

Ancient history: withdrawal

The oldest recorded contraceptive method is withdrawal of the penis just before ejaculation so that no sperm enter the vagina. The Bible, in the Book of Genesis, reports that a man named Onan "spilled his seed on the ground" when he did not want to conceive a child by the woman he was with. Withdrawal is probably the contraceptive method still used by more people the world over than any other. It is also one of the least effective methods. This is because some semen often escapes before noticeable ejaculation, and because it is very difficult for a man in the midst of such an emotional process to be sure to act at just the necessary time. This method deserves to be historical only.

Permanent obstructions

The most dependable method of preventing conception, except total abstinence from sexual intercourse, is to tie off, surgically, the ducts that convey the gametes to a place of fertilization (Fig. 4-3). In a man the duct is the vas deferens (between the epididymis and the seminal vesicle), and the operation is called a **vasectomy**. In a woman it is the oviduct,

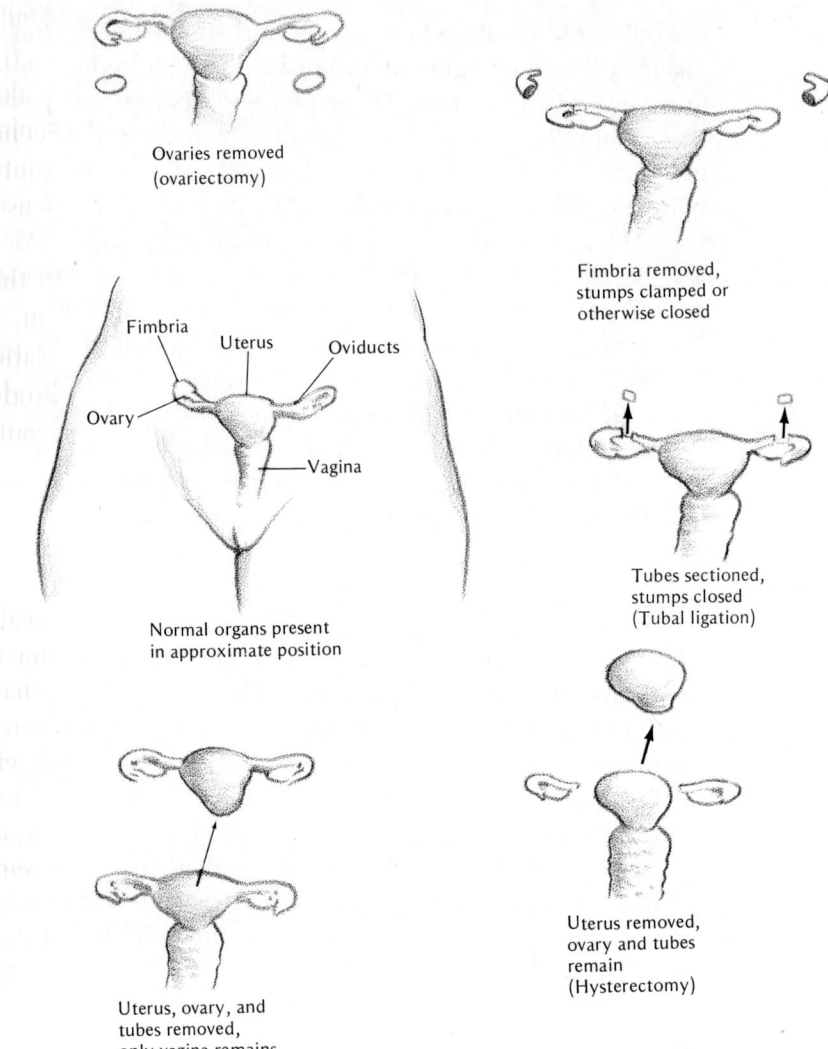

Fig. 4–3: A. Operations for sterilization of the female.

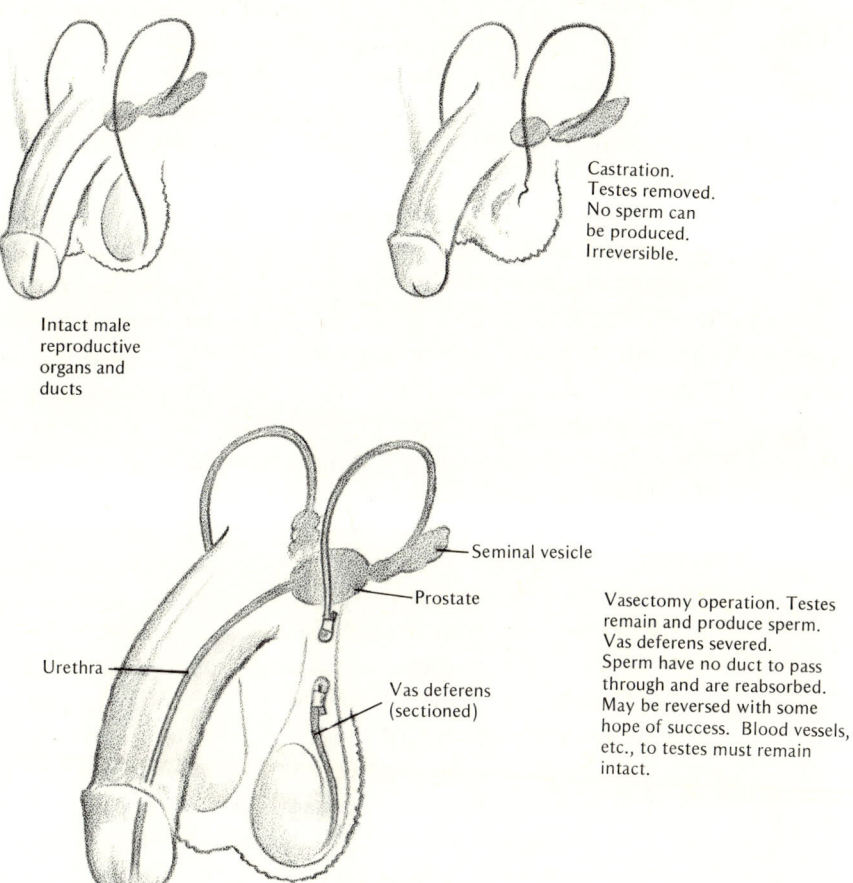

Fig. 4–3: B. Operations for sterilization of the male.

sometimes called the Fallopian tube; the operation is called a **tubal ligation**. In both cases the procedure is to expose the duct, cut it, and tie off each end. A vasectomy is simple, because the vasa (the two vas ducts) are easily accessible in the upper part of the scrotum. Two small incisions are made under local anesthesia; the operation is usually performed in the doctor's office. The man will probably not feel like working for a day or so, but complications are rare and minor. Since sperm may be stored in various parts of the ducts, it is necessary for the man to use other types of contraception for several weeks; after that time, no other method of contraception need be used again. The operation for sterilizing a woman is more difficult because it is necessary to cut through the abdominal wall. In the past, it has been desirable to perform tubal ligations just after childbirth, because the uterus with the attached oviducts is then large and easy to find. Recently, it has become possible to tie off the tubes with only two very small incisions, and with a special viewing device to guide the procedure. It is still hospital surgery, often with the use of general anesthesia, but further technical advances may make the operation even more easily done.

It is not the difficulty or pain of the operation, since these are slight, that are the major obstacles to sterilization of either sex—it is the permanence of the change. It is possible for a skilled surgeon to sew together the cut ends if a person wants to be fertile again, but the duct often does not re-form properly, and even when there is a tube again, it may not work well. Reversal rates for sterilization surgery range from 50 percent down. The unfavorable side effects of each operation are very rare, and they seem to be entirely psychological. Some people worry that they might change their minds; others feel that they are maimed and less of a man or a woman. A tubal ligation has no effect on hormone cycles, vaginal cycles, or uterine cycles. A vasectomy does not change the ability for intercourse—the volume of ejaculated fluid is not reduced significantly, and hormones are produced as before. A few years ago it was very difficult in most places to get a vasectomy; it required the consent of a whole panel of doctors, who would agree only if there were four or more children. Now the operation is becoming more common every year. One survey recently indicated that vasectomy and tubal ligation combined represent the means of contraception for a larger group of couples in the United States than does any other single method. Many doctors and most non-Catholic hospitals will now perform a sterilization operation if the patient appears to know what he is doing, and if the spouse signs an agreement. The failure rate of these methods is almost zero; only a few unsuccessful operations have come to notice. For someone who is confident that "this many children is enough," it is the simplest and best contraceptive. Many researchers are trying to develop reversible vasectomies and ligations, but their success is not yet assured. One promising approach would involve dissolvable plugs, or even valves.

The IUD: Why does it work?

An old contraceptive method that has been recently improved is the **intrauterine device**—the IUD. This is a small piece of flexible plastic, or sometimes a piece of durable wire spring, that is placed in the uterus and thereafter prevents pregnancy. IUD's come in many shapes and materials (Fig. 4-4). This variety is found because no one has any good idea of what would be the best shape, since no one knows how the device acts to prevent pregnancy. The best guess is that, like a few of the new hormone preparations, it upsets the muscular activity and the lining of the uterus just enough to expel the ovum before fertilization, or to prevent implantation even if fertilization does occur. The major problem with the IUD seems to be that although once inserted it is supposed to stay indefinitely, an occasional one comes out without being noticed. The newest types of IUD seem less likely to be dislodged. Side effects are few; some women experience cramping and heavy menstrual flows for a few months after insertion, and in a few the IUD must be removed because of these effects. The presence of such a strange foreign body permanently in place might be expected to cause ulcers, inflammation, or even cancer, but there has been amazingly

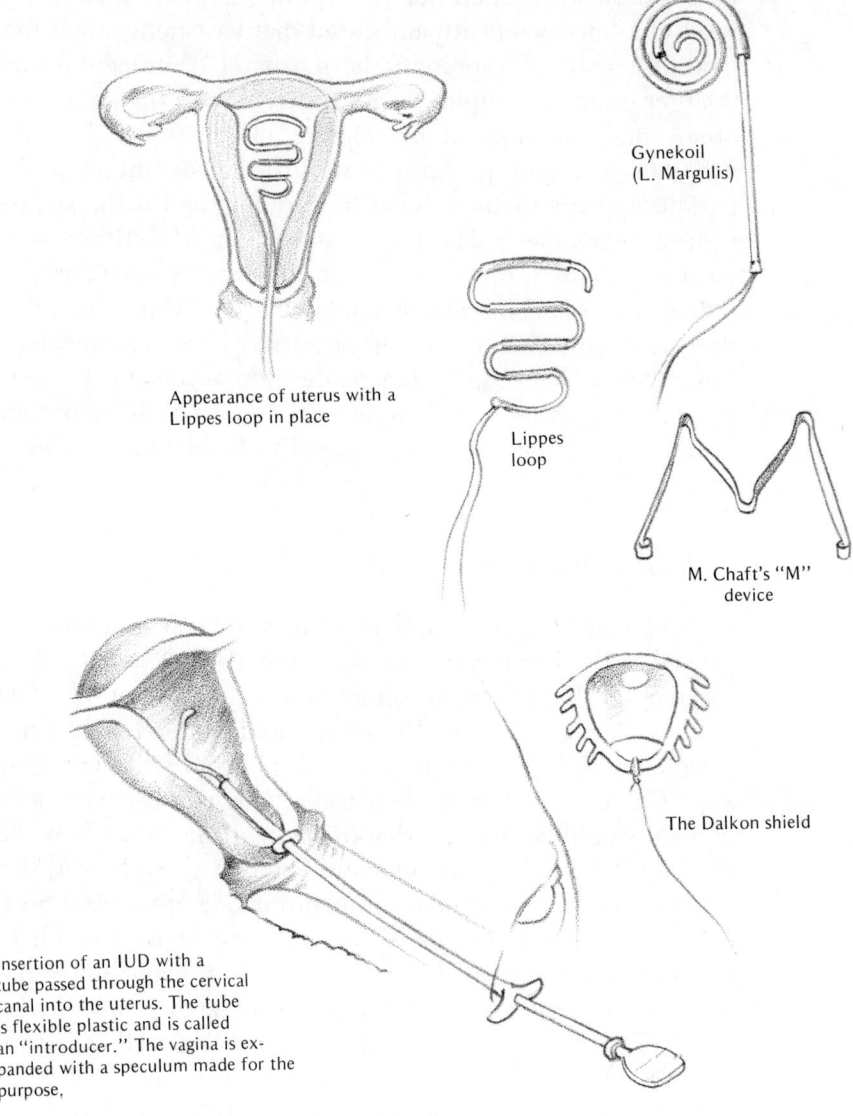

Fig. 4–4: Various intrauterine devices for contraception.

little evidence of any of these. It is reasonable to suppose that IUD's will continue to be improved, especially if their mode of action becomes understood. Their popularity is growing.

When it's almost too late

Several contraceptive methods under study would not require preparation or caution before intercourse but would prevent pregnancy by action afterward. Given the impulsive tendencies of humans, they are probably needed. For some time, victims of rape have been treated with large doses of synthetic estrogen to prevent implantation. The procedure has not always worked, however, and has had some unfavorable side effects. More suitable synthetic hormones and schemes of treatment are being devised. There is some probability of a "morning after" pill, not to be used as an emergency measure but as the standard contraceptive, to be taken the morning after intercourse. Another possibility is the once-a-month pill, which would be taken at the time the menstrual flow was expected. If there had been no zygote, it would have no effect. If there were a blastocyst implanted, menstruation would occur anyway, and the blastocyst would pass out with the cast-off uterine lining. Such a pill might be a chemical relative of progesterone that would block its action. Still other new methods are being considered, but none of them is very close to actual development and use. Research and development in this field is slow, since everything must be checked both for effectiveness and absolute safety.

Which method is best?

When comparing the various contraceptive methods, it is necessary, for most methods, to take into account the failure rate both under ideal conditions and under conditions of normal use. Most contraceptives depend on the dependability of the user. The approximate figures shown in Table 4-1 give the probable number of pregnancies in a group of 100 married women using the method for a year. Reported failure

Table 4-1. Comparison of effectiveness of contraceptive methods

Method	Birth rate with ideal use	Birth rate with actual use
Vasectomy or tubal ligation	0	0
Oral contraceptive (pill)	0.1	1
IUD	1	2
Diaphragm	1	10
Condom	2	15
Foams or jellies	4	10
Rhythm	10	40
Withdrawal	15	30
None	90	90

rates for the same contraceptives vary greatly from study to study. The figures given here may be quite different from those given elsewhere. No one knows with certainty which are right.

The population explosion

Contraceptives are important for more than the convenience and well-being of individual couples. Widespread control of births is the only action that can possibly avert the tragic consequences of the population explosion—one of the most serious of the threats that presently confront the human race.

When considered on the time scale of known human existence, the population growth rate in the last few generations is seen to be totally unprecedented and definitely abnormal. It took several hundred thousand years for the human population to reach 1 billion—in about 1800 A.D. It reached 2 billion by 1920, and it is now 3 billion (Fig. 4-5). Whereas the total world population must have been relatively stable for thousands of years, it probably doubled in the 200 years from 1400 to 1600. It doubled again in the 100 years of the 1800's, and now the doubling time is 40 or 50 years. This doubling was not just the *addition* each time of numbers equal to the population in 1400. Rather, in each doubling period the population increased as much as in all

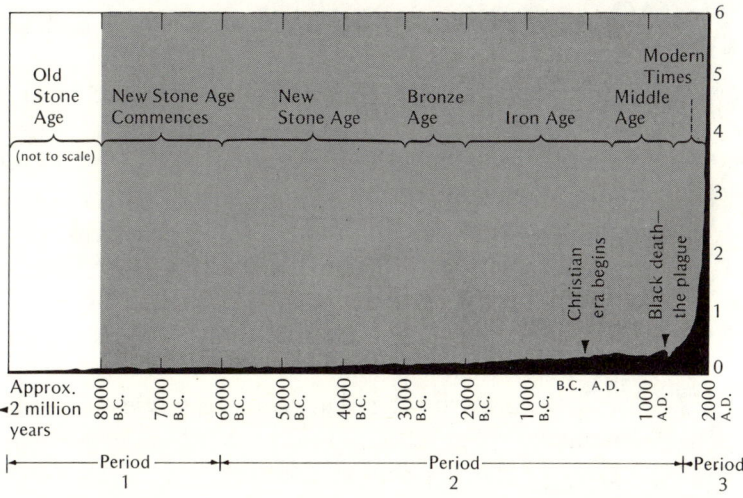

Fig. 4–5: Population through history (*Population Bulletin*, 1962, Vol. 18, No. 1.)

previous history. In 50 years our population seems likely to grow from the present 3 billion to 6 billion people on this finite earth.

The reason for this explosive growth rate in modern times is easy to understand, and it is quite ironic. Humans multiply slowly compared to most other mammals, since they usually have no more than one child a year, and there is a wait of 15 or 20 years before sexual maturity is reached. The species has continued only because humans have always been able to raise to maturity a larger proportion of their children than most mammals do. But in human terms, the noticeable thing once was that almost any family could expect to lose several of their children—always a tragic occurrence, even though not unusual. Now, in contrast, we expect that all our children will survive to maturity without having illnesses serious enough to threaten their lives. This is the most unquestionable advance in human happiness that can be credited to modern civilization. It can easily be doubted that cars, lawnmowers, and TV are better and make for a happier life than horses, goats, and ceremonial dances. But to be spared the tragedy of seeing your children die is a real gain, not an illusion.

It is just this greatest human advance that is responsible for the great problem of the population explosion. Relatively simple public health measures have reduced infant mortality and early death in most societies. Children survive to maturity and have children of their own. However, most of these societies still have the birthrates that were suitable for family survival when many of their children died. Only easily available contraceptives, and the will to use them, can solve the population problem.

Even growth rates that seem insignificant can lead to phenomenal population growth: With a 2 percent per year growth, the population doubles in 35 years. We are finally beginning to realize that we live in a limited environment—"Spaceship Earth." Growth of the human species, no matter how slow, must stop sometime. The longer it takes to reach that stopping point, the worse off we will be. Why not stop soon?

The price we pay for the exploding population differs according to the part of the world in which it occurs. In the underdeveloped countries, any agricultural, educational, and technical gains are reduced and often negated by an even faster population growth. Overpopulation causes dependence of these countries on the developed countries for emergency aid, since any crop failure can mean immediate famine. Where there are poor, hungry people, there is political instability; unstable governments are ineffective in promoting progress.

Developed countries are paying a direct price for their own population growth in their unstable economies, the costs of their government programs, and the necessity for changing life styles to fit the crowded scene. The whole world suffers from the pollution of the environment, and from the excessive use of irreplaceable natural resources. By trying (whether successfully or not) to feed a growing number of people, man uses more and more fertilizers and pesticides, which pollute the water and upset the balance of nature.

Technology, by itself, will not solve this problem. There is no magic scientific method that will make the difficulties go away without our having to worry much about them. People will have to change their ways of doing things to ways that are more appropriate for the situation that actually exists. Specifically, birth control programs, in the form of family planning, will have to be introduced and used almost universally as an essential step in the control of population. Hoping for voluntary action will not be enough; it will not happen fast enough in most of the world. Some kind of collective (governmental) action will be necessary. An essential first step is educating the public about the general population problem, the benefits of family limitation for the family itself, and the means of family planning. But governments need to go far beyond this. Sterilization and other contraceptive devices should be legalized and made readily available and free on a worldwide basis. Monetary incentives for sterilization and for birth control should be considered. Such an approach—helping families to limit the number of children they have to those they actually want, and applying gentle pressure to change the number of children they decide to have—will go a long ways toward alleviating the problem. Already the rate of population growth in the United States has slowed dramatically because couples have decided, for a number of reasons, that they will be happier with fewer children and that they can control their own destinies in this respect. The governments and the inhabitants of many overpopulated countries, such as India, desire to control their population growth and would be eager to accept more birth control aid if we would only give it. Voluntary and incentive programs will work wonders.

But will voluntary birth control programs solve the problem forever? This is a much tougher question. Suppose that most couples in this country decided to limit their families to replacement numbers (two children per family), for the well-being of the children themselves. Suppose further that those few who are not concerned about children's welfare had no children, because they realized this would save them a great deal of trouble. We would still be left with several groups that believe positively in large families and population growth. These groups would include the Hutterites, the Amish, the Mormons, and some conservative Roman Catholics. Such groups generally cause no direct social problems; most of them do a very good job of looking after their children and helping one another. But in a very few generations during which most of us had families of two children while Hutterite couples each had ten to fifteen, a significant proportion of the population would be Hutterites. Where would our stable population through voluntary action be then? Would we have to pass restrictive legislation against those genuinely virtuous people—for example, requiring their sterilization after two children. Surely we would want to do nothing like that! But what *would* we do? Perhaps that won't be our problem; maybe we can just help the people who want to limit their families to do so easily, and leave the splinter groups for our grandchildren to worry about. The problem might look different to them.

We may have the same problem much sooner on the world scene. Some countries of the world that have exploding populations are not yet ready to take serious steps to limit this growth. Yet those countries may reach famine conditions very soon. If we send food to keep people from starving, won't there be twice as many to starve in the next generation? We may not have extra food for the next generation. Won't we, by our lack of action, have caused the starvation of those extra people? Is this moral? Maybe we should consider insisting that when we give aid to any overpopulated country, the contraceptive aid should at least equal in dollar value (or maybe even in tonnage!) the food and economic aid that we send. How can we require such a thing of other countries if we are not applying strict measures to limit population here at home? Contraceptives and other aids from science and technology are just the means; societies must implement them. The following is a proposed program for dealing with the population explosion.

National program
1. Subsidize massive research in contraceptive technology and possible population programs.

2. Conduct massive education and propaganda campaigns in favor of smaller families.
3. Provide free contraceptive information, advice, and supplies to anyone who requests it; provide free sterilizations on request. Offer subsidies to groups such as Planned Parenthood.
4. Legalize and improve abortion as a "backup" measure for contraception.
5. Offer a subsidy of $1000 to anyone between the ages of 20 and 50 who will be sterilized.
6. Eliminate income tax deductions after two natural children; institute a positive tax on more than three children, with sliding scale for income and number.

The following proposals for a national program are worthy of debate, which might produce more satisfactory ideas:

7. Require compulsory sterilization of any mother on welfare who has two children.
8. Require compulsory sterilization of *any* mother who has two natural children.
9. Offer incentive payments for late marriage, delayed childbearing, and small families.

International program

10. Offer massive contraceptive aid (on the general financial level of recent military aid) to any overpopulated country.
11. Insist that to any overpopulated country, the contraceptive aid at least equal (in dollar value, or even better, tonnage) the food and economic aid.
12. Encourage private agencies such as churches to change their relief budgets to make population-control aid at least equal to food aid.

5 The early embryo

The trophoblast

When the shapeless mass of cells in the early embryo begins to assume a definite form, the first structures formed are those necessary for obtaining nutrition. The 200-cell blastocyst that becomes implanted in the uterine wall has become a hollow sphere made up of small cells. At one place on this sphere there is a larger mass of cells that is several layers thick. The thin shell is called the **trophoblast**; the thicker portion is the **inner cell mass** (Fig. 5-1). "Trophoblast" is a Greek word meaning an embryonic layer that nourishes; its function is to absorb nutrients from the surrounding uterine tissue. (It is sometimes called the **chorion**.) The trophoblast will develop into the placenta, which at birth will weigh about $\frac{1}{8}$ as much as the fetus.

The external membranes

The inner cell mass will eventually give rise to the embryo proper, but the first developments from it are two additional temporary structures useful during embryonic life only. Some of the cells in this inner cell mass seem to change a little in appearance, so that those next to the cavity of the blastocyst are distinguishable from those closer to the outside. The cells adjacent to the cavity are now called **endoderm** (meaning inner layer) while those toward the outside

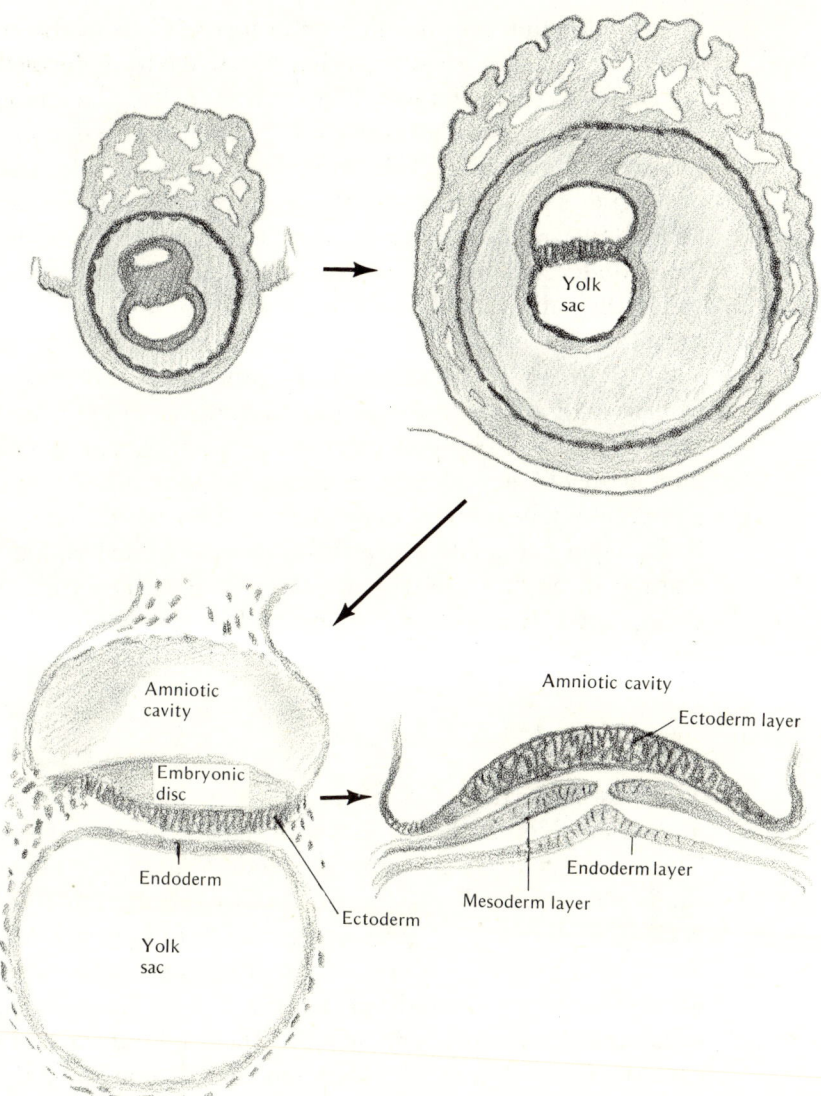

Fig. 5–1: Development of the early embryo: formation of the primary germ layers.

are called **ectoderm** (meaning outer layer). Cells in the ectoderm begin to move apart so that a new cavity is formed in the center of that cell mass. This cavity is called the **amniotic cavity**, and the cells that surround it will become a membrane known as the **amnion**. The amnion will later be a fluid-filled protective sac surrounding the embryo. After the amnion forms, still another cavity appears, this time in the midst of the endoderm. This new cavity and its surrounding cells are called the **yolk sac**. (The yolk sac is a relic of our evolutionary ancestors the reptiles, who have—like birds—eggs with large yolks. Part of reptile and bird embryology involves the formation of a layer, the yolk sac, that extracts food from the large yolk mass. The yolk sac became a part of the developmental machinery of our ancestors, in that other parts of embryonic development depended on it for more than nutrition. When mammals evolved the process of nourishing the embryo through a uterus and placenta, the large yolk mass disappeared. Human embryonic development retains a small but recognizable yolk sac for its other functions.)

Primary germ layers

The most important parts of the ectoderm and the endoderm are not involved in the formation of the amnion and yolk sac. These two layers, ectoderm and endoderm, lying between the amnion and yolk sac, are the embryo itself. Some of the cells of the embryo, at the junction of ectoderm and endoderm, change slightly into still another cell type, making what is to be the **mesoderm** (middle layer). The mesoderm cells multiply rapidly, and some of them migrate out of the embryo proper to form a covering for the amnion and a lining for the trophoblast. The three layers, ectoderm, mesoderm, and endoderm, are called the three **primary germ layers**, and certain definite parts of the later fetus can be traced to each of them. The endoderm gives rise to the absorptive cells lining the intestinal tract; the ectoderm forms the outer layer of the skin (among other structures); and the mesoderm forms the bulk of the body, including the circulatory system and all the supportive structures such as bone and muscle.

A starting point in common

The elaborate sequence of events just described seems almost random and pointless to anyone studying it for the first time. The structures formed as described thus far give no hint of the final product. Yet, what has happened sets the stage for everything that is to occur later. It is the standard way for all vertebrate embryos to start. The major variations from species to species are in the shapes of the various sacs formed, the **extraembryonic membranes** (trophoblast/chorion, amnion, yolk sac). These structures differ considerably from fish to frogs to reptiles according to the size of the eggs. Even among mammals the process by which the chorion and amnion are formed is so amazingly variable that we may even be mistaken in using the same terms in all species.

The primary germ layers of the embryo itself are not so variable. Even though the steps by which the three germ layers form and develop may depend a great deal on the amount of yolk in the egg, the final arrangement of the three layers in this early embryo is amazingly similar in mammals, birds, reptiles, amphibians, and fish. The three-layer stage is a common starting point for all. Events leading up to it differ, and the embryos to follow will slowly become increasingly distinctive, but this stage is the same in all vertebrates.

Formation of the placenta

The trophoblast of the blastocyst sends out root-like projections into the surrounding uterine tissue. These projections digest their way in among the uterine cells just as the whole blastocyst did earlier. As heart and blood vessels develop in the mesoderm layer of the embryo, they project into the trophoblastic extensions (sometimes called chorionic villi). This brings the blood vessels of the embryo into close contact with the blood vessels of the uterus (Fig. 5-2). The blood of the embryo and of the woman bearing it *do not* intermingle; but there is an all-important exchange of materials between them. Cords of embryonic tissue, containing blood vessels, are bathed in pools of uterine blood. Any materials

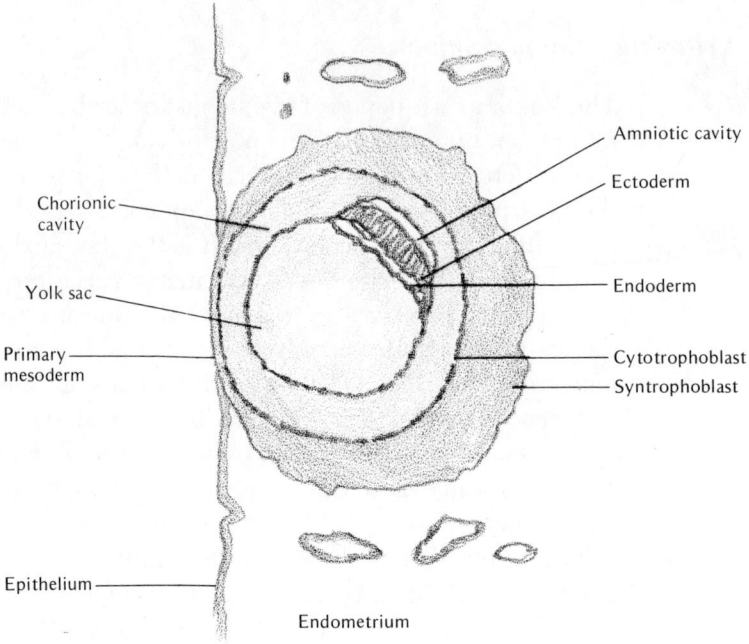

Fig. 5–2: The early embryo just after implantation in the uterine wall.

must pass through two or three layers of embryonic cells to move from one circulation into the other.

As the embryo grows, the projections of the trophoblast become concentrated in a disc-shaped area on one side—projections here elongate, branch, and increase in number, while the projections from the surface elsewhere on the trophoblast fail to increase in size and eventually disappear. This concentrated organ of combined embryonic and uterine tissues is called the **placenta** (Fig. 5-3).

Placental function

The primary function of the placenta is simple exchange of materials. As nutrient materials in the embryonic circulation are used up, they diffuse from the uterine capillaries into the capillaries of the embryonic part of the placenta. At the same time, waste materials that build up in the embryonic blood diffuse out.

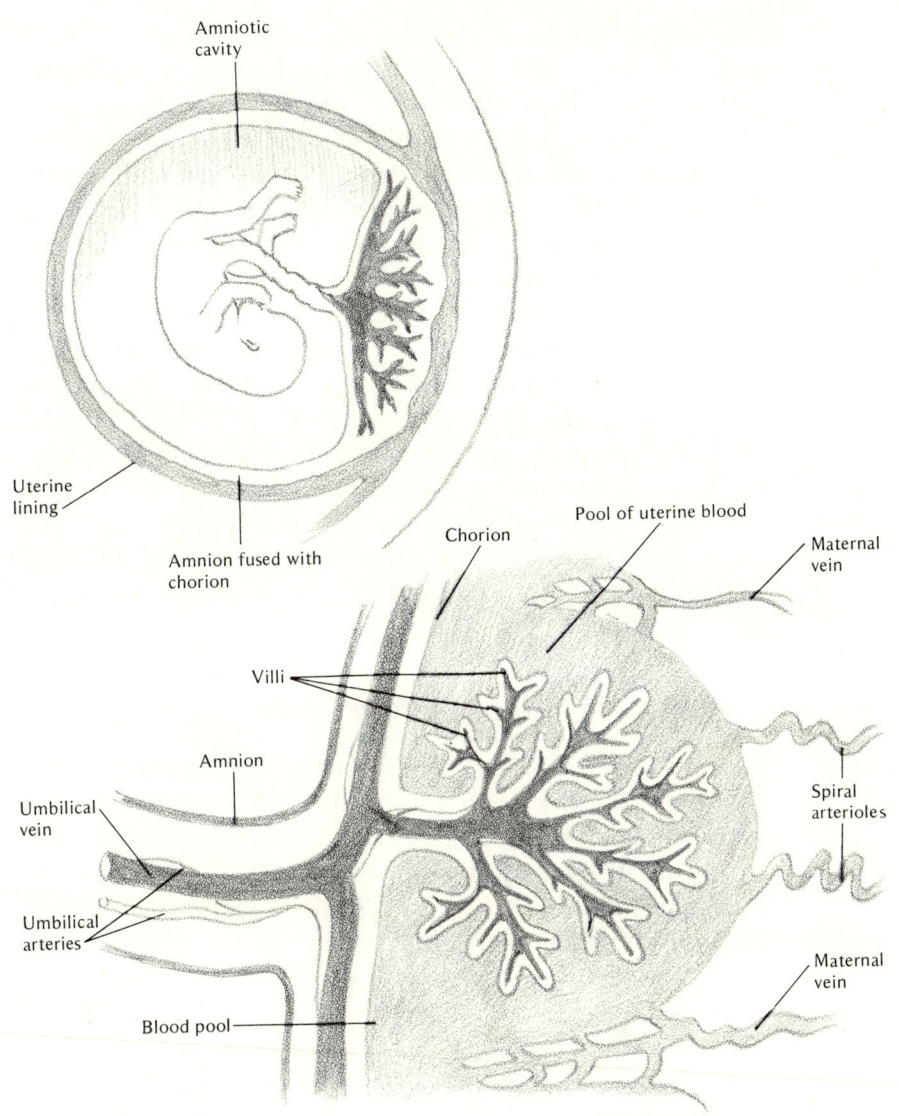

Fig. 5–3: Later in embryonic life the fetus is sustained by its connection to the maternal uterine circulation through the placenta and umbilical cord.

A second function of the placenta, less well understood, is control of some of the materials passing from one set of blood vessels to the other. Some of this passage of materials goes on faster or slower than might be predicted from the size and nature of the materials. It thus appears that the cells that separate the two circulations prevent the passage of some materials and actively move other materials in one direction or the other.

A third vital function of the placenta is as an endocrine gland, secreting hormones. In humans and other primates, and in a few other species that have been studied, the trophoblast secretes, almost at the very first, a hormone (called **chorionic gonadotropin**) that sustains the corpus luteum, which in turn secretes progesterone to support the maintenance of the uterine lining so that menstruation does not occur. Later in the life of the embryo, the placenta itself secretes progesterone and a variety of other hormones, many not well studied.

Fetal-maternal relationship through the placenta

In many ways, the fetus acts as a parasite, existing at the expense of the pregnant woman. Although the uterus must ordinarily be in the proper state for implantation and formation of the placenta to take place, the entire process is primarily an action of the embryo itself. There are occasional instances in which the ovum fails to enter the oviduct when it is released from the follicle, and is fertilized in the abdominal cavity. Or the zygote may become lodged in the oviduct and grow to such a size that it ruptures the oviduct and is deposited in the abdomen. In either case the embryo attaches to some abdominal organ, forming a placenta there. In almost every such case, there is insufficient blood supply available to support development past the third or fourth month, and the embryo dies and must be removed surgically. In rare instances the embryo happens to attach to the rich blood supply of the intestines and grows to maturity. It will then, of course, need to be delivered by abdominal operation. It is clear from the evidence of this abnormal kind of happening that the formation of a placental attachment may be helped by a receptive uterine lining but does not completely depend on it.

Interaction between woman and fetus

The embryo can be thought of as a "do-it-itself-kit." The uterus is simply a warm, protected source of nutrient that permits the embryo to grow. A mammalian embryo is nearly as independent of outside influences as is a developing bird in an egg. The embryo is such an efficient parasite that it is normally assured of a sufficient supply of nutrients, even at the expense of the woman's health. (This is the result of the active transport of chemicals by the cells of the placenta from the uterine blood supply into the embryonic vessels.) The function of the uterus, as indicated, is largely a supportive one. It is complicated by the fact that some hormones probably pass between the two circulations. The many hormones produced by the placenta are secreted by the layer of cells in contact with the uterine blood; they must pass through two layers of cells to reach the embryonic circulation. Because the embryonic blood volume is so much smaller, the placental hormone concentrations are probably about equal in the two circulations. These hormones have important effects on both the development of the embryo and the physiological adjustment of a woman to her pregnancy. Specific influences of the woman on the fetus are few, but the pregnant woman's moods may have some effect on the embryo, by affecting levels of certain hormones (such as adrenalin). However, there is no nervous connection between the two systems, and any complicated nervous influences are impossible.

Dangerous influences

In most cases, the uterine environment is satisfactory for embryonic growth. A number of circumstances, however, can cause injury to the developing embryo. If the pregnant woman is subjected to actual starvation, the embryo may be aborted. In less extreme cases, the fetus seems surprisingly resistant to poor nutrition.

Oxygen deprivation may be an important danger to the embryo; women with emphysema or asthma may have an increased likelihood of bearing injured offspring. Still, such relationships have been hard to establish clearly. Other dangerous influences, to be discussed in the next chapter, are certain drugs and a few viral infections.

The amnion

As the embryo grows larger, the yolk sac does not keep pace and becomes less prominent by comparison. The amnion, in contrast, grows at the same rate as the embryo or even a little faster. Meanwhile, the shape of the embryo changes so that its attachment to the placental disc elongates to become the **umbilical cord**, which contains principally the large blood vessels carrying fetal blood to and from the placenta. Embryonic shape changes so that the embryo comes to be encased within the amnion, which now constitutes a fluid-filled covering around it. The fluid-filled amnion serves an obvious protective function—the embryo floats in a fluid bath and is little affected by jerks and jars. After several months of fetal life, the space between the amnion and the trophoblast/chorion is obliterated, and the amnion fuses with the chorion.

Amniocentesis

There is now a procedure for studying the embryo while it is still developing within the uterus. There are always a few cells that are dislodged from the outer surface of the embryo or from the inner surface of the amnion; these cells remain alive for some days, floating in the amniotic fluid. In a procedure known as **amniocentesis**, a needle is inserted through the vagina or the abdominal wall into the amniotic sac, and about 20 ml of fluid is withdrawn (Fig. 5-4). The cells in this fluid can either be studied directly, or they can be induced to grow and divide in tissue cultures. Any fetal

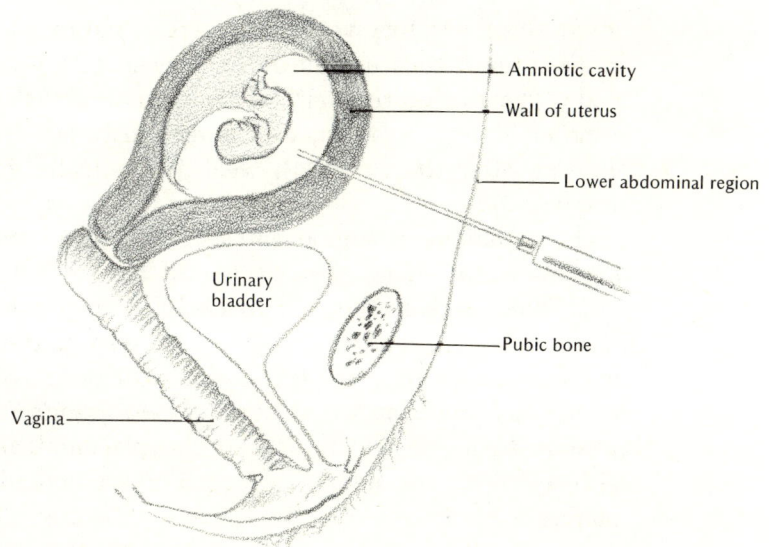

Fig. 5–4: Amniocentesis, a procedure for sampling the fluid that surrounds the fetus.

disorder that can be diagnosed from random single cells can be studied by this technique. It is a new method—about 500 tests had been done up to 1971, with no noticeable injury to either the pregnant woman or the fetus. Since that time it has become much more common, and the incidence of complications is still low. Early in pregnancy the procedure requires vaginal puncture, since the uterus is then surrounded by the pelvic bone. The risk of infection is somewhat greater then than it is later in pregnancy, when the enlarged uterus projects higher in the abdomen and puncture can be done more simply through the abdominal wall. The volume of amniotic fluid increases as the embryo grows, adding to the greater ease of the procedure at later embryonic age. At 15 weeks there is 150 ml of amniotic fluid, and it is fairly easy to withdraw some. Improvements in the technique are also improving the feasibility of performance of amniocentesis early in pregnancy.

The problem involved in amniocentesis when the embryo is advanced enough to make the procedure easy (15 weeks or more) is that 3 to 4 weeks may be required to grow enough amniotic fluid cells in culture for the necessary tests to be made. This pushes the date for therapeutic abortion, in case a serious defect is found, to a dangerously late time, since abortion after the twentieth week is medically very undesirable.

The technique of amniocentesis is still very much in the developmental stage, since it is still not known just how much fluid is really needed to enable the best procedures for culturing to be used. It will be necessary to do still more investigation to get a realistic estimate of the risks involved for various procedures and times of performance. It is now possible to locate the fetus and the placental attachment with an ultrasonic device (which works much like radar), making it easier to avoid the fetus when doing a puncture. If desired, skin samples can be taken from the fetal surface or samples of fetal blood from the placental vessels. These different sampling techniques make different kinds of diagnostic study possible.

Chromosome studies

The best-developed application of amniocentesis is in the prenatal diagnosis of chromosomal abnormalities, especially trisomy 21 (Down's syndrome). Cells from the amniotic fluid are grown in tissue culture and then killed and stained during mitotic division. From photomicrographs of these preparations it is possible to count the chromosomes, and, for more accurate diagnosis, to arrange the photographs of all the individual chromosomes into rows and groups according to size and shape. (Such an arrangement is called a **karyotype**.) A karyotype allows the identification of individual chromosomes. If a woman is a carrier of the rare translocation form of Down's syndrome, it is possible to detect from a karyotype whether or not a fetus has Down's syndrome. If the prospective parents are willing to have therapeutic abortion in case an abnormality is detected, a woman who is a translocation carrier can have children without having any with Down's syndrome. (Whether or not a fetus that is itself a translocation carrier should also be aborted is a subject of disagreement.)

Routine screening with amniocentesis

The probability of chromosome abnormalities is enormously greater in children conceived by older women, over 40. Routine screening of such pregnancies to detect fetuses with trisomies is wise, on the basis of family happiness and even of public financial policy. Assuming that amniocentesis continues to be a safe procedure when done routinely, it is possible to make some estimates of its financial advantages as a public health measure.

At least 1 percent of the offspring of women over 40 are afflicted with trisomy 21. Amniocentesis done on 100 such pregnancies would be expected to diagnose one afflicted fetus, on the average, which could then be aborted. Amniocentesis plus a karyotype costs about $200, while a therapeutic abortion (somewhat late in the pregnancy) might cost about $500. Thus the total cost of avoiding one case of trisomy 21 would be 100 × $200, plus $500, for a total of $20,500. This sum can be compared to the estimate that a person afflicted with the mental incapacity of Down's syndrome costs the state an average of $60,000 during his lifetime, in addition to the expense and personal tragedy for the family. It would pay for the state to do such screening free of charge for all older pregnant women who would be willing to choose abortion if a fetal abnormality were found.

Other chromosome abnormalities

A very considerable problem of judgment would arise if, during a program of screening for trisomy 21, some other trisomy should be found. Trisomies of other autosomal chromosomes would be no problem, because the deformities are always great, and the fetus would be unlikely to survive to birth. But the commonest of the other trisomies that might be found in a several-months-old fetus are those involving the sex chromosomes. These trisomies cause such mild conditions that it is likely that a child born could lead a normal life. The monosomy (XO, Turner's syndrome) causes sterility and poorly developed ovaries, but the girl is usually fairly normal in appearance and within the normal range of intelligence. The XXY and XYY trisomies both produce men at the taller end of the normal range in height, but this is no handicap. A man with XXY is sterile and of somewhat feminine appearance, but not more so than many men with normal chromosomes. The XYY man is fertile and does not seem to pass on his chromosome abnormality. Either of these two conditions seem to make the man several times more likely to end up in a prison or institution for the retarded than the average. But even so, the great majority of such men lead a normal life. Should fetuses with sex-chromosome abnormalities be aborted, then? If not, should the parents even be told of the abnormality? (They might treat the child differently and cause trouble by such treatment.) These questions are now being debated.

Single-gene defects

A much wider application of the technique of amniocentesis is developing, as there is now the possibility of diagnosis of single-gene defects. In order to understand why it is even possible to detect a single abnormal gene by studying the cast-off cells of an embryo, it is necessary to consider what a gene is and how it works.

It has "always" been known that children inherited physical traits from their parents. What was actually inherited and how the transfer came about through the gametes was altogether vague and mysterious. In about 1900 the unit of inheritance, the gene, was found to be localized in the chromosomes, but the physical nature of the gene itself remained a puzzle. The physical and chemical description of the gene was worked out in the 1950's, but that simply pushed the mystery on further: How does the gene work to bring about the inheritance of visible traits? Some of the answers to that question are now very clear—others are far from complete.

The nature of the gene

The crucial research publication on the nature of the gene was a paper by James Watson and Francis Crick, who proposed a credible structure for the chemical DNA that would fit the required specifications for the genetic material. DNA is the abbreviated designation for a long, much-coiled molecule (properly called deoxyribonucleic acid), made up of four different kinds of subunits. (Nucleic acids do not fit into any classification of the commonly discussed foodstuffs, such as carbohydrates, fats, and proteins. There is some resemblance to proteins, but they are not proteins.) Molecular biologists commonly speak of the genetic information as being stored in the form of a "code"; that is, as messages written in the sequences of those four different subunits found in the long DNA molecule. It is as though a language were written with only four characters. (European languages generally have about 26 characters; Morse code has only two—dot and dash.) The total length of the coiled DNA molecules found in any cell is about 4000 times as great as the diameter of the cell. The subunits (code letters) of DNA are small, and therefore much information can be stored in the great lengths of the DNA molecules within the chromosomes. It has been calculated that if the patterns of the four different kinds of subunits were used to represent letters of the alphabet, all the information written in a large set of encyclopedias could be stored in the DNA contained in the nucleus of a single cell!

Action of the gene

The information in the chromosomes of course does not code for words; it actually codes for protein molecules, and there are probably a thousand or so different kinds of protein molecules in the average cell. Proteins are vital and powerful agents within the cell; they are also large, complex molecules. A protein molecule is a long chain made up of different kinds of subunits, called amino acids. It is now understood that what a gene (a sequence of subunits in a DNA strand) does is to provide the pattern for the cell to manufacture a specific kind of protein molecule. Any one protein (such as the hemoglobin found in the red blood cells, carrying oxygen) is made up of a specific sequence of amino acids. There are twenty different kinds of amino acids, and each protein chain contains several hundred of these subunits. In a gene (one part of a strand), a group of three adjacent subunits is the "code" for a specific one of the twenty possible kinds of amino acids. The sequence of subunits in the DNA comprising the "hemoglobin gene," for example, gives the cell a pattern, or instructions, for putting specific kinds of amino acids together in the right order to make hemoglobin. (The cell has an elaborate set of machinery for reading the code and manufacturing the protein.)

It is not too difficult to see how having a gene that gives a pattern (or a blueprint) necessary for making some constituent of the cell such as hemoglobin, would make it possible to inherit "human hemoglobin" rather than the slightly different "rat hemoglobin," for instance. This direct understanding of gene action also applies to the genes that code for enzymes. Most of the protein molecules that are found in any cell are enzymes. Enzymes are different kinds of protein, each of which controls the rate of one specific chemical reaction; together they direct the chemical activity of the cell. For every enzyme that is found in any cell of the body, there is, on one of the kinds of chromosomes, a DNA pattern for making the enzyme.

Some genetic traits like those previously discussed can be described as the presence or absence of the ability to make certain proteins and therefore can be directly understood as the result of gene action. But this is not so for most of the human genetic traits that we know about. You may have inherited "attached" rather than "free" earlobes, the ability (or lack of it) to roll your tongue, or a big nose rather than a small one. At present it is not possible to understand any of these traits simply as the obvious result of having the pattern to manufacture one more kind of protein molecule. Yet all the evidence that has been accumulated in hundreds of active laboratories for the last 20 years leads to the conclusion that this is the *only* thing that a gene can do: *direct the manufacture of a specific kind of protein.* The big task for research workers in molecular biology and embryology for the next 20 years is to learn to understand how single genes (coding for single proteins) can specify structural traits that seem very far removed from the amino acid sequence of a protein.

Gene products and practical actions

The complicated and highly theoretical research that has taken place in the last 20 or 30 years concerning the mechanism of gene action has made possible the first real understanding of several inherited diseases. Because of this understanding, a few genetic disorders can now be treated, and some can be avoided by genetic counseling or by amniocentesis and selective abortion. This application is newer than amniocentesis and karyotyping for elimination of chromosomal abnormalities, and it has different problems and potentialities, which will be discussed later in the chapter.

What are alleles?

Geneticists would have nothing to study if everyone's heredity were alike. It is generally impossible to know that a locus for a certain trait even exists unless there are at least two alleles at that locus. It is now understood that alleles (the alternative genes that may be found at a certain spot on a certain chromosome) are sequences ("code words") of DNA that differ in just one or a few "letters" of the code, so that the protein produced under the directions of one allele differs in just one or a few amino acids (out of the hundred or so in the whole sequence) from the protein produced under the direction of the other allele. The usual gene for hemoglobin and any one of its several alleles, for example, both give instructions for making hemoglobin (red, oxygen-carrying protein), but the hemoglobins produced are slightly different in amino acid sequence and in physical properties.

The origin of alleles

Any set of allelic genes (the alternative sequences of DNA subunits that may occupy a certain locus and influence a certain inherited trait) has come into being by the process of **mutation**. In almost every case where genetic information is duplicated and passed on to other cells, the duplication is done very accurately, so that the new cells have exactly the same DNA sequence (genes) as the original cell. But in a very small fraction of these many gene duplications, a mistake is made. It may be because some unusual chemical in the environment of the cell "confuses" the copying machinery; it may be because injury from isotopes or x rays throws this copying out of kilter. Some mutations just "seem to happen"; no one yet knows why. Mutations represent small random changes in a sequence of subunits in a DNA strand, and thus in the amino acid sequence of the protein it codes for. A typical simple mutation causes a certain amino acid somewhere in one kind of protein molecule to be replaced by another of the twenty possible amino acids. In some cases this makes no difference at all to the function of the protein molecule. In other cases the **mutant gene**, the new allele, directs the formation of a protein that does its job but not very well. In still other cases even this small change happens to have such a vital effect that the whole functional capability of that protein molecule is destroyed. Most mutant genes are either unimportant, slightly harmful, or very harmful to the cell. In the exceptional circumstance, however, the mutant gene may happen to code for a protein that does a *better* job than the original, or that performs a new and advantageous function for the cell. If so, it is retained in the successful descendants of that cell, and in the descendants of that individual—it becomes a part of the process of evolution.

Chemical genetics: enzyme deficiencies

Any genetic enzyme deficiency is the result of a mutant gene coding for an ineffective form of the enzyme. People afflicted with the more serious of such deficiencies are not likely to have many offspring, so the mutant genes—the defective alleles—are not common in the human population. They will be found only in occasional families, showing up in a few people over the generations. The fact that really deleterious alleles can be found at all is due to one of two reasons. First, there is always some mutating occurring (although the chance of its happening to any one gene is small in any one person), and therefore new examples of the mutant gene will occasionally be found in someone's gamete. The other reason is that most of the alleles for enzyme deficiency are harmful only when they are in double dose; such recessive genes may remain hidden for many generations, actually appearing in an individual only when two carriers (heterozygous persons) meet, marry, and reproduce.

Phenylketonuria

The best known of the inherited enzyme deficiencies is known as **phenylketonuria** (commonly called PKU). The defective enzyme in this ailment is one that normally acts on the amino acid **phenylalanine**, not to incorporate it into a protein, but to modify it into one of the hormones, pigments, or other special products manufactured and used in the body. In the absence of this particular enzyme, extra phenylalanine not being used to manufacture protein accumulates and may be transformed by other enzymes into various related chemicals not normally found in any amount in the body. (Phenylketonuria takes its name from the chemical description of the substances formed and excreted in the urine. Babies afflicted with the homozygous form can be detected by simple tests on their urine.) The chemicals produced from phenylketonuria seem to have a poisonous effect on the developing brain, and the child has some degree (often very serious) of mental deficiency. Ever since the disease has been thoroughly understood, it has been possible to feed afflicted infants a special (and expensive) diet containing very little phenylalanine—just enough for use in making proteins. If this is done for a few years, the brain develops fairly normally, and the person seems to be able to eat a normal diet thereafter, functioning well and normally.

Almost any recessive gene can be detected even in single dose in the heterozygous person, when there is careful study at the chemical level, although, of course, the overt symptoms do not appear. In a person heterozygous for the gene for phenylketonuria, the cells actually contain only about half the normal amount of the enzyme. (Presumably, they contain an equal amount of the mutant or defective form of the enzyme, but since this has no enzymatic powers, it cannot be detected by the usual tests for enzymes.) The reduced amount of enzyme seems to do all that is necessary most of the time. In special tests where people are fed large concentrations of phenylalanine, however, those heterozygous for the defective gene cannot assimilate as much of the amino acid, and it stays in their blood longer. This is of no direct significance to the carrier's health, but it is useful as a genetic test to detect the heterozygous person. If a man had a brother afflicted with phenylketonuria, he might have such a test done. If he proved heterozygous for the PKU gene, he would want his wife, when he married, to take this test also. If (as might happen, although it would be unlikely) she turned out also to be a carrier of PKU, they could take some kind of precautions to avoid the problem of homozygous PKU in their children. There are several other genetic defects similar to PKU in which, for lack of an enzyme, some abnormal product accumulates and causes damage (Fig. 5-5).

Albinism

A genetic enzyme deficiency that has been known for a long time is **albinism**. An albino lacks the dark pigment **melanin**, more or less of which is normally found in everybody's skin. Melanin is manufactured from the amino acids tyrosine and phenylalanine by a series of enzyme actions. An albino lacks one of the enzymes in that series. (In this condition, it is the lack of an enzyme product that causes the trouble, in contrast to PKU, where the problem is caused by the presence of an abnormal chemical product.) Albinism is recessive, since a cell with a single gene for the normal enzyme seems to be able to make enough enzyme to manufacture a nearly normal amount of pigment. Presumably, some biochemical test could be devised to detect a heterozygous person, but very little work has been put in on the problem because

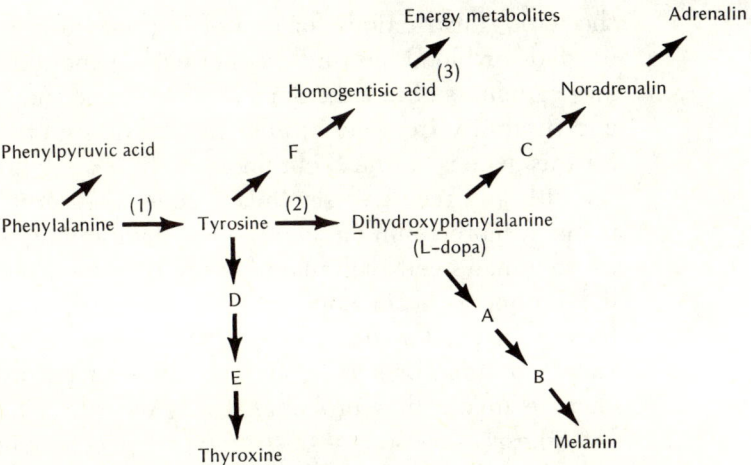

Fig. 5–5: Inherited deficiencies in the metabolism of the amino acids phenylalanine and tyrosine. Each arrow represents a specific enzyme that transforms one kind of molecule into a slightly different one. The letters A, B, C, and so on, stand for intermediate compounds. Three genetic metabolic deficiencies are listed, and the enzyme affected in each is indicated. Other, rarer, deficiencies are known that involve others of the enzymes in this series.

albinism, although troublesome, is neither fatal nor disabling. An albino cannot suntan and so must avoid any great exposure to sun; the most serious handicap is an excess sensitivity of the eyes to bright light, because the eyes, too, lack melanin, which normally absorbs some of the stray light. Albinos thus have poor vision.

Hemophilia

The gene for **hemophilia** is famous because it showed up several times in the last century and a half among the royal families of Europe. It is due to a recessive gene that contains the wrong instructions for one of the enzymes in the blood needed for blood clotting. People with hemophilia may lose large amounts of blood from a small cut, and may suffer pain and disability from very minor bruises or sprains. The locus for this gene is on the X chromosome (it is a sex-linked gene). As with any recessive sex-linked gene, the actual condition is much more common among men than among women, because a man's cells will manufacture only ineffective enzyme if he inherits the hemophilia gene from his mother; he can have no normal allele. A heterozygous woman will normally show no symptoms even though the concentration of the enzyme involved is probably low. (According to the Lyon hypothesis—see pp. 55–58—some of her cells will manufacture the normal enzyme and some the useless one.)

Favism

The condition sometimes called **favism** is one in which it is possible to see how the lack of a single enzyme can lead to complicated symptoms. This condition is due to another X-linked gene, one that codes for an enzyme called glucose-6-phosphate dehydrogenase. This enzyme seems to be relatively unimportant in most cells, but it is important in the rather unusual energy metabolism of the red blood cell. If this enzyme is ineffective, the red blood cell will not have enough metabolic energy to keep its cell membrane in good repair. Consequently, a red blood cell with a defective gene for G6PD is fragile and easily broken. If a man has inherited a gene for a defective enzyme, he will have no normal allele present, and all his red blood cells will be fragile. A woman homozygous for the mutant allele will be that way, too. If a woman is heterozygous for the normal and abnormal genes, some of her red blood cells will be fragile and others normal, because one of the X chromosomes is inactivated in each cell, but not always the same one.

The condition of G6PD deficiency was first discovered as a sensitivity to the eating of fava beans, a native European legume. For some unexplained reason some normally harmless constituent of these beans destroys the red blood cells

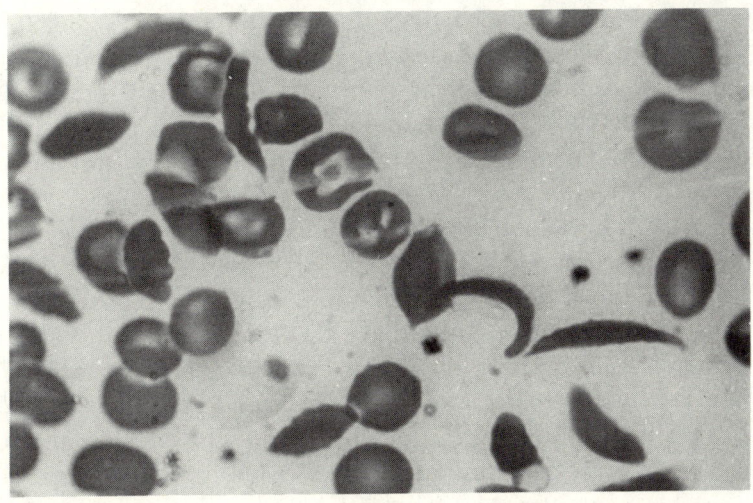

Fig. 5-6: Sickle-shaped red blood cells of a person with abnormal hemoglobin. (Courtesy of Michael Myszewski, Drake University.)

that are weakened by the enzyme deficiency, causing anemia and even resulting in blockage of the kidney by the hemoglobin released from the blood cells. Probably a more important, and advantageous, effect of the gene is that red blood cells weakened by this enzyme deficiency burst when invaded by malaria parasites, preventing the parasite from building up in the person's body. This makes the person nearly immune to malaria, although in some danger of anemia.

Sickle cell anemia

The most famous of all the chemical genetic diseases is **sickle cell anemia**. In this disease the gene in question codes not for an enzyme, but for the protein hemoglobin that makes up most of the bulk of the red blood cells and carries the oxygen. The mutant gene for sickle cell causes only a tiny change in the hemoglobin molecule—the substitution of just one of the 287 amino acids by another amino acid. This abnormal hemoglobin still carries oxygen, but the presence of the wrong amino acid changes the way the molecule can fold up and makes it much less soluble. For some reason, it also fails to pack well into the red blood cell, so that these cells come to have a strange shape, which resembles a hand sickle—hence the name (Fig. 5-6). These misshapen cells clump easily and then break up as they are tumbled about in the blood stream. The afflicted person suffers from chronic anemia, and has occasional acute attacks of pain caused by the clogging of blood vessels with clumped cells. A person heterozygous for the gene is said to have the "sickle cell trait." This ordinarily causes no problem, but the hemoglobin in the blood cells is about half the normal variety and half abnormal. When oxygen to breathe is in short supply (as at high altitudes), even these cells assume the sickle shape and may break down and cause trouble. The red blood cell with about half normal hemoglobin is rather fragile, and like the cell with G6PD deficiency, will disintegrate if invaded by a malaria parasite. In malaria-ridden parts of Africa, it has been a real advantage to have the sickling trait, and the gene is fairly common there, despite the fatal effects of the homozygous condition. The gene is found in about 8 percent of Americans with African ancestors; to them it is of no advantage, since they live in a malaria-free environment.

Two parents with the sickling trait (heterozygous for the mutant gene) have one chance in four that a child will have the crippling sickle cell anemia. The only thing that it has been possible to do to prevent the disease has been to counsel those with the trait (heterozygous) not to marry each other, or not to have children if they do marry. Now amniocentesis with selective abortion is a possibility, although detection is still very difficult and uncertain. Unlike conditions such as PKU, this condition cannot be solved by diet, and the only treatments have been supportive measures like multiple blood transfusions. The ailment is usually fatal before maturity unless there are repeated massive transfusions. There seems now to be a faint hope for effective treatment. In recent research, two or three chemicals have been found that seem to prevent sickling and breakdown of the blood cells. If these chemicals can be harmless to the rest of the body when given in effective doses, they could prevent a great deal of suffering and tragedy.

Prenatal detection of chemical diseases

It is estimated that nearly a third of the serious birth defects that occur are due to single-gene effects, generally abnormal enzymes. It would prevent a great deal of human suffering if such serious problems could be detected in the early embryo by amniocentesis and therapeutic abortion performed. But this is a much more difficult problem than screening for trisomies, for several reasons. The basic problem is that there are more than 200 such genetic chemical diseases known, and probably several hundred or even thousands more yet to be discovered. Although there are many people afflicted with these diseases, any one of the diseases will be found in only a very few individuals. This means that it is not sensible to consider screening all pregnancies for most of the conditions: There are too many different ones to look for. So study, counseling, prevention, and treatment must generally be done with families in which the condition has already occurred. Luckily, it is usually possible, or will eventually become so, to detect, among the relatives of afflicted individuals, the heterozygous carriers for these usually recessive genes. It would then be feasible to test whoever marries one of these carriers to find out if the spouse, too, happens to carry the gene. In that rare event, amniocentesis might make sense when the woman becomes pregnant.

Another problem in such prenatal detection of rare chemical diseases is that no hospital is equipped to deal with more than a few of the hundreds of possible problems. Detecting an enzyme deficiency (or even worse, a reduced concentration of an enzyme) is far more difficult than preparing a karyotype and noting any trisomies. An enzyme assay requires a special set of reagent chemicals (which probably spoil in a few days or weeks if they are not used) and some kind of procedure that the technicians must learn. No hospital laboratory can be tooled up to perform immediately any one of 300 enzyme assays, each of which might be needed only once in every 2 or 3 years. It costs a hospital laboratory only about $10,000 to set up for chromosome analysis, but the cost of labor, supplies, and special equipment to maintain universal enzyme capabilities would bankrupt any establishment. The laboratories that work with such diagnoses are working to set up a system of "regionalization," by which centers all over the United States and Canada cooperate in solving these problems. Each of the centers is staffed and equipped to perform the biochemical or chromosome analyses necessary to diagnose the most common genetic diseases. In addition, some of the centers have the specialized personnel and facilities required for diagnosis of one or more rare genetic defects. All the centers are to be staffed to provide genetic counseling and follow-up to any diagnosis. Through a central clearing house all requests for diagnostic assistance are channeled to the appropriate center. Blood, tissue, or urine samples taken at one place can simply be mailed to the center equipped to study them. Hopefully, it will soon be possible to ship growing cultures from amniotic cells in the same way.

Genes, active and inactive

The greatest obstacle to solving chemical genetic problems with amniocentesis and selective abortion is tied in with the intricate cellular events during embryology. In no one cell of the body are all of the thousands of genes stored in the nucleus actually active in giving instructions for proteins. Most of the proteins for which there are patterns are just not being made. During embryonic development the cells become different from each other (a process called "differentiation"). The basic events in differentiation are the activation of some genes and the suppression of others, so that each final cell type has only the set of active genes that are appropriate to it. (The inactive genes are probably chemically bound up with some specific repressor proteins.) The first evidence of differentiation in the embryo is the formation of the three primary germ layers: endoderm, mesoderm, and ectoderm. The cells in each of these three primitive tissues already have sets of genes that are active and sets of genes that are repressed (inactive), along with some genes that could become either active or repressed. As embryonic development proceeds, more and more different kinds of specialized cells are developed, with the inactive genes more thoroughly suppressed and the active genes more efficiently at work. Each of the 100 or so different kinds of cells in the adult body has a different set of active genes. No doubt there are many "overlapping" genes that are active in several or many cell types, but the overall pattern of active and suppressed genes is unique for each kind of cell.

The cells that can be obtained from the amniotic fluid when the amnion is large enough for withdrawal of fluid are already highly specialized. There are several types of cells: amnion lining, skin, blood. But each of these few types of cells has only some of the total genes in an active state. An embryo can be homozygous for an unfavorable gene that causes an enzyme deficiency in some adult organ, and this gene might still not be acting at this stage of embryonic life or in the particular cells available for study. It requires long and difficult analysis to be sure whether or not a certain condition can be diagnosed in cells from the amniotic fluid, and some ailments will never be able to be studied that way. There are two theoretically possible methods to obtain a prenatal diagnosis of such conditions. One would be to take a sample (called a "biopsy") from the appropriate organ of the embryo, to study conditions affecting that organ. Even now, in rare cases, a sample of the fetal blood is taken from the placental vessels, or a small bit of skin may be scraped off. These procedures have caused no harm, but the obtaining of biopsy specimens from embryos is a risky business and may never become an important means of diagnosis. Even if biopsies someday become safe, there will still be the problem of the gene that is not active before birth. Someday in the future it may be possible to treat a tissue culture of amniotic cells with some special agent (so far not available) that will activate the genes to be studied. Cell biologists would generally agree that such a procedure should be possible eventually, but they do not know when.

Cystic fibrosis

There are a few chemical-genetic diseases that are common enough to make it feasible to have screening programs for them. One of these, as mentioned, is the sickle cell gene among American Negroes. Another possibility is **cystic fibrosis**. This is caused by an enzyme deficiency that leads to abnormal, viscous mucus in the body fluids, so that the lungs and pancreas, among other organs, become clogged and inoperative. The ailment may cause death in childhood or may require hospitalization and expensive care for many years. It is estimated that each case of cystic fibrosis costs someone (the government, the insurance companies, the family) an average of at least $80,000 during the patient's lifetime. Some 5 percent of the general population of the United States are heterozygous carriers for this defective gene. If screening tests (already technically possible) were carried out on the cells of 800 women (before they became pregnant), one would expect an average of 5 percent of these 800—40 women—would prove to be carriers of the gene. If the husbands of these 40 women were tested, two of the families, on the average, would have both wife and husband heterozygous for the gene and would have the danger (one chance in four) of having a child afflicted with cystic fibrosis. If each family planned to have two children, amniocentesis would be done at each pregnancy, with the probability of detecting one afflicted embryo, which would be aborted. Another amniotic test would be done when a pregnancy occurred after an abortion. If we assume that screening tests cost $20 each (probably a high estimate), the 840 screening tests would cost $16,800. If each amniocentesis with tests costs $200, the five required would cost $1,000. If a late abortion costs $500, the single projected one would come to that amount. Therefore the total cost to prevent one cystic fibrosis sufferer can be calculated at $18,300. This figure for preventing one cystic fibrosis sufferer compares with the $80,000 cost of caring for such an unfortunate. These figures, if correct, indicate that such a program would make sense strictly from a financial viewpoint, even without taking into account the possibility of avoiding suffering and heartache.

The social morality of abortion: a biological view

The new possibility of preventing many tragic birth defects by therapeutic abortion adds to the already large demand for abortions from those who did not intend to get pregnant. Many people are troubled by the feeling that abortions are wrong and too closely akin to murder. A few are convinced that the embryo is a full human being with all rights and feelings from the moment it is conceived; some of these people are trying very hard to persuade others to agree with them. The crucial question of the abortion problem is the following: "Is abortion an act that destroys a human life?" Biologists, as their name indicates, are professionally concerned with considering the meaning of the word life and so are especially qualified to speak on this topic.

Inducing abortion

The medical procedures used to induce abortion depend on the stage of the pregnancy. Pregnancies in the first 3 months are commonly terminated by a new procedure called **vacuum aspiration** or "suction curettage." A narrow flexible tube is inserted through the vagina into the uterus, and the embryo and its associated membranes are removed by suction. This is the method most used in the special abortion clinics that were set up in New York State after abortion was legalized there. The woman requires only a local anesthetic at the cervix. The technique takes just a few minutes, and after an hour's rest and a safety recheck, the woman can return home. There is little discomfort and very little danger.

An older method of abortion that can be used in the early months of pregnancy is **dilatation and curettage**, or "D & C." Tubes are inserted in the cervix, starting with a small one and gradually increasing in size. This dilatates or opens the cervix so that a spoon-like "curette" can be inserted in the uterus to scrape out material. D & C is often required to remove remnants of placenta after a spontaneous abortion, or to remove troublesome noncancerous growths. The operation requires either local or general anesthesia and usually involves a day or so in the hospital. As a means of abortion,

D & C is still used by some private physicians, but it is being replaced for that purpose by the vacuum procedure.

After the third month the embryo and associated membranes are so bulky that abortion is a much more difficult procedure. The method currently most used is **saline infusion**, in which a strong salt solution is introduced into the anmiotic cavity, killing the embryo and inducing uterine contractions to expel it. This operation requires about 2 days of hospitalization, considerable discomfort, and some danger to the woman. There is still not as much discomfort and danger as with a full-term delivery. Some very late abortions are performed by a surgical opening of the uterus to remove the fetus, but this operation is being replaced by saline infusion. When fetuses are near the age (about 6 months) when they might be able to survive outside the uterus, such late abortions have serious legal and moral problems. They are done only when medically necessary for the safety of the woman.

What is life?

While many introductory biology books are unwilling to venture into an attempt to define such a mystical entity as "life," this situation is changing. More and more authors are realizing and explaining that life, which was once thought of as something like a mystical fluid that could reside in a body but flows out at death, can best be defined as a collection of wonderful but quite observable properties that are found in cells and in organized collections of cells. These properties can be described in several different ways, but they include the following: (1) possession of an organized structure, with wide but still limited range of variation from species to species; (2) metabolism, an orderly and regulated set of chemical reactions that break down and synthesize chemical molecules and transform energy from one state to another; (3) homeostasis, the ability to respond to environmental and internal changes and stresses to preserve the general form of the organized structure; (4) response, the ability to react to stimuli in such a way as to preserve the individual and the species, including behavior for feeding, self-protection, and reproduction; (5) growth, the ability to extend the organized structure to a greater volume of space; (6) reproduction, bringing about the creation of separate, organized structures in increasing numbers.

These properties are usually (although not always) found together; so it is convenient to refer to them with a single word, "life," which is always associated with cells or their equivalent, or with groups of cells. These are the common properties found in all kinds and grades of cells and are the necessary basis of the other properties that various cells have. Any particular cell does have other properties in addition to those included in this definition of life. The other properties might be considered part of the "life" of that cell.

There are two grades of organization of cells. The less complicated group includes the bacteria and the blue-green algae, which have cells of comparatively simple structure. All other cells are in the more complicated group, which includes many one-celled plants and animals, fungi, and all cells of higher plants and animals. Except for this division, it is not really possible to speak of higher and lower grades of cells. If anything, the free-living single-celled animal or plant (such as an ameba or a green alga) is of a higher level than a single cell of a multicellular organism, since the free-living single cell can carry out all of the life processes itself, while the cell in a larger organism is usually specialized in such a way that most capabilities have been reduced while one or a few capabilities have achieved increased efficiency. On the cellular level, human cells are not any more alive, or even "higher," than other cells.

The special nature of a human being lies not in the types of chemicals in his makeup (we share those with all life, including bacteria) or in the special properties of his cells (we share those with all animals). Even most of human tissues and organs are not very different from corresponding structures in other mammals, with the exception of the brain, and even that only at the higher level of organization; the human brain has the same kinds of cellular interaction as the brains of lower animals. What is distinctively human is the way all these ordinary biochemicals, cells, and tissues are arranged and organized into a body that has dexterities and capabilities different from those of any other creature. In the case of the brain, our capabilities are clearly superior to other animals by most objective criteria. However, there is no evidence of important "humanness" at any lower level. (To be sure, a human cell can be recognized by its chromosome set and by immunological tests, but it is no more distinctive than any of the other millions of animal species.) If we examine our own common feelings, we realize that we do not think that human life, as distinct from animal life, exists in tissues. There are a number of human tissues growing in tissue culture in various laboratories and subjected to various kinds of experimental operations. Almost no one feels that this is wrong or objects to these cultures being destroyed when necessary. It is the organized, functioning human being that is valuable, not the cells or tissues.

The life of the zygote

The human zygote is a single cell with no observable capabilities to distinguish it from the zygotes of hundreds of other animals. It can't even care for itself as an ameba can. What it does have is a possibility—that of becoming a human being if certain necessities for survival and development are available to it. The most useful way of approaching this question that I know of was provided by the biologist Garrett Hardin, who suggested that the zygote bears almost the same relationship to the newborn baby as a set of blueprints bears to a house. Each provides the information for something of value but contains very little substance and can easily be replaced by copies or equivalents. The human zygote does indeed have a unique set of genetic instructions, but there is no reason to think that any one zygote from a certain set of parents will be any more or less interesting or valuable than any of the other possible ones. It is tear-jerking propaganda and unclear thinking to contend that some zygote that was unable to continue to develop might have been another

Beethoven, Einstein, or Bunche. By such thinking, it would be necessary to have all the thirty or so children theoretically possible to a couple. And even then one would have to mourn the thousands of oocytes that never got a chance, the millions of unused sperm (each unique), and to mourn as well all the many billions of possible combinations of unique sperm and unique eggs, each of which might be equally likely to have turned out to be a Nobel prize winner!

The life of the embryo

As with the house under construction, the developing embryo steadily gains in value, from the low value of a replaceable blueprint to the full value of the complete structure. We can realize that people generally agree on this fact (a zygote seems to be equated with a person *only* when considering voluntary abortions) if we think of how people regard a natural, spontaneous abortion—a miscarriage. The fact, now becoming more widely known, that a significant fraction of zygotes fail to implant, or otherwise fail to develop in the first few weeks, does not cause any widespread consternation. I have heard no proposal that we mount a major effort to "save" all of these "innocent little babies" that are being lost by early miscarriage. An early miscarriage that shows up as a missed period is not a tragedy; at the very most it may be a disappointment. There is a sense of serious loss only if the couple has been trying for years with no success to have a child. The loss even then is one of withdrawn opportunity, not of substance. The loss that a couple feels with a spontaneous abortion increases steadily with the extent of the duration of the pregnancy, so that a stopping of fetal movements or a stillbirth is a major sorrow. Finally, if the parents have had a chance to see and interact with a newborn baby, its loss is tragic.

A house being built increases in value mainly because of the material and labor that goes into it—the structure (organized substance) that results. But if someone is building the house from his own design for his own use, it may mean a great deal to him emotionally and personally—if it were to burn down, he would feel more than a financial loss. The embryo also increases in value to its prospective parents as it develops. The mass and developed capabilities of the embryo increase only slowly. What gives it the greatest amount of its increased value (resulting in the sense of loss if it is destroyed) is the amount that the parents have invested, not financially but emotionally. From the time that a set of parents know that "a child is on the way," they are anticipating it and planning for it. They speak of the child-to-be as though it already existed; they may even pretend to speak to it or give it a name. This emotional identification with the developing embryo increases as pregnancy advances, especially after fetal movements can be felt. By the time of birth the parents have invested a large store of love, hope, anticipation, and commitment—and the value of the fetus is high. Even for an unwanted pregnancy, the emotions of the pregnant woman are inevitably tied up with the developing embryo. Whether or not she feels more hate than love, dread than anticipation, rejection rather than commitment, she cannot ignore the situation and its implications. So the loss (either spontaneous or planned) of even the unwanted fetus fairly late in pregnancy is of some consequence emotionally.

Since the value of the embryo increases steadily through its development, the ethical status of abortion changes similarly. Destruction of the zygote or very early embryo by preventing implantation or causing menstruation will usually occur without anyone's knowing that anything unusual happened that month. It is just one more added to the large number of natural abortions and is not an ethical problem at all. After the first month or so, an abortion is not something desirable to plan for, and its undesirability increases with the passing months. As the fetus nears the age when it could survive if removed from the uterus, its value is such that abortion is normally not acceptable, except for some such overriding reason as grave defect or danger to the pregnant woman.

Compulsory pregnancy

Dr. Hardin has suggested that when we are considering the acceptability of abortion, once a specific decision of an individual case is upon us we should not think of the alternatives as "abortion versus no abortion," since saying "no abortion" tells nothing of the situation involved—one cannot do *nothing*. Neither is it "abortion versus contraception"; once conception has taken place, that alternative is out. (Looking ahead, of course, contraception is infinitely to be preferred.) But when it comes down to it, the choice is between voluntary abortion versus compulsory pregnancy.

Making a woman carry for 9 months a fetus she passionately does not want is a kind of involuntary servitude—slavery—of a uniquely intimate and destructive kind. Looked at this way, compulsory pregnancy is an obscene suggestion. As of now, the ethical problem of abortion late in pregnancy is taken care of by the fact that late abortion is uncomfortable, expensive, and somewhat dangerous. For these reasons, a woman who does not decide until the fifth month of pregnancy that she doesn't want a baby will usually wait out the second half of pregnancy, too. This state of affairs will not continue very long into the future, for late abortions are sure to become less troublesome to perform than they are now. Then we will be back to the problem of having to decide. It would be convenient to be able to say "After such and such a time, an abortion is not acceptable." But there is no point in embryonic development that can logically be taken as a cut-off point—we are faced with a gradually changing situation. Each case of desired voluntary abortion must be decided on its merits. The need for abortion may be the result of irresponsibility—the failure to use contraceptives—and abortion may still be the best alternative available.

No matter what we take as an arbitrary cut-off point, decisions—painful ones—will always have to be made. It will become possible some day to remove a defective embryo and treat it in an artificial uterus. Will we want to do that?

I am convinced that the abortion laws in all states should make abortion strictly a medical rather than a legal matter. Great effort should be put into making contraceptives available (perhaps it would be best if they were free to everyone), and people should be urged to practice contraception, reserving abortion for contraceptive failure and for therapeutic elimination of defective fetuses.

6 | Embryonic development: a commonplace miracle

The embryo begins as a single cell, the zygote, and ends as an organism containing millions of cells of at least 100 different cell types, arranged in at least four hierarchial levels of structural organization—tissues, organs, organ systems, and whole animal. The human zygote can survive only in a special protective and supportive environment, it has no defense against temperature changes, drying out, or other risks that might befall it. The adult human that is the final product of development has a finely tuned system for maintaining an optimum temperature throughout the body, and other systems for keeping optimum concentrations of water and salt. The zygote shares with all single-celled organisms a lack of sensitivity to many characteristics of its environment; it cannot distinguish in any but the most general way among different patterns of sight, sound, or touch. The zygote even lacks the motility that so many single-celled organisms have with which to respond to whatever stimuli it does distinguish. The final adult product of development can respond to many kinds of signals from the environment with a complicated array of different behaviors appropriate for survival and health. In many cases his response matches the stimulus closely.

All these wonders are products of the developmental process (Fig. 6-1). The human zygote has no obvious features to distinguish it from other zygotes, and it goes through almost exactly the same set of developmental steps as zygotes of other mammals do. Yet the human being produced is capable of understanding and control of the environment, and of a degree of complexity in relationships with other humans and animals to a degree that is not approached by any other animal.

Embryonic development, whether of mouse, whale, or man, is an authentic miracle in the original sense of the word: something worth wondering at. It will remain a miracle—an object of wonder and awe—even when the process is thoroughly understood. We are still, however, a long way from total understanding of the complexities of embryonic development.

Our established patterns of thought, both popular and scientific, are generally concerned with stable, dependable, objects that endure to be measured, to be described, and to become familiar to us. We are not adequately equipped intellectually to deal with an object that changes while we are observing it, not just in color, size, or position, but into a different kind of object—for a zygote is, in truth, a very different kind of creature than a man, for all that it eventually develops into one.

Another fundamental problem that we all have in comprehending embryological changes is that the changes proceed so clearly from less organization to more organization. We are used to the idea that a complicated system gives rise to something simpler: a man moulds a vase, a lathe cuts out a bolt. But a bolt doesn't change into a whole machine! In fact, we know that it is a fundamental law of nature that orderly, organized systems tend to get out of order and disorganized—whether expressed in a popular maxim or as one of the laws of thermodynamics in physics. We know that machines break down instead of getting more complex; a room gets littered and it takes a special effort to straighten it up; we each face death and dissolution in the end. Yet, the embryonic process is one in which the (relatively) simple

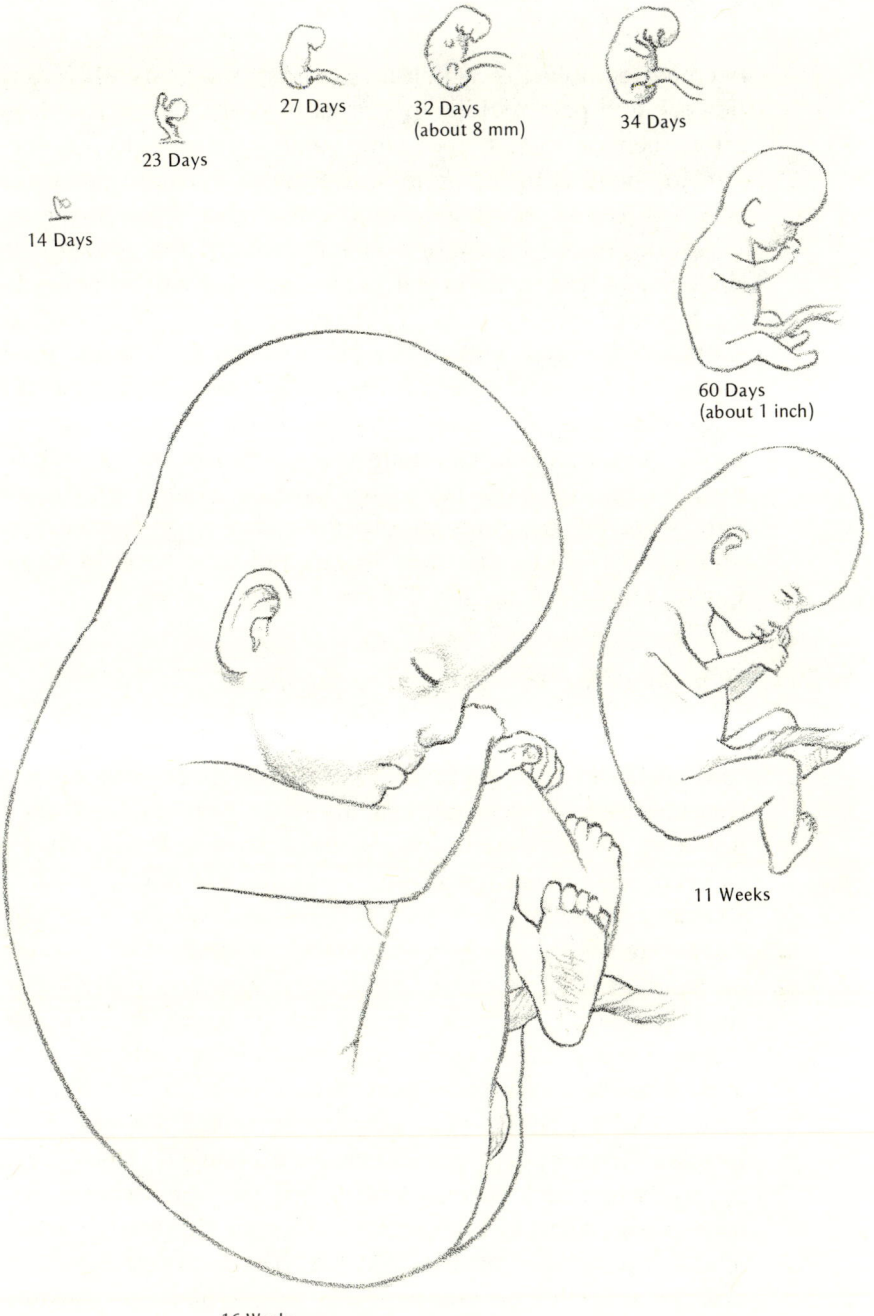

Fig. 6–1: Some stages in early embryonic and fetal development, shown at approximately actual size.

becomes complex, in which a system increases in attributes, powers, and potential. Such a process as embryonic development seems almost to be going against nature. Embryonic development is, in fact, one of a group of creative processes that use similar strategies to create "more" out of "less."

Adding to the fascination of embryology, but also to the difficulty of understanding it, is the fact that it is so unfamiliar to us. Even though we all share a common set of experiences before birth with other people (and with most other mammals, too), we have no memory of them. And the early development of any animal that we might pay much attention to, and that is large enough to see clearly, occurs within a shell or inside a uterus. So we have no common, ordinary names for the structures we observe—they seem to be exotic objects, outside of our experience, and it is hard to know how to start thinking about them.

Preformation: An early theory of embryology

The early embryologists, in the 1600's, had the same intellectual problems that we face today in thinking about objects that change their very natures, and that even change from a lower to a higher condition. Using only magnifying glasses to examine tiny embryos, they were not restricted by much real data to which their speculations needed to conform. So they suggested a simple theory of embryology that avoided philosophical difficulties. This theory was called **preformation**. According to preformation, the earliest embryo has within it a miniature, condensed person (often referred to as a "homunculus," or little man) which more or less unfolds during development. This was the dominant theory of embryology for two centuries. There were many vigorous arguments as to which parent contributed the homunculus, and as to what, then, was the contribution of the other parent. The dominant opinion was that the male seed contained the tiny person, and the woman was just a place of nurture. When primitive microscopes became available for observing sperm, at least two investigators with better imaginations than eyesight drew sketches of sperm with a little man huddled up in the head of each (Fig. 6-2).

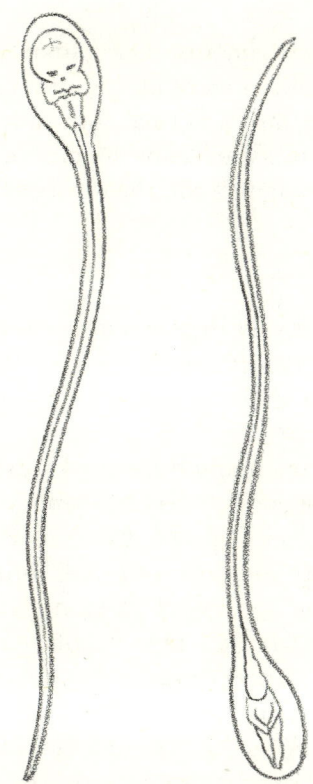

Fig. 6–2: The homunculus in the sperm. Seventeenth-century observers thought they could see small individuals (called homunculi) in the heads of sperm cells. These were considered to be preformed embryos. Drawings are from Hartsoeker in 1694 (left), and Dalampatius in 1699 (right).

Aristotle, 2000 years previously, had broken open developing hens' eggs at different periods of incubation and had seen clearly that the early embryonic stages bore little or no resemblance to the final product. The more scholarly preformationalists read Aristotle and did similar experiments, with similar results. But this did not immediately scuttle the preformation theory. Its adherents suggested that the puzzling, vague shapes in the embryos simply masked the real, underlying, preformed, miniature adult. It was not until experimental embryologists studied the effects of injury to early embryos that the preformation theory fell out of favor.

In the simplest of these experiments, it was found that if the first two cells of a frog embryo were moved apart from each other, each could develop into a normal (but initially small) tadpole. Other more complicated experiments made it impossible to continue believing in a full theory of preformation.

Twinning

It is not really necessary to do experimental embryology to see that the preformation theory cannot be true. It is evident from a consideration of human twins that the zygote is not equivalent to a baby.

Twins are of two fundamentally different kinds. Fraternal twins develop from independent zygotes produced by the fertilization of two different eggs by two different sperm. Fraternal twins are therefore just ordinary brothers and sisters of exactly the same age. A set of identical twins, on the other hand, develops from a single zygote, just like the identical frog embryos experimentally produced from a single fertilized egg. The mode of formation is a little different, since the coating of the oocyte remains around the zygote until the time of implantation and would prevent any separation at the two-cell stage. Identical human twins are formed within the blastocyst by a splitting of the inner cell mass into two bunches of cells (Fig. 6-3). The early embryos share a single trophoblast, which develops into a single placenta. Genetic instructions are identical in the two embryos formed in this way; but the twin children are definitely two separate individuals, even though very much alike. The small accidental differences in the course of their embryonic development and in their postnatal experience produce their individuality.

Mosaics

The evidence against preformation supplied by identical twins is supplemented by recent discovery of the opposite phenomenon. For over 10 years it has been known that some people are mosaics, made up of contributions from more

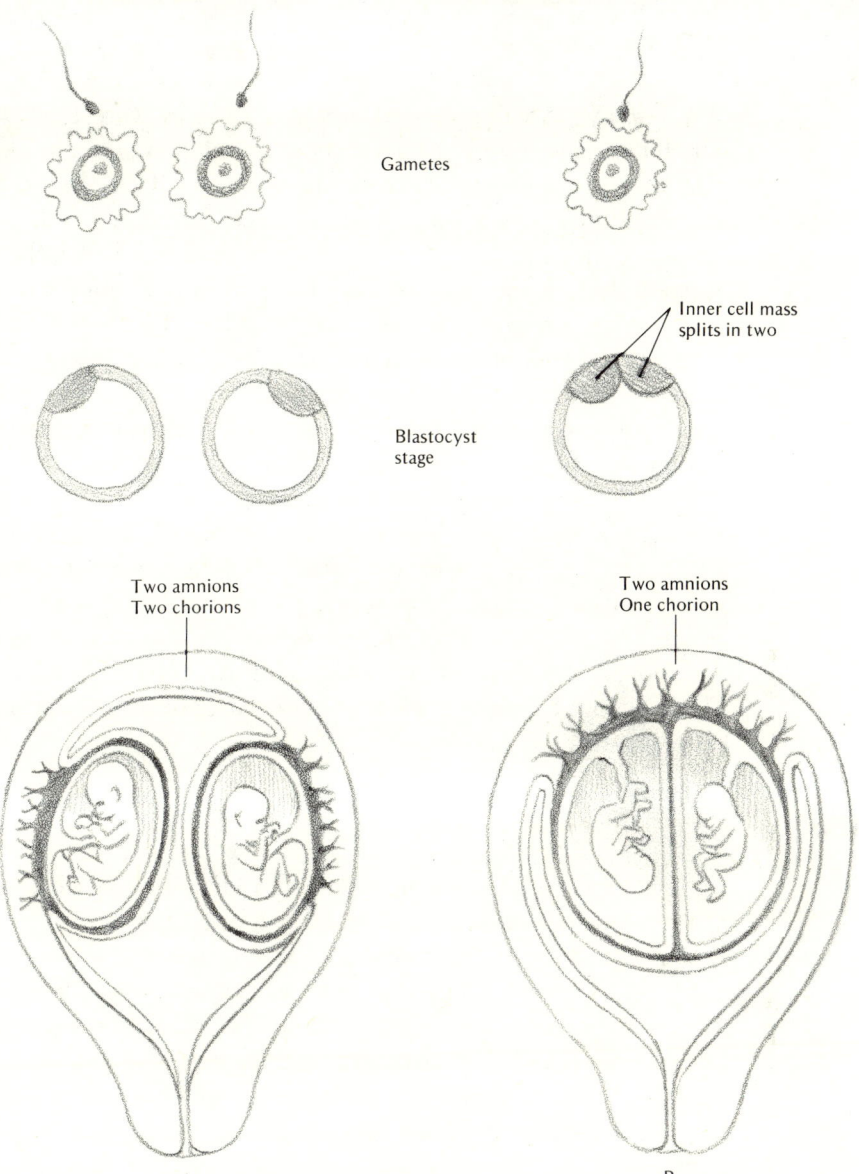

Fig. 6-3: Two kinds of twins. A. Twins that result from the separate fertilization of two eggs by two sperm are called fraternal twins. They develop individual placentas and chorions. (Chorions are represented in the diagrams by the middle shaded layer). Fraternal twins may be both male, both female, or one of each. B. Identical twins result from the fertilization of a single egg by a single sperm, and the subsequent splitting of the inner cell mass within the blastocysts. Identical twins share a single chorion and placenta. Identical twins are always of the same sex.

than one zygote. The most dramatic examples were cases in which two zygotes had apparently adhered in such a manner that each gave rise to the cells on one side of the body. It happened that the zygotes had received different combinations of genes for degree of skin pigmentation. The mothers noticed that in the summertime their children tanned to a darker color on one side than on the other. When the children were brought in to a research center for study, the mosaic condition was discovered. One detectable result was that the blood types of some of these children were mixtures of types that were possible from their parentage but impossible for a single genotype; the children had both kinds of blood cells. But they were not "two people each"; they did not have "split personalities." Each was a perfectly well-integrated individual. In one case a child, one of whose two zygotes had contained XY and the other XX, was a nearly normal boy, with only slight abnormalities of the genitalia. These examples make it clear that the zygote *is not*, cannot be, a pre-formed person.

Epigenesis

Embryologists now clearly understand that the zygote does not contain a preformed, folded-up adult ready to be unfolded or blown up. The zygote is simply not the same kind of object as a man, a baby, or even a 2-month fetus. Instead, it is a very large cell with certain special features. As it grows and divides, it changes into a mass of cells with slightly different properties. The processes going on within and among these cells act to change the properties of the new cells, too. Embryonic development can be thought of as a series of steps, each step based on the state of the embryo when its action starts, but each one changing the embryo in this state into something different from what it was. This process is called **epigenesis**. The word is from the Greek and simply indicates that what happens in each step is "built on" what has already happened. Even though the shape of the early embryo is most nonhuman (and is very similar to some simple kinds of animals), these early embryonic stages are important, in fact absolutely necessary, for the ultimate development into a fetus and then a child.

The genes found in the zygote are sometimes spoken of as constituting a set of directions for development, rather like a blueprint. This is not quite accurate. Nothing in the zygote has the shape of the final product, in the way that a blueprint outlines the structure of the house. Even when, someday, scientists have described all the genetic loci in all the chromosomes, and have the ability to create pictures of these genetic instructions as they look at the molecular level, there will be no sketch of a person to be seen. The genes give the instructions for only one step at a time. All the embryonic processes that go on are quite similar in rats, cows, and people. The differences are only in such characteristics as rates of growth and final proportions of organs. Most of the embryonic structures are similar in all mammals, and embryos only gradually become recognizable as belonging to one species or another. The baby produced by the commonplace miracle called human development is truly a product of the *process*, not the fluffing up of a condensed homuluncus already there.

Everyone who cares to study human development can find out that the human embryo at its early stages is most unlike a baby, and it only gradually comes to resemble a functional human. Yet people find it difficult to accept the idea of epigenesis: that something is there at the later stages of embryonic development that *was not there* at the early stages.

Popular preformationism today

There is a naive, imprecise kind of preformationism at the heart of all the emotional campaign against legalized abortion. There are many good reasons why abortion can be deplored or opposed, or perhaps even why the government should not allow it. But the bulk of the propaganda against it calls early embryos "babies" and persists in ascribing hopes, fears, and virtues to an embryo that is at a stage where the nervous system is only minimally functional. This is pre-

Table 6-1. Developmental stages of the embryo

As the embryo grows, it goes through a regular and quite well-known series of developmental stages. Some of these are listed below, in rather sketchy form. It is possible to compare the various early stages with different levels of organization found within the whole body or in different groups of relatively simple animals. Thus for the stages up to 8 weeks there is an organizational unit of the human body and a kind of lower animal, both of which correspond in a very rough way to the level of organization found in the embryo.

Embryonic age (weeks)	Size	General appearance	Equivalent organization level	
			Anatomy	Lower animal
1 day	0.135 mm	Single cell	Single cell	Ameba
1 week	0.5 mm	Mass of similar undifferentiated cells	Single tissue	Sponge
2	1.5 mm	Three kinds of differentiated cells in three layers	Organ	Hydra
3	2.5 mm	Body assumes shape of tube-within-a-tube; some indications seen of most major organs	Organ system	Flatworm
4	5.0 mm	Organs being actively formed, heart beating, but with only two chambers		Earthworm
8	23.0 mm (1 inch)	Most organs formed, some human appearance to fetus, sexual differentiation can be distinguished	Full vertebrate form	Lower vertebrate
12	56.0 mm (2 inches)	Obviously human form, but few human capabilities		
16	112.0 mm (4 inches)	Spontaneous movements can be felt		
20 to 28		Many small but vital changes in form; fetus becomes capable of independent existence		

formational thinking—that because an embryo can, usually, develop into a newborn, it *is* a baby (just like the newborn). And that is incorrect. It does not fit observations that can be and have been made.

The periods of embryonic development

The development of the embryo is a continuous process, and it is appropriate to refer to "the embryo" at any stage before birth. (Table 6-1 summarizes the developmental stages of the embryo.) However, for the sake of convenience the developmental period is often divided into three periods: the period of the blastocyst; the embryonic period; and the fetal period.

The period of the blastocyst lasts from fertilization to about the fifteenth day of development. As described in the previous chapter, during this time the structures for nourishment and the protection of the embryo (chorion, amnion, yolk sac) are formed, and the three primary layers of cells (endoderm, mesoderm, and ectoderm) that make up the embryo proper have been established. More than 90 percent of the embryonic tissue at this stage is in the chorion, amnion, and yolk sac—the temporary structures.

The embryonic period is the time of **organ building**. It includes the third through the eighth weeks of development. During these 6 weeks there is a rapid initiation of all the organs of the body, at least in "outline form." At any given time during these 6 weeks, several of the organs will be forming. The formation of any particular organ usually takes only two or three weeks, and the various organs begin formation at different times during the embryonic period.

The fetal period is a time of **maturation**. The organs that were established during the previous period gradually mature into functional form. These changes are less dramatic than the earlier ones, but are just as necessary.

The sequence of development

Although most organ systems are developing more or less simultaneously during the embryonic period, it is possible to see some pattern or sequence in which the development occurs. The very first structures, as already indicated, are nutritional. The trophoblast/chorion actually absorbs most of the nutrient for the blastocysts, but the yolk sac may have a little of this function even in primates like man. Since part of the yolk sac persists as the gut, it must be said that the digestive system is the earliest to start, simply for purposes of nutrition. The first new structures to appear after formation of the three primary layers are a part of the skeletal system, the **notochord**, and the forerunner of the nervous system, the **neural tube**. They are important to the early embryo not for support and integration (for which they become important only far in the future) but because they serve to define the midline of the embryonic disc, and to orient all the other structures that will be formed.

Formation of notochord and neural tube

On about the eighteenth day after fertilization, a strip of mesoderm (between ectoderm and endoderm, of course) along the middle of the embryonic plate begins to have a changed appearance. The structure formed is called the **notochord** (Fig. 6-4). (A similar structure of "stiff" cells provides the structural support for some tiny relatives of the vertebrates. The notochord of vertebrates, including man, later becomes imbedded in the bones of the vertebral column.) The presence of the newly formed notochord stimulates a strip of ectoderm above it to thicken along the sides, to fold or roll up into a tube, and to differentiate into cells that will become neurons, or nerve cells. The edges of this **neural tube** meet to form a complete cylinder by day 22, but the ends are not closed off until much later. Along the side of the neural tube, but gradually becoming detached from it, are cords of cells called the **neural crests**. These will migrate to other parts of the body to form the nerve cells outside the

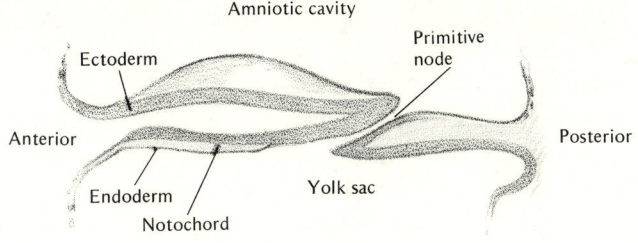

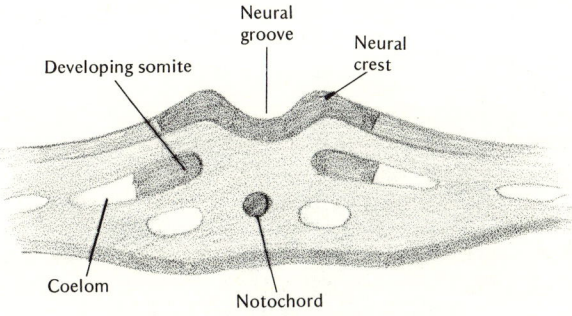

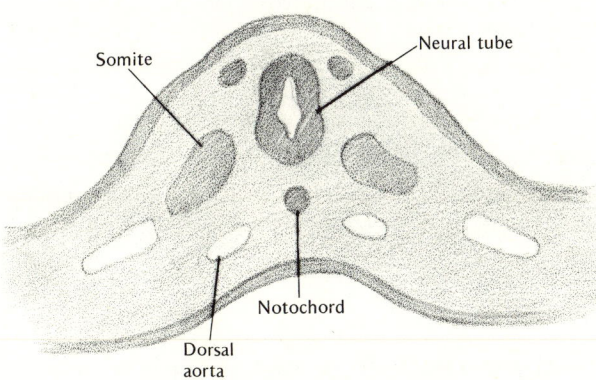

Fig. 6–4: Formation of the first embryonic organs: the notochord and the neural tube.

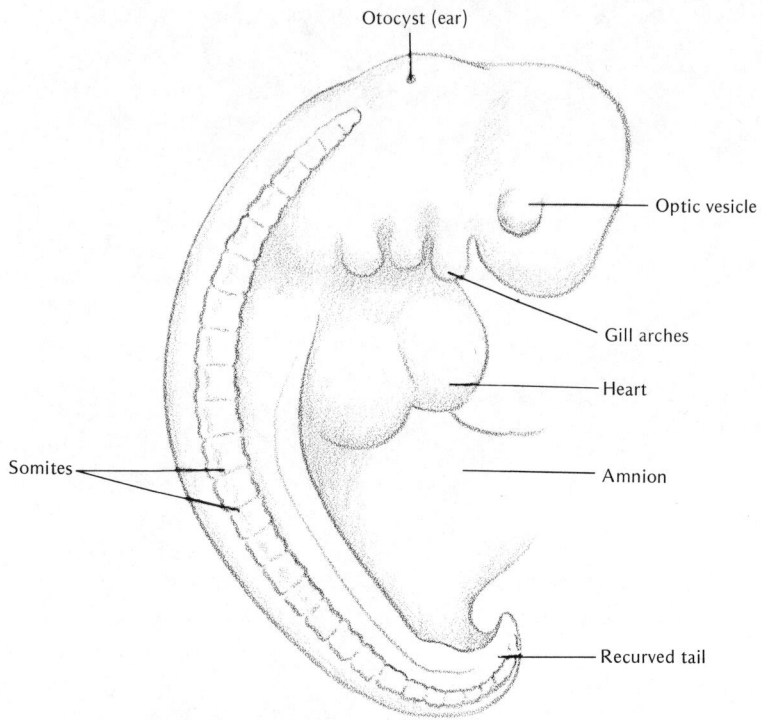

Fig. 6–5: Somites and gill pouches, structures that become something else. Human embryo of about 27 days. Actual size is about ⅙ of an inch.

central nervous system, while the neural tube itself will become the central nervous system (brain and spinal cord).

Somites

The mesodermal cells on each side of the notochord become grouped into dense clumps called **somites**, so that from day 20 to day 30 there is a row of somites, like a string of beads, along each side of the notochord-neural tube (Fig. 6-5). This is one of the most conspicuous features of embryos of that period, and embryologists count the number of somites to determine just how old the embryo is. Such rows of somites are important in the development of some lower kinds of animals in that they establish a series of nearly identical body segments making up much of the length of the body (as in an earthworm). Segmentation is not apparent in most parts of our bodies, except that even in human development each somite gives rise to one of the vertebra, and to the muscles associated with it. The somites also contribute to the formation of other organs, but not in a segmental way. So they are not very important for a simple understanding of development.

The circulatory system

The embryonic structures described so far are important as a kind of superstructure on which further development is based. The first system to become functionally important in the ongoing life of the embryo is the circulatory system. As soon as it is formed, it begins its vital role of carrying materials from one part of the embryo to another. It continues this role throughout life, even while the structure of the body and the composition of the circulatory system are changing. Not only do the other organs of the embryo develop later than the circulatory organs; they never assume as much importance before birth. As long as fetal circulation is adequate, the fetus can survive even grave defects of most of the other organs, because the pregnant woman's own organs, by the exchange of materials through the placenta, perform most of the functions necessary for the life of the fetus.

At the same time that the notochord and the neural tube are forming (days 18 to 21), there are cords of cells in the mesoderms of the yolk sac, chorion, and embryo proper that simultaneously develop cavities within the cords, making them into loosely organized vessels. These primitive vessels rapidly join to form a network, connecting the vessels in all parts of the embryo and its surrounding structures. One of the larger vessels located on the midline near the anterior end of the embryo will become the heart; the connecting vessels will be arteries and veins. The heart begins to beat on day 22; the first few beats cause just an ebb and flow in the adjacent vessels. But in a day or two the circle of vessels is complete enough so that the blood is being pumped throughout the chorion, distributing food and oxygen.

The blood cells are derived from cells that got "trapped" as the vessels were formed; they proliferate and give rise to cells that can manufacture hemoglobin. Some of these stem cells for the production of circulating blood cells remain in special widened parts of the circulatory system. These blood-forming tissues are found first in the yolk sac, in the liver during fetal life, and in the spleen and bone marrow of the adult.

Development of the heart

By day 26, when the fetus is less than 5 mm in size, the circulatory system is complete and functional, but still does not have any of the complications of the adult system. The heart of any air-breathing mammal is divided into four chambers, to make propulsion of the blood more efficient, and to connect to the two different circuits of blood: to the lungs and to the rest of the body. These features begin to form in the early embryo, even though they will not be functional until birth.

On day 26 a septum between the right and left sides of the heart begins to form, and it is completed in the following 12 days (Fig. 6-6). This septum separates the two atria from each other, and the two ventricles from each other. In addition, the atrium on each side becomes divided from the ventricle into which it empties by a constriction of the tubular heart, by the enlargement of the cavities relative to this constriction, and by the growth of valves between the chambers. An opening through the septum between the two atria persists until birth as part of the mechanism for shunting the blood away from the nonfunctional lungs.

The blood vessels

The heart of the early embryo connects with the aorta by several pairs of vessels called **aortic arches**. In a fish, similar embryonic structures would become the vessels supplying blood to the gills. In the mammal, most of the aortic arches disappear, but parts of them persist as sections of the final circulation. The front end of the final aorta is the left side of

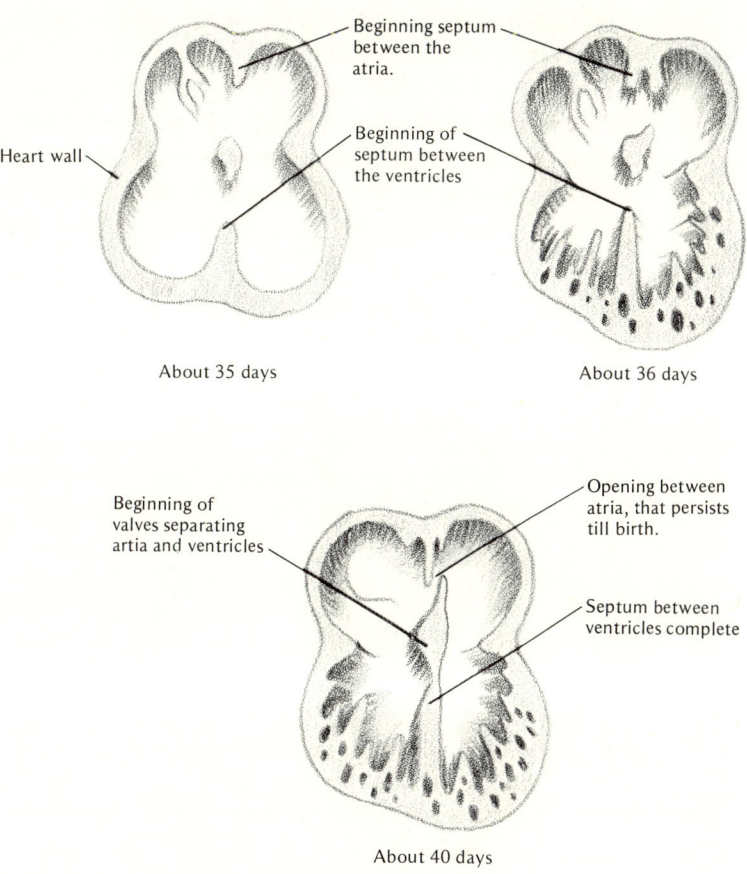

Fig. 6–6: Formation of the heart chambers.

arch 4. The carotid arteries are divided from arch 3, and the pulmonary artery from arch 6. Another part of arch 6 forms the **ductus arteriosus**, which shunts the blood that enters the pulmonary artery over into the main circulation. (At birth, the ductus arteriosus constricts and grows shut, and a flap closes over the hole between the atria, gradually growing into a permanent closure.

As the arms and legs grow out from limb buds, buddings from the central arteries and veins grow with them, to join in the extremities and supply their circulation.

The circulation of the embryo and fetus, though complete

and functional, is not the same as that of the child after birth (Fig. 6-7). The most important difference is that of the umbilical vessels that carry blood to and from the placenta. The umbilical arteries branch off from the large arteries supplying the legs. The umbilical veins drain through the liver into the vena cava. These umbilical vessels degenerate

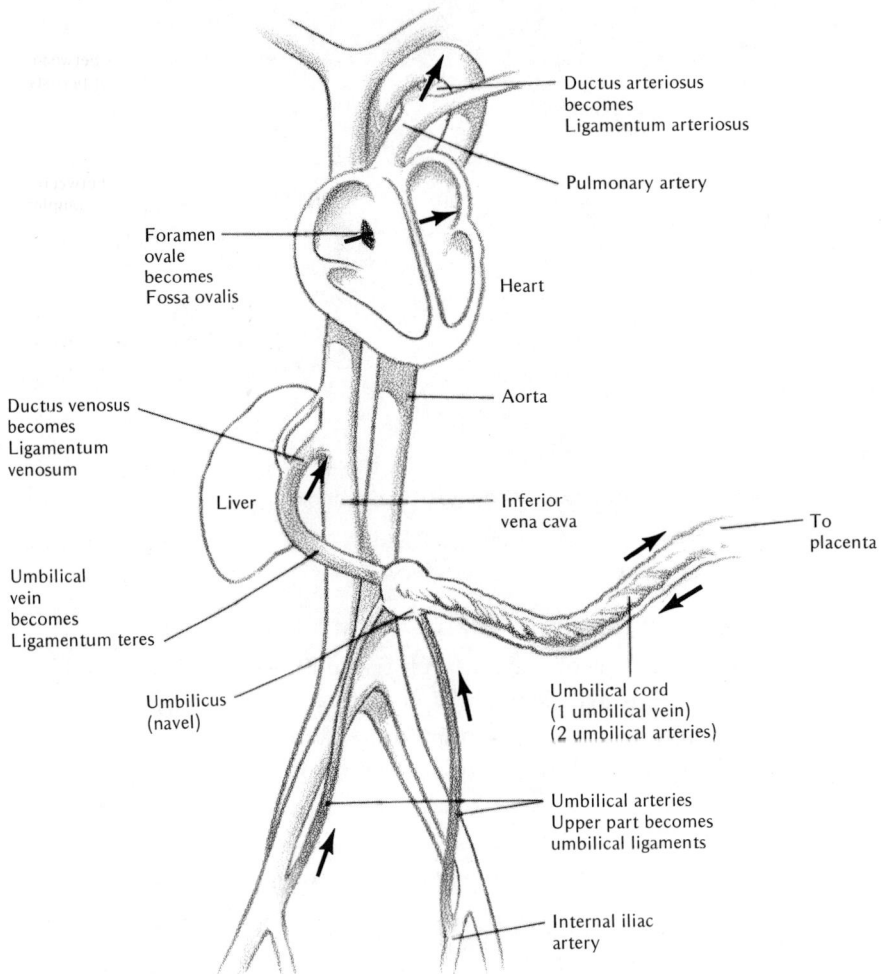

Fig. 6–7: Temporary blood vessels in the fetus. Those in the umbilical cord are lost; others become ligaments.

after birth. In the earlier embryo, much of the body wall is drained by "cardinal veins," but by the eighth week, they have largely disappeared to be replaced by another set of vessels. Less profound changes occur in later fetal life. Throughout all these changes, the flow of blood is not interrupted.

Because of the many changes that take place in the development of these vessels, and because the usual final pattern is not the only one that will work well, there are many small variations from person to person in the exact way the veins and arteries are hooked together. The functionally serious variations in structure of heart and arteries can sometimes be corrected by surgery soon after birth or in later life.

The digestive system

At about the time the neural tube has been formed (day 22), the posterior portion of the embryo, where there is no notochord or neural tube, begins to fold under the rest of the embryo. A few days later the anterior portion of the embryo, including some of the neural tube, also begins to fold under, so that what was on the upper surface becomes the top of the head and part of the face surface.

Before this folding starts, the cavity of the yolk sac is bounded on one side by the future "inside" of the embryo. After the folding has started, part of the cavity of the yolk sac is enclosed by the folds of the body wall into blind pouches in front and back, called the **foregut** and **hindgut** (Fig. 6-8). The folds of the body wall continue to grow together, gradually pinching off the yolk sac and leaving it outside the embryo. While this is going on, parts of the gut wall begin to give rise to organs associated with the intestinal system.

Organs derived from the gut

By 35 days, the parts of the gut that will become throat, esophagus, stomach, and intestines are distinguishable, though far from their final shapes. But several other organs

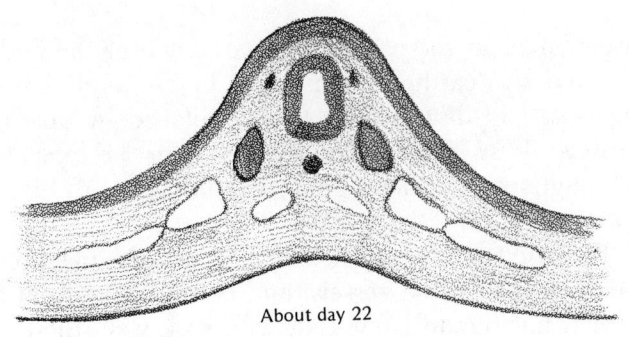

About day 22

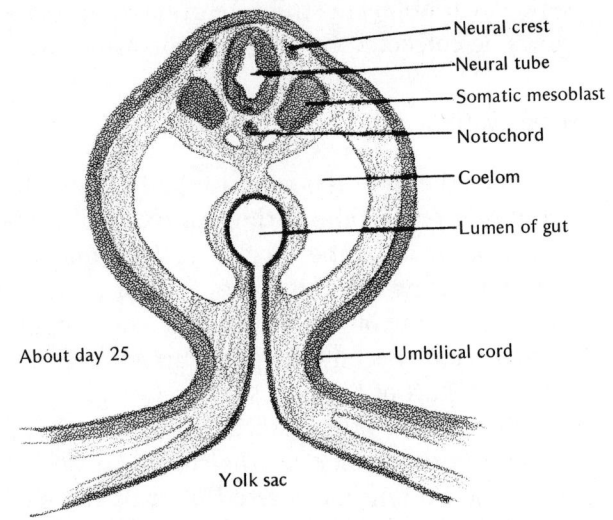

About day 25

Neural crest
Neural tube
Somatic mesoblast
Notochord
Coelom
Lumen of gut
Umbilical cord

Yolk sac

Fig. 6-8: Formation of the gut from the yolk sac.

develop from this early tube (Fig. 6-9). The pancreas and liver (which will discharge secretions into the later intestine) develop as outpouchings from the early embryonic intestine. Each outpouching branches and grows, finally becoming a compact, multibranched gland. An outpouching of the gut nearer the head end becomes the respiratory system. At 35 days the trachea is evident, but each lung is still a small and simple bud. In the next few weeks the lungs are elaborated, but they are folded up and nonfunctional until birth.

Derivatives of the gill arches

The front part of the foregut is associated with structures called **gill arches**, with **gill pouches** between them (Fig. 6-5). In a fish embryo, similar structures would turn into the functional gills. In the human or other mammals, they turn into a variety of organs, including the jaw, the middle ear, the parathyroid glands, the thymus gland, and the larynx.

Development of the face region

The development of the face is one of the most complicated processes observed in embryogenesis. Perhaps that is the reason that defects such as cleft palate are relatively com-

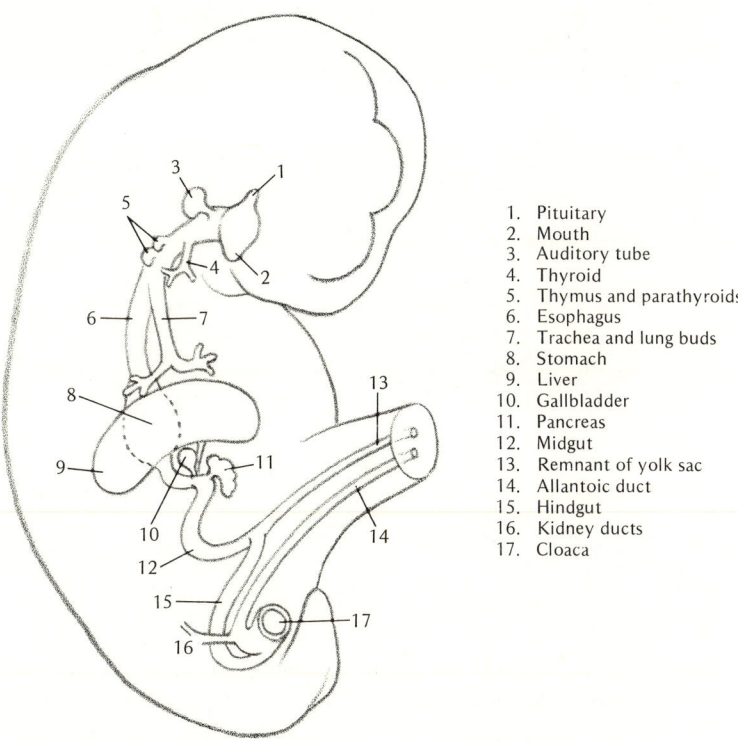

1. Pituitary
2. Mouth
3. Auditory tube
4. Thyroid
5. Thymus and parathyroids
6. Esophagus
7. Trachea and lung buds
8. Stomach
9. Liver
10. Gallbladder
11. Pancreas
12. Midgut
13. Remnant of yolk sac
14. Allantoic duct
15. Hindgut
16. Kidney ducts
17. Cloaca

Fig. 6–9: Organs that develop from the gut. Developing gut of a 35-day embryo showing the various parts derived from it.

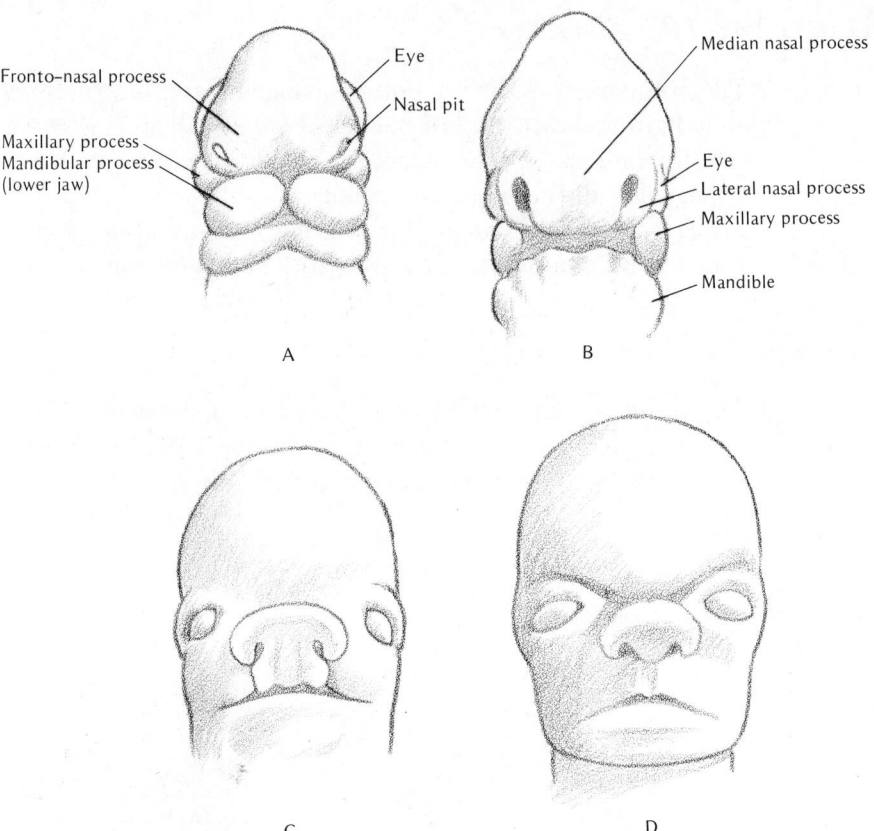

Fig. 6–10: Facial development. Total embyro length at these stages of development: A. about ½ inch; B. about ⅓ inch; C. about ⅗ inch; D. about ⅘ inch.

mon. The face is formed by a growing together of the top of the head with parts of the gill arches from the throat region (Fig. 6-10). There are five projections, called "processes," that are usually identified. Growing down from the top of the head is the **front nasal process**. At its base it is split up into the **median nasal process** (which will form the center of the nose and the middle of the upper jaw), and two **lateral nasal processes** (destined to form the outside margins of the nostrils and the cheeks and to fuse with the upper jaw). Derived from the gill arches are the two **maxillary processes**.

These grow in from each side; the line of growth extends from the front surface into the back of the mouth-nasal cavity. When the maxillary processes meet in the middle, they will have formed the upper jaw and the palate (which separates the nasal passage from the mouth cavity). All five of the growing processes must enlarge at just the proper rate and then fuse properly when they meet to form the final structures. By the eighth week the structures are all there, although the face does not yet have a human appearance.

The arms and legs

Small buds that will grow into the limbs are visible in the embryo at 28 days. These gradually elongate and begin to develop further structure. By day 40 it is just possible to distinguish upper arm, lower arm, and hand, and the fingers are beginning to form. By day 47 the fingers and toes are fairly well formed (Fig. 6-11).

Bones

The skeleton, as a structure of hard, rigid bones, is slow to develop. Some parts of the skeleton are established first as cartilage, which is replaced later by harder bone. Other bones are deposited directly in softer membranes. **Ossification** begins in many centers and spreads outward from each until there are multiple meetings of bone. Thus the fetal skeleton contains many more bones (isolated hard spots) than are found in the adult. Ossification is only partial even at birth; ossification of the long bones of the arms and legs is complete only when growth stops late in puberty.

Urinary system

The kidneys and their associated ducts seem to develop almost independently of other organ systems. The kidneys start to form about day 25, going through several stages that can be thought of as three individual kidneys, one replacing the other. Each "new" kidney uses some of the ducts and

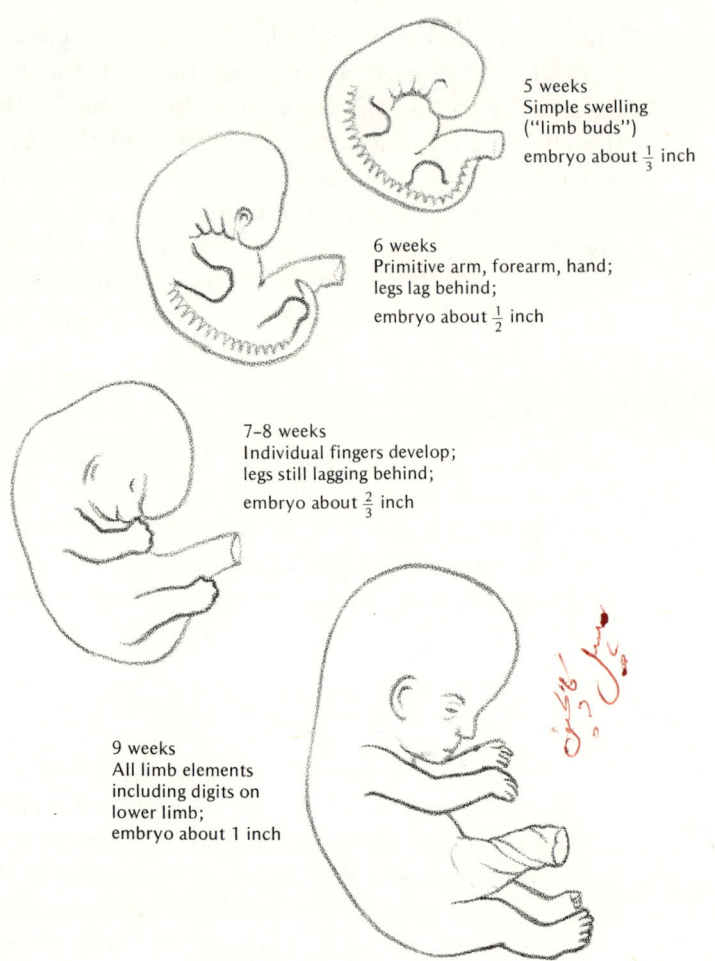

Fig. 6–11: Growth of arms and legs.

structures of the previous one. The third and final kidney begins to develop on day 30, budding off of an earlier duct. The urinary ducts drain the kidneys into the hindgut of the embryo until about day 35, when a septum separates the **urogenital sinus** (receiving the urinary ducts) from the rectum. The urogenital sinus is later enclosed by a growing together of the walls at its outer limit, and it becomes the urinary bladder. The passage through the body wall from the bladder to the outside is now the urethra.

Genital system

The gonads develop from some relatively undifferentiated endoderm; they are associated at the beginning with two sets of genital ducts (Fig. 6-12). The one that will persist in the male, derived from a "discarded" duct of an early kidney, is called the **Wolffian duct**. The female duct, called the **Mullerian duct**, forms out of the mesoderm by itself and

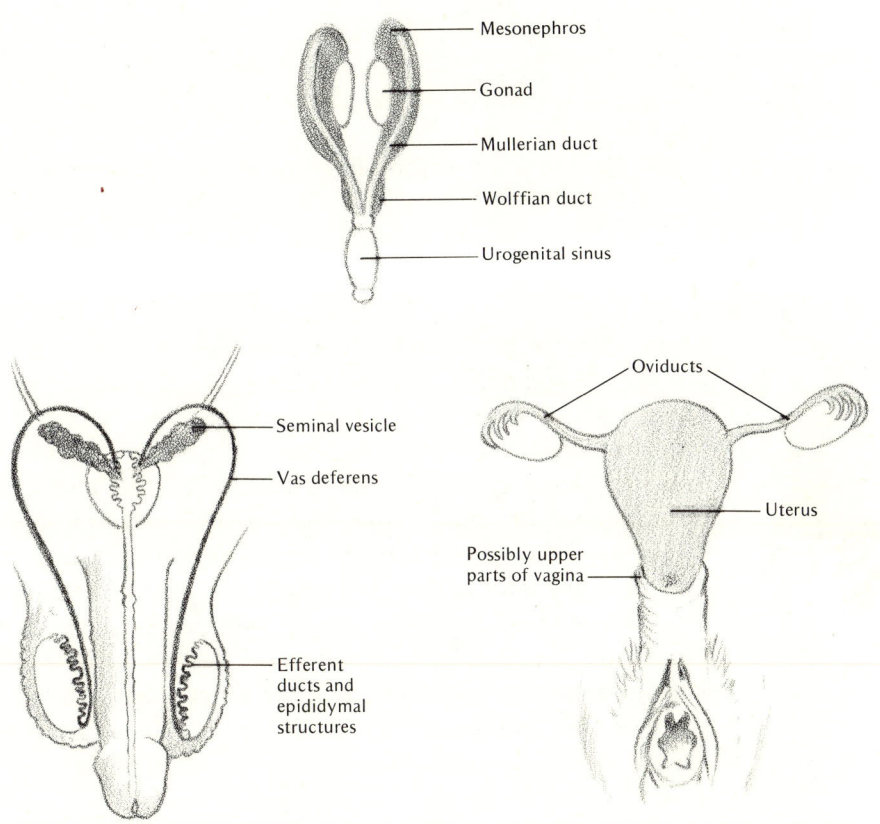

Fig. 6–12: Development of the internal sexual organs of the two sexes has a common starting point.

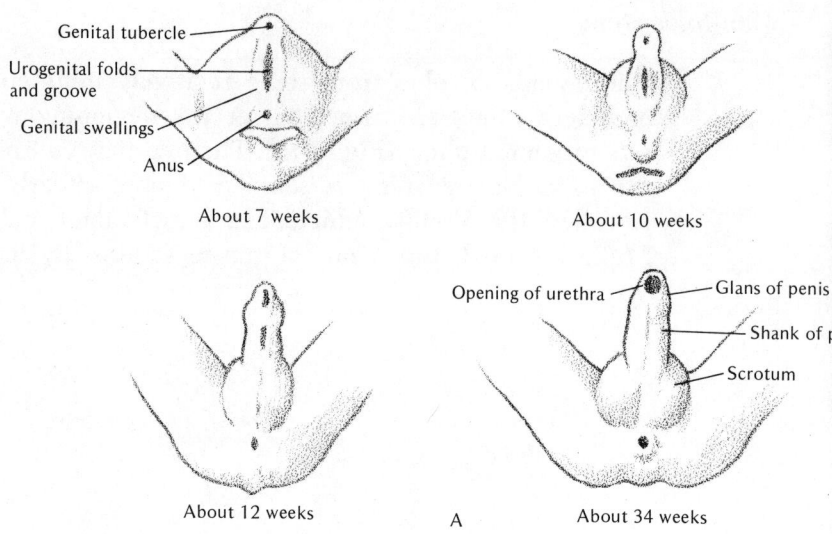

Fig. 6–13: The external genitalia develop from a common stage. A. Male. B. Female.

leads from the body cavity to the outside. At about day 37 the gonad of an XY embryo begins to differentiate into a testis and to secrete testosterone. Under the influence of this hormone, the Mullerian ducts degenerate, while the Wolffian ducts grow, become attached to the testes, and develop into the epididymis and vas deferens. In the XX embryo, on the other hand, the gonad becomes transformed into an ovary later in development. Early in the third month oogonia are forming and beginning their proliferation. At the same time, the Wolffian ducts degenerate, while the Mullerian ducts become oviducts at their inner ends. The lower parts of the Mullerian ducts fuse into a single uterus and vagina.

External genitalia

Previous to the seventh week, the external genitalia of embryos of the two sexes are indistinguishable. By the end

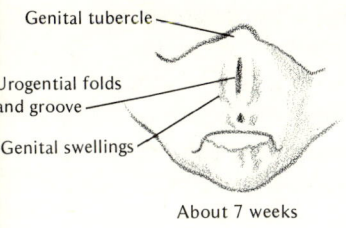
About 7 weeks

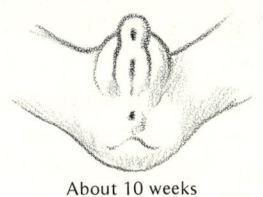
About 10 weeks

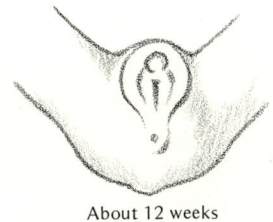
About 12 weeks

B

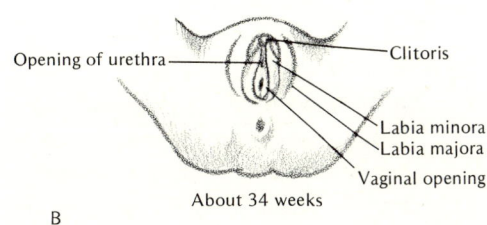
About 34 weeks

of the eighth week it has become relatively easy to distinguish them. The undifferentiated structures include a protrusion called the **genital tubercle**, a slit dorsal to the tubercle called the **urogenital groove**, a pair of **urogenital folds** bordering the groove, and a pair of **genital swellings** outside the folds (Fig. 6-13).

In the male, the tubercle enlarges to form the glans on the end of the penis. The urogenital folds form into a tube (connected with the urogenital sinus) which is the major part of the penis. The genital swellings enlarge and protrude, becoming the scrotum, into which the testes descend. The testes pass through an opening in the abdominal wall called the inguinal canal. (This remains a weak spot through which a loop of the intestine may herniate later in life.) The penis is formed by the eighth week; the testes descend during the eighth month.

The same original structures become the external female genitalia, but in the female they do not assume their form until the third month. The female condition represents less change from the undifferentiated condition than does the male. (For this reason, several early authors overestimated the numbers of females among early aborted embryos.) The tubercle becomes the clitoris, the groove the vaginal opening, while the folds turn into the labia minor and the swellings into the labia major.

The nervous system

The original nature of the neural tube persists in the spinal cord, although the cavity of the tube becomes very small in comparison to the thickened wall. In the brain, though, the tubular nature of the structures is difficult to recognize after development is over. The front end of the neural tube, destined to become the brain, fuses much later than the posterior parts of the tube, becoming complete about day 30. The brain part of the neural tube becomes bent on itself, and the various sections grow to such different sizes that their origins and structural relationships are hard to see from any simple inspection of an adult brain (Fig. 6-14). The most anterior part of the brain, the cerebral cortex, comes to cover most of the rest of the brain, except for that covered by one of the more posterior parts, the cerebellum. Most of the recognizable parts of the brain have already started their development by day 30, but cells are still increasing in number and making their functional connections until after birth. Although the nervous system is one of the earliest ones to become visible in the early embryo, it is probably the last, in the human, to come to its final form during fetal growth.

The brain and spinal cord do their work of integrating events throughout the body by virtue of receiving impulses from sense organs throughout the body, and sending impulses to muscles and glands in other parts of the body. There is a standard pattern in that any specific sense organ is connected to cells in a certain spot in the brain, and each muscle is supplied with fibers from its special spot in the brain. These connections have to be established by the elongation of nerve cells (based in the brain or spinal cord)

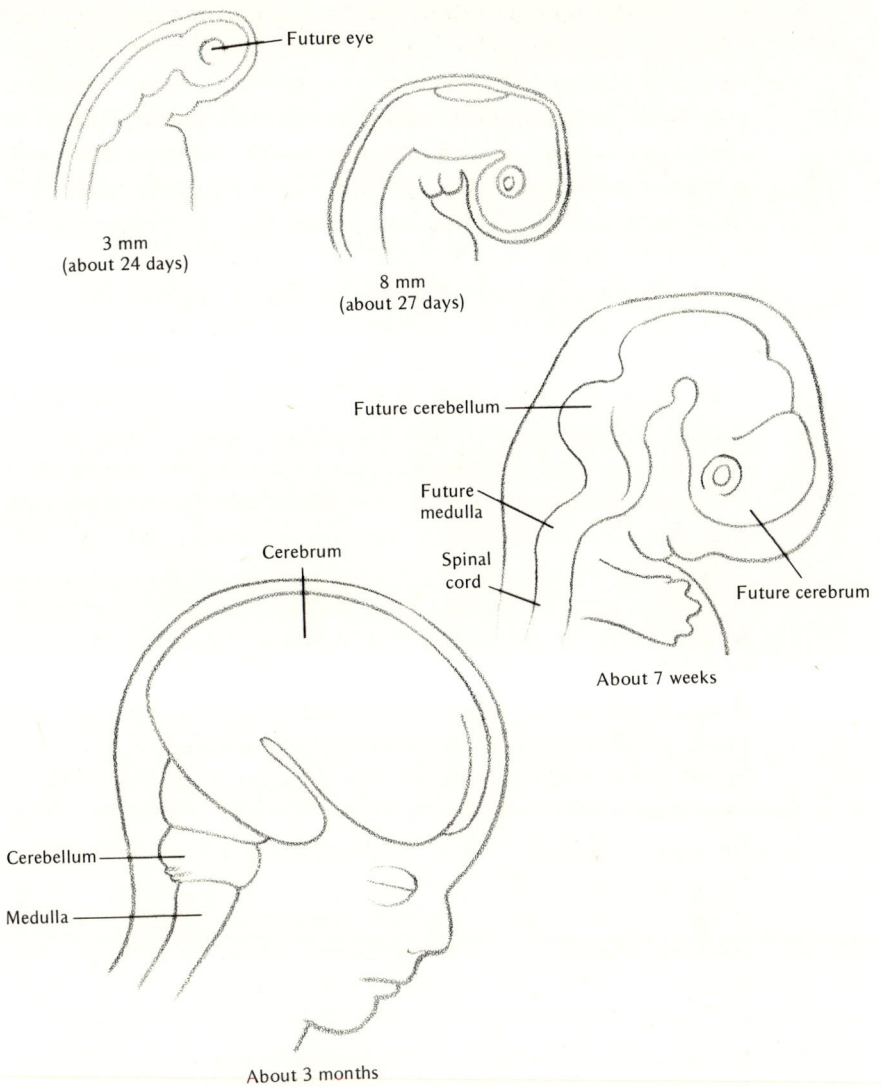

Fig. 6-14: Development of the brain.

so that processes or extensions from them come in contact with the sense organ or muscle. The growing nerve extensions from many brain centers grow out in the same bundles; yet, each growing nerve fiber seems to "know" where it should go to make the proper connection. What guides these specific connections? The tentative answer seems to be that there is simultaneous but independent development of the nerve centers on the one hand and sense and action organs on the other. Simply by the interaction of parts of the brain with each other, certain cells become "set aside" to make connections with, say, touch receptors in the pointer finger. Meanwhile, the fingers have developed. Extensions of those "pointer-finger–touch" brain cells grow out in the arm as it develops and can somehow recognize the touch receptors in the pointer fingers, growing into their vicinity. This description does not remove the mystery; it merely defines what needs to be explained.

While the connections to peripheral organs are being made, there are even more nerve fibers growing from point to point within the brain and spinal cord. Any instincts, reflexes, tendencies, and sensitivities that humans have are the product of whatever connections are established. There is little understanding of the mechanisms of any part of embryonic development; it is safe to predict that development of the brain will be the last to be understood, since it is certainly the most complicated.

The human being's most complex sense organ, the eye, does not develop on the periphery to receive outgrowing extensions of brain cells. The retina of the eye is formed by an outpouching of the brain itself, so that the cells of the retina are really a piece of the brain displaced a few centimeters. The retina induces the overlying ectoderm to become the lens. The lens cups inward, and pinches off so that the ectoderm grows over the spot again. The new coating of ectoderm is then induced (perhaps by the lens) to become the cornea.

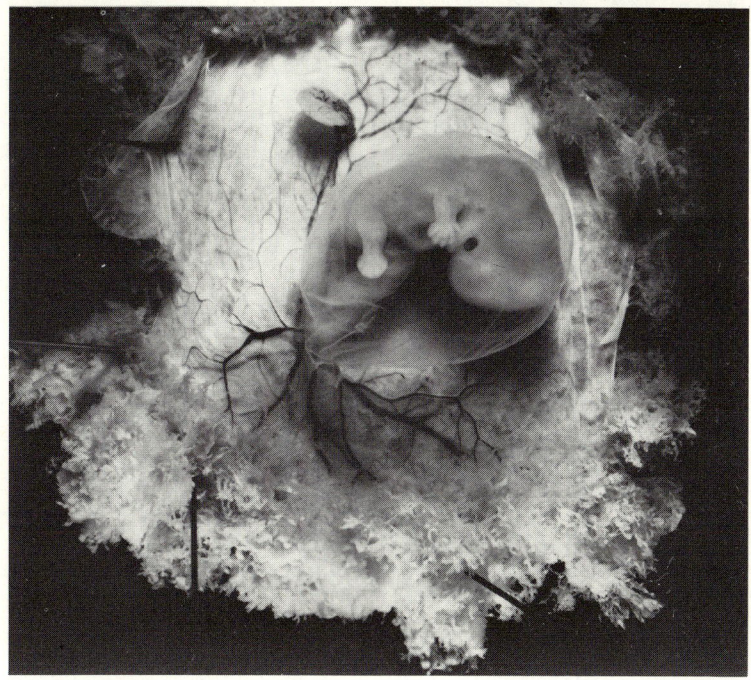

Fig. 6–15: Fetus enclosed in the embryonic membranes. (Courtesy of Carnegie Institute of Washington.)

The fetal period

The changes in the fetus from the end of the second month to the time of birth are slower and less dramatic, but they are still profound (Fig. 6-15). Few of the organs are really functional when formed. They still require a great deal of internal elaboration, growth of tissues, establishment of ducts, and all sorts of other details. In the nervous system, most of the multitudinous nerve connections are probably not established at the beginning of the fetal period. The relative size of the parts of the brain does not become "human" until birth. The brain of a 6- or 7-month fetus looks quite a bit like the brain of a monkey; the distinctly human indentations and foldings are established later.

Fetal preformation?

It might be tempting to think that the old preformationalists were just wrong in their timing; that the 2-month embryo, at last, is a preformed person who needs simply to expand and grow. But there is nothing simple about the development during the fetal period. If the 2-month fetus were merely to increase in size, it would be a useless monstrosity at birth. A fetus has the appearance, the visible pattern, to suggest that it is a "person." But it does not have a person's capabilities.

Most of the fetal organs, although partially formed, are not capable of maintaining homeostasis for the fetus. It is obvious that the fetus depends on the placenta and the systems of the pregnant woman for most body functions, since malformations in most organs do not cause death until after birth. The uterus is a support system that takes care of most problems, as long as the fetal circulation is working. Yet a support system is just about all the uterus is—it does not "shape" the embryo, does not direct its development. The fetus itself is capable of developing, using its own internal direction, into a baby, if only it is given proper support.

The time in development at which babies are normally born is not an accident, even though it may vary a few weeks without serious harm. Rather, the time of birth is at the stage in development at which the fetus is likely to be able to survive under normal conditions on the outside. The attainment of about the maximum possible size is timed to coincide with that stage of development. It is at present possible to use extraordinary measures to support a premature fetus. Originally, this support was only controlled temperature and special feeding formulas; now it can include intravenous feeding. Some day it will be possible (whether practical or not) to use an artificial uterus to support an embryo for its full time of development. However, it will still be a very different object at 9 months from what it was at 2.

Earlier, it was suggested that the zygote might be compared to the blueprints for a house, and the newborn baby to the house itself. (It should be pointed out that this was a little misleading on two counts. First, the zygote actually turns into a baby, whereas the blueprints do not become a house. Furthermore, the zygote, if given the proper support, does the developing by itself, but building a house requires a crew of workers to do the actual building. The second difference is that the blueprints, unlike the zygote, contain a representation shaped like the final product.) If the zygote is at least something like a blueprint, the 2-month-old fetus is somewhat like the house at the stage when the roof is on, the wall studs are up, the sheathing is on the outside walls, and the subflooring has been laid. It looks quite a bit like the final house will look, but it is far from livable.

Recapitulation: an overextended theory

The study of embryology was strongly influenced by the publication, just over 100 years ago, of Darwin's theory of evolution. It was observed that embryos of different vertebrate groups resemble one another in their early developmental stages but become less similar as they grow older. The characteristics that can be detected first in an embryo are those that mark it as a vertebrate; later it is obviously a mammal rather than a reptile, and still later the embryo is obviously that of a human rather than of another mammal. There arose a general theory about these resemblances. The theory held that each animal, as it develops from the zygote to birth, passes through a series of stages in which it resembles, in sequence, the evolutionary ancestors of the species. This is the theory of **recapitulation**. The theory holds that the development of the embryo recapitulates the evolution of the animal: "Every animal climbs up its family tree." A more pompous version of this statement is "Ontogeny recapitulates phylogeny." (Ontogeny is embryonic development; phylogeny is evolutionary history.)

The recapitulation theory is attractive to someone who has seen, for example, the following: A human embryo of 4 or 5 weeks has, on each side of the neck region, a row of gill slits, complete with blood vessels. These look just like the gill slits in fish embryos that later develop into functional gills. The gill slits in mammalian embryos never function in oxygen exchange, but they are there anyway. According to the recapitulation theory, such mammalian gill slits are "hangovers" from the time when our ancestors were fish.

The theory of recapitulation demands that the order in which features appear in embryonic development be the same as the order in which similar features were present in the series of our evolutionary ancestors. This is not always so, and in many particulars the full-blown recapitulation theory does not fit the facts very well. Then, too, it is not really an explanation, since our evolutionary ancestors are not here forcing our embryos to have certain characteristics. So contemporary embryologists have a rather unsettled feeling about the theory. They generally feel that it calls attention to some interesting resemblances but that the explanation of these observations requires a little different approach.

Why do embryos sometimes resemble evolutionary ancestors?

There are two general explanations for the fascinating resemblances of embryos to evolutionary ancestors, as described by the idea of recapitulation. First, the early embryos of any vertebrates are necessarily constructed on a simple plan; the earliest ancestors of the vertebrates must also have been simple. Building up a complicated organism like a vertebrate from something simpler is bound to be somewhat similar, however it is done—whether by embryonic development or in evolutionary change. No cause and effect is needed to explain the basic resemblance. The second explanation is that the structures in the embryos of the ancestral form were retained for other developmental functions when the species evolved to a different sort of animal. Thus the gill slits of the mammalian embryo become the thymus gland, the parathyroid glands, the ear canal, and perhaps some of the tonsils. In the process of epigenesis, the mammalian embryo builds something different than does the fish on the similar "foundation" provided by the gill slits.

How can embryonic development be explained?

The marvelous growth and change occurring in the embryo are based on information contained in both the nucleus and the cytoplasm of the zygote, but development is not simply the unfolding of a preformed pattern. Many structures that appear in the course of development have their particular form because such a structure was present in an ancestor, but the developing embryo is not just a jumble of ancestral remnants. Further explanation is needed. The structures that appear during the course of bird and mammalian development are known in great detail, and enough work has been done with monkey embryos so that, when the knowledge thus gained is added to the study of the occasional human embryos that become available from accidents and operations, the structural changes in human development can be thoroughly described. Some of these were summarized earlier, but in themselves they mean little. There are two fundamental questions that need to be answered about developmental change: What physical processes give rise to the transient structures of the embryo and culminate in the intricately balanced adult? What gives direction to these processes?

Embryonic processes

As discussed in the previous chapter, the genes in the zygote and in each embryonic cell act by providing the patterns for manufacturing specific necessary enzymes. No one cell can use all the enzymes for which its nucleus has patterns, so one of the fundamental processes of embryonic development is the **differentiation** of cell types by the activation of some genes and the inactivation of others. At the same time, the embryonic mass of cells is dividing into **fields** that will develop into specific organs; this development is gradual, with the field for a limb later being subdivided into fields for arm, wrist, and hand. The visible structures of the embryo are formed by such processes as **differential growth** of one part as compared to another, by the formation and **spreading of**

layers of cells, by **migration** of groups of cells, and by **rolling of layers** into three-dimensional structures. These processes do not go on independently of each other; there is overall control of development by hormones, as well as nervous signals, so that everything fits together, especially in the later stages of development.

Directive forces in development

Several different approaches need to be used to explain *why* all these things are happening during embryonic development. To begin with, every event that occurs during the development of an embryo is necessarily governed by the general rules of physics and chemistry, acting upon the molecules of the zygote. Zygotes are, however, too complicated and unique for detailed physical explanations to be very practical. Some features of the embryo exist because they serve to sustain the ongoing life processes of the developing organism. Other structural and chemical details of the embryo at any one stage are there as precursors of the later structures that will develop from them. Finally, and most important, the embryo is not just the sum of its history, the inevitable workings of chemistry, and the necessities of everyday metabolism. It is a "creative system," responding to changing conditions in often unexpected ways. These responses usually culminate in a remarkably effective final individual, considering how many different ways something might have gone wrong at some point in the process of its development.

Explaining everything by physics and chemistry

The idea that something as astounding as embryonic development can be completely explained by the laws of physics and chemistry arouses considerable opposition from many people. Some of this opposition stems from a desire to preserve an aura of mystery, or a feeling that there is something mystical in human development that is beyond any explanation. Others distrust a purely physical explanation because they feel that standard physics and chemistry fail to answer many questions concerning embryonic development. Most biologists would agree with the last statement. But biologists, like all other scientists, know that there cannot be deviations from the laws of physics and chemistry—they are *descriptions* of what things are and how they behave, without exception. Embryologists perform their investigations under the assumption that nothing they find will or can violate the laws of physics. Further, it is assumed that the detailed, minute-to-minute events in the embryo can eventually be explained as the sum of the expected actions of the molecules that are in the cell and in its environment. The laws of physics and chemistry are expressed so that they describe the properties and interactions of simple molecules. Each such molecule, once specified, can be confidently described. One individual molecule of a certain compound has the same properties as any other molecule of that compound; it does not matter whether a molecule of sugar came from a maple tree or from clover via a honeybee—its chemical properties are just the same. Furthermore, molecules can be acted on by reagents to form predictable combinations; they are not individuals.

The importance of history

Precisely because cells, and multicellular embryos, are such elaborate constructions of complicated molecules, all the zygotes or embryos of even one species are certainly not alike. They cannot be explained as simply as standard types of molecules. The nature of embryos depends on what has happened previously in their histories. The evolutionary history of the species may provide some clues to the reason certain structures occur in the embryo; but the genetic instructions contained in each zygote depend on a different kind of history—that is, on the genotype of the parents, and on how the genes happened to be assorted in the meiotic divisions forming the two gametes. The yolk and other chemicals that accumulate during the maturation of the ovum have a profound influence, too (only dimly understood) on how the zygotes will develop. Even after the zygote is formed, the particular environmental conditions during all stages of embryonic development have their influences, and it is rarely possible to predict what directions the influences will take. During development there are some randomly determined events that influence the outcome. One of these is the inactivation of one of the X chromosomes in each cell of a female mammal. (This is the Lyon hypothesis discussed in earlier chapters.) The cells making up any one organ or tissue might happen to have the genes active in just one or the other of the two X chromosomes, since the "choice" is made in early development, and all descendants of each early cell will have the same X chromosome inactivated. So even chance plays a role. All these various historical explanations need to be added to the simple physical ones in understanding embryology.

Organs in the process of becoming

A special "biological type" of cause and effect relationship is seen in embryonic events that can be understood because of their importance to the *future* needs of the embryo. In no lifeless system can we expect an event to take place because of the future. Causes precede events in lifeless systems. It would be nice if an automobile could grow a larger gas tank because it had to cross a desert, but we know that inanimate things cannot react this way. This sort of future-caused event is found in industry, however. Because some people want to drive across a desert or do other special things with a car, a factory turns out a number of different models, adapted for different kinds of tasks.

When the nervous system begins in the embryo, it is a mere strip of slightly unusual tissue, running along the future back of the organism. It has no function in the early embryo; but it will become something essential in the future. In this same sense, much of what goes on in a developing embryo is related to other future needs of the embryo or of the adult. Of course, on a second-to-second basis, such events as the starting of a nervous system take place because structures and chemicals interact in a regular manner. The embryo has the genetic information allowing this constructive interaction because of evolutionary selection—only animals with good nervous systems, for example, survived to reproduce. Thus events in the past resulted in present events that are best understood because of what will be needed in the future!

The immediate needs of the embryo

An entirely different biological kind of cause and effect is provided by the present needs of any embryo. A half-completed mechanical device can just lie on the work bench until someone finishes it. An embryo is no different from any other living creature, however, in that it must respire, obtain food, discharge its wastes, and transport important compounds throughout the body to wherever they are needed. So some of the things that occur at any given stage of embryonic development are related to the paramount necessity of staying alive right then. (These physiological processes are difficult to study, since physiologists are not accustomed to working with a system whose very nature changes while they are studying it.) Embryonic development could not proceed without embryonic maintenance. Throughout the past millions of years, hereditary factors have been selected to maintain viability during the course of development.

The kinds of physiological processes that maintain the present health of the embryo are generally those that provide for homeostasis of the organism. That is, they act to maintain some characteristic of the embryo in a steady condition. This means that some organ of the embryo has to perform some activity to counteract an outside influence. If there is a moderate but disadvantageous increase in the saltiness of the environment, the kidney may secrete some salt. This works for internal changes, too. If one tissue uses up glucose, some organ, perhaps the liver, will manufacture more glucose to replace it. The same sort of internal adjustment occurs within each cell, too. In such a case, the embryo, or even the individual cell, acting as a competent organism, is not simply acted upon by some influence; it actively responds to the influence, and usually counteracts it. Furthermore, in responding to a challenge and resisting it, the embryo *changes itself*. The kidney or liver that restored a steady state in the above examples probably had to grow larger to do so, and so had increased capability for the next challenge. Such active adaptation to meet problems is important even in adults (where the hard-working man develops stronger muscles, for example), but it is of singular importance in the embryo. It is the means whereby embryonic development is a "creative process."

The creative process

Growth and development constitute a creative process. By this we mean that the embryo is an organized system that responds in two ways to stimuli and influences. First, the general health and steady condition of the system are maintained (homeostasis). Second, in the course of responding to influences the embryo itself is changed so that it responds slightly differently to the next stimulus. In this respect, embryonic growth is similar to the other creative processes, which include (1) biological evolution, (2) human development from a child into an independent adult, (3) historical change of societies, (4) individual invention, problem solving, innovative thought, and artistry, and (5) interpersonal relationships.

All creative processes have in common the fact that they represent an active response to a stimulus (not just receiving its impact) and that the result is a new and still more competent system because of the changes brought about by responding. Creative processes contrast with plasticity, which is passive change directly caused by some influence, and also with random change (which usually makes the system less capable). Creative processes also contrast with homeostasis, which is stability or lack of change because the response nullifies the outside influences. (But as described above, the creative process includes maintaining homeostasis as a part of the response.)

Whereas the other creative systems listed perform their creative changing largely in response to random challenges from the outside, the embryo is unusual in that it is usually responding to internal stimuli that are built in and standardized. The combination of active genes in the various cell types is such that the proper stimuli are given to one group of cells by another group of cells, and at the proper time, so that each group of cells will react creatively. For this reason, the creative growth of the embryo usually takes place at a standard pace; it is responding to a standard set of stimuli.

Each individual embryo, of course, has (except in the case of identical twins) a unique set of genes. The interaction of these genes during development in such a way that the whole embryonic system responds creatively works so well that the result is usually a harmonious whole unless there is a serious gene defect. Even a moderately serious gene defect may be compensated for by some action of the embryo, so that its impact is much less than might be expected. An embryo also responds creatively to stimuli or injury from outside itself and repairs, at least partially, many kinds of defects caused by injury.

Stimulus and response in the embryo

Many examples of stimulus and response in the embryo can be given. The lens of the eye is formed when an outgrowth of the brain comes into contact with the surface ectoderm. The influence of brain cells on ectoderm cells causes those ectoderm cells to differentiate into lens cells; they become flattened and transparent.

In the very early embryo, a temporary structure, the notochord, is formed from part of the mesoderm (middle layer). The notochord acts on the ectoderm that is right above it, to produce embryonic nervous tissue. The notochord has provided a trigger stimulus, inducing the nervous system to start forming. Because of the location of the notochord, the nervous tissue of vertebrates lies on the back rather than the belly side of the animal.

When a sheet of tissue is rolling into a tube (as in the formation of the penis) and the edges meet, the surfaces fuse, and growth slows or stops in response to the contact of the surfaces.

The development of the placenta to a certain stage allows

production of chorionic gonadotropins, which stimulate the gonad of a male embryo to produce testosterone, which in turn induces the development of other male characteristics.

Continuation of the creative process

Development and growth after birth are a continuation of the creative processes that operate during embryonic life. The only difference is that many new and variable outside influences are added to the internal stimuli to which the system responds. Throughout development the system is responding creatively to whatever situation it is in. In the embryo, the situation (given a healthy uterine environment) is largely internally determined by the embryo itself. It is therefore intrinsically difficult to direct or even alter embryonic growth in useful directions from outside the embryo. But in the child, the developmental process (especially development of behavior) responds to many environmental stimuli, including the parents, as well as to internal ones.

The child is not just a plastic lump to be shaped by directive influences, but a creative system that responds actively to stimuli according to its individual nature. Therefore, the results of a stimulus may at times not be accurately predicted—and may even be the opposite of what the parent intended. The difference between "inborn" and "learned" behavior is, for this reason, not an absolute difference. It is just that there is a large contrast between the time before and the time after birth in *how much* of the learning is in response to internal, genetically programmed stimuli and how much is in response to stimuli from the outside.

Developmental defects

Since every individual (with the exception of identical twins) has a unique combination of genes, a combination that has probably never directed the development of an embryo before, the amazing thing is that most babies that are born are competent and functional—essentially perfect. Only 1 to 5 percent of all babies born have important defects. (The uncertainty of the number reflects disagreement as to how serious a deviation from the usual must be for it to be called an important defect.) At least one-fifth of the deaths of very young infants is known to be due to defects present at birth, and the real proportion is probably even higher. So it is of great interest to understand how such defects can occur, and how they can be prevented. Unfortunately, prevention of most defects is not likely to be possible in the near future; the necessary experimental work has hardly begun when compared with the large amount of research that will be needed to solve the problems. It is possible, however, to see how some birth defects can be understood as defects of the standard processes that shape the embryo. For each of the basic embryonic processes discussed below, some defects that result from disturbances of the process will be described.

Gene action

Single-gene defects that can cause such problems as albinism, sickle cell anemia, and phenylketonuria have been discussed earlier. Another example, in which it is possible to see how a single-gene defect could cause a complicated developmental defect, is one of the conditions known as the "adreno-genital syndrome." In one form, a female fetus develops normal internal organs, but the external genitalia are of the male type. Here the defect is in a gene that normally codes for an enzyme acting in the adrenal cortex, as a part of a sequence of enzymes leading to the metabolic hormone cortisol, a steroid related to the sex hormones. Because of this block, the precursors of cortisol go into alternative pathways, including one for the production of testosterone. There are feedback mechanisms (involving the pituitary) that tend to keep the level of cortisol constant; they cause the total production of steroid hormone to increase so that the little fraction appearing as cortisol maintains the normal concentration of this hormone. The side effect of all this is an enormous amount of testosterone, causing the masculinization of the fetus, a state that continues throughout life as a sterile male.

Many other birth defects of all sorts seem to be caused by the chance presence of mutant genes at several loci working together. The presence of just one of these genes seems to cause no trouble, perhaps because the work of the defective enzyme can be taken over by another, similar enzyme specified by a gene at a different locus. But if a zygote inherits defective genes at more than one locus, a defect results. It will probably take some time before such complicated problems can be untangled and understood.

Tissue formation: differentiation of cells

The zygote does not start out as just a uniform bag of cytoplasm; it has considerable internal structure, much of it not well understood. Most evident is the fact that there is more yolk in one half of the zygote than in the other. The first cell divisions that form the embryo take place rapidly, one after another, without any growth or even much readjustment between divisions. The nonuniform cytoplasm of the large zygote is thus parceled out into 20 or so nonuniform cells. These cells have already started toward differentiation. In the usual course of events, those cells that received the yolk-rich cytoplasm will become endoderm, the rest ectoderm. The differentiation into endoderm and ectoderm layers is the first step in embryonic differentiation. The next step is apparently due to the action of these two types of cells on each other: The cells in the boundary layer between them change into the third type, the mesoderm.

By a complicated system of interaction among cells, and by reaction to environmental stimuli, different cells of the early embryo establish different patterns of gene action, which persist through cell division into the daughter cells. Because different sets of genes are active, the shape, appearance, and reactions differ from one cell type to another. Further differentiation may occur in the daughter cells.

The action of one tissue on another to cause a further differentiation is called **induction**; it is thought to be the influence by which most of the embryonic differentiation takes place. Two examples have already been given: The notochord induces the differentiation of ectoderm into nervous tissue, and nervous tissue later induces the differentiation of another area of ectoderm into the lens of the eye.

Exactly how the influences from one kind of cell can cause another type of cell to change into still a third type is unknown, despite a great deal of research that has already been done on the matter. Each type of differentiated cell probably contains some specific proteins within its nucleus that combine with the unnecessary genes and keep them from operating. Perhaps proteins produced in one kind of cell can diffuse into another and act on its genes to change it. Whatever the exact mechanism, it is evident that any failure in the induction of differentiated cells would have serious repercussions, since the failure to produce one type of cell would snowball, preventing further inductions. For instance, if the notochord failed to develop, it could not induce differentiation of ectoderm into nervous tissue. That by itself would be more than enough to doom the embryo; but among other effects, there could be no lens, because its inducer would be lacking. Therefore there are few birth defects that can be blamed on defects in differentiation; most of them would be fatal in early embryonic life. One probable exception to this rule seems to be the condition called "thallasemia." This is similar to sickle cell anemia and G6PD deficiency, in that the red blood cells are fragile and do not support the growth of malaria parasites. Thallasemia is due neither to a defect in the gene for hemoglobin structure nor to a problem with the energy supply for maintenance of cell structure. Normal hemoglobin actually consists of a combination of two different kinds of protein chains, which are normally produced by the cell in similar amounts. In thallasemia, the controlling factors that keep both required genes working at equal rates have gone wrong, and the hemoglobin produced has too much of one kind of subunit. This defect in differentiation is minor enough so that the embryo may survive at least until birth.

Formation of the organ fields

At the same time that the different tissues or cell types are becoming distinguishable from each other, the whole embryo is being divided into fields (initially invisible) that will gradually form into visible organs. When first formed, the fields overlap. That is, a cell in the region of overlap between the abdomen field and the leg field could become a part of either the abdominal wall or of a leg muscle, depending on the circumstances. As visible structures begin to form, possible fates of cells become more and more mutually exclusive; each of the cells is irrevocably headed toward one outcome. The original fields become subdivided into smaller units: Finger fields develop at the end of the arm field, for instance. If some injurious agent (a drug, a virus, an oxygen shortage) affects the embryo when it is in the limb-field stage, the whole limb might fail to develop. An injury at that time could not affect only one finger or one part of the arm. But if the injury came after the arm field was divided into subfields for upper arm, lower arm, wrist, palm, and each of the five fingers, an injurious agent might affect any one or a combination of them. The injuries caused by thalidomide (to be discussed later in this chapter) appear to have been brought about by an interference by the drug with the process of field formation. Those cases in which the whole arm was missing have been injured at the early arm-field stage. Those with recognizable hands at the ends of very short arms might have been affected at a later stage, with injury done to the arm fields, but not so much to the hand and finger fields.

Differential growth

The various parts of the embryo do not all grow at the same rate. In fact, some parts may stay the same size for a time, while others grow rapidly. The obvious result is that the whole shape of the embryo changes. Such **differential growth** is important throughout development, from the establishment of the earliest structures to the attainment of the final form of the adult. For instance, when the limbs are forming, they are at first nothing but small bumps on the side of the embryo. But the cells in those protruding limb buds divide rapidly, so that the area grows much more quickly than the bulk of the embryo. As the limbs elongate, their ends develop more projections, which will elongate into fingers or toes. Sometimes such differential growth is furthered by the regulated death of certain cells. In the development of the final shoulder joint, for example, some of the cells in the armpit area die rather suddenly, despite good supplies of food and oxygen. This deletion of cells helps to narrow the arm at its point of junction with the body, and it is an important part of the attainment of final form. Similarly, cells between the finger buds die and so help separate the individual fingers.

Differential growth occurs throughout later embryonic development and through all of childhood until growth stops. The inherited differences among us in body shape must work by encouraging differential growth to a greater or lesser extent. Someone with genes for long arms must have a faster elongation of the arms, or this growth will continue longer, than in a person whose genetic instructions call for short arms. The control must be rather precise, since it is often possible to see strong family resemblances in the subtle differences among people's facial features. Perhaps a gene for a certain length of nose actually specifies, in some unknown manner, how many cell divisions will take place in the cells that form the nose.

Formation and spreading of layers

Formation of layers of cells is especially important in the earliest stages of embryonic development. One of the earliest developmental events is the formation of the three primary germ layers and the spreading of the mesoderm to cover the outside of the amnion and the inside of the chorion. Since all organs are composed of several layers of different kinds of cells, a great deal of formation and spreading of layers must go on in the course of organ formation during the first 3 months of embryonic growth. Injury to any of the layers being formed may prevent the proper development of the whole organ. This is the reason viral infection of the embryo, especially by **rubella** (3-day or German measles) can cause such an array of defects in the heart, inner ear, eyes, and less often, in many other organs. Generally, the earlier such an injurious agent affects the embryo, the more serious is the damage.

Importance of the growth rate

In order to cause a defect, it is not always necessary that the agent of injury kill the sensitive cells. It is often sufficient to slow their growth. It has been suggested, for example, that the common defect of cleft palate is caused by such slowing of growth. The palate, separating the mouth cavity from the nasal cavity, is formed by the growth of palatal shelves from each side (part of the growth of the maxillary processes). The shelves meet in the middle and fuse into the solid palate. If some injurious agent merely slows the growth of the shelves while the head (and the cavity in it) continue to grow, the space to be covered becomes too wide for the shelves to establish contact when they come to the natural end of their growth period.

The whole group of septal defects of the heart may be caused by just the same kind of problem. If the growth of the septa between the two atria or between the two ventricles is slowed, the heart cavity may enlarge so much that later growth of the septa does not establish a complete wall.

Birth defects: the importance of timing

As indicated in the preceding discussions, it is possible to deduce what kind of developmental accident might have led to any given birth defect. However, the causes of the majority of these accidents leading to birth defects are not known. Some are known to be genetic, some caused by radiation, some by chemicals, some by viruses. But no one knows whether the large number of unexplained defects is also caused by these same kinds of agents or by something else not yet discovered.

One of the things that has been rather well established by the experimental injury of embryos in experimental animals is the importance of timing in determining what harm will be done. If the injurious agent acts during the period of the **blastocyst**, up to about the fourteenth day, the embryo will either recover without visible damage or else will spontaneously abort at an early stage. If the injury is during the **embryonic period** of rapid organ building, the embryo will probably survive to term but will then appear as a neonatal death or as a birth defect. If the injury occurs during the **fetal period**, it probably will not cause a visible defect (the structures are already formed), but may weaken some organ so that illness will occur later in life. Within the organ-building period, the exact timing of the injury will determine what kind of defect is caused, depending on the injurious agent and on what is being formed at the time.

Radiation

Since the use of x rays in medicine began 75 years ago, it has been found that embryos are especially sensitive to them, being injured by doses one-tenth as great as those that would cause permanent injury to an adult. The nature of the injury depends on the time of irradiation. If the gametes or zygotes receive a large dose of x ray (or of any other ionizing radiation), there may be chromosome aberrations such as translocation of a part of one chromosome to another, or loss of parts of chromosomes. These will be likely to be fatal to the cells, or if not, to cause abnormalities in the embryo. Usually the zygote either dies or recovers. If the embryo is irradiated during the early divisions, the likely result of a small dose is loss of one chromosome to produce a monosomy. Apparently, all types of monosomy are formed, but the only ones that survive are, as usual, the XO condition—Turner's syndrome. Irradiation with somewhat higher doses during the period of organ building can cause any of a great variety of defects, depending on what is being formed at the time. It seems, though, that since it is such a potent agent of injury, radiation would be expected to cause even more structural defects than it does. Perhaps because the injurious effects of radiation are felt by all cells, it would not be likely to cause more retardation of growth in one tissue than in another. Thus it would not induce structural defects, like the cleft palate, caused by slow growth of one part while other parts get disproportionately large. During the fetal period most organ systems are not easily injured, but substantial dosages of radiation (as in therapy of the woman or even in fluoroscopy), can interfere with the maturation of the brain and cause serious mental defects.

Chemicals that cause defects

All drugs have their effects by stimulating or inhibiting some specific physiological functions of the body. Thus it would be expected that drugs could cause specific inhibitions of growth in an embryo and bring about congenital defects. Experimental studies with animals have shown that almost any drug or chemical can cause birth defects if administered under some special conditions. Therefore, it is surprising that only a few drugs used in human medicine are known to cause birth defects. Actually, a drug is likely to cause serious damage to the embryo only if it has three properties: (1) low toxicity to the pregnant woman, so that she will be likely to take it in high dosage; (2) a very high dose required to kill the embryo, which otherwise will be aborted instead of showing a defect at birth; (3) a low dose required for causing developmental defect. Most drugs that can cause defects are either toxic to the woman or fatal to the embryo and so do not cause significant numbers of malformations.

Thalidomide

Unfortunately, there has appeared one drug that fulfills all the requirements for a **teratogenic agent** (one that will cause birth defects). In 1958 thalidomide was introduced in European medical practice as a mild sedative, with unusually few side effects. As such it was prescribed for many pregnant women. Within the next year there was noted a great in-

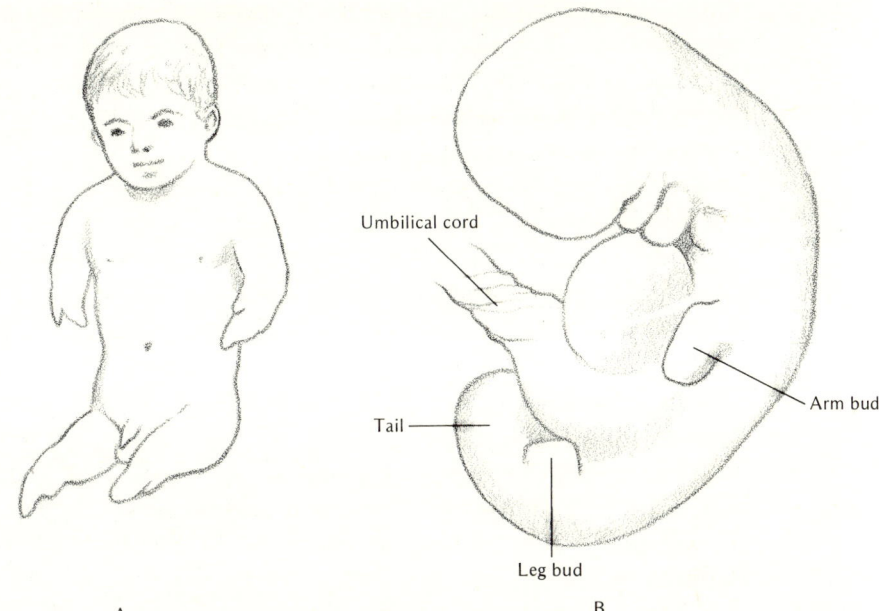

Fig. 6–16: The thalidomide abnormality. A. Chemical interference has prevented normal limb development. B. An embryo at about 1 month shows the beginning "buds" of what should normally develop into limbs. The embryo must initiate and complete this process within a genetically prescribed time or the process cannot continue. The thalidomide drug prevents the completion of the process if administered during the crucial period and thus results in undeveloped or only partially developed limbs on an otherwise normal body.

crease in the numbers of babies born with severe defects of arms and legs, or even their absence. The correlation between the drug and the defects was detected, and the drug was withdrawn from use—but not before thousands of severely malformed children had been born. (By a happy and unusual combination of wisdom and bureaucratic delay, thalidomide was still awaiting approval for use in the United States when its effects were discovered.)

Thalidomide interferes in some unknown way with the process of field formation in the growth of the limbs (Fig. 6-16). If taken on day 37 of pregnancy, it can cause absence of the arms, but if taken any time from day 38 to day 45, it will cause the development of short, flipper-like arms. From days 40 to 45 it can cause absence of legs; from days 42 to 47, severely shortened legs. After day 50 it causes few if any visible defects.

Are there other "thalidomides"?

The blame for defects was properly placed on thalidomide only because it produced a previously rare, dramatic, and wide-spread birth defect. Other drugs that simply cause occasional embryonic death, or increase slightly the incidence of a common defect like cleft palate, would be extremely difficult to detect as being dangerous for pregnant women to take. After the thalidomide tragedy, that drug and many others were tested for their effects on development in a variety of experimental animals. On the commonest laboratory animals, rats and mice, thalidomide had no harmful effects at common dosages. Another drug called meclizine (used to reduce vomiting in pregnant women) *was* found to cause many serious defects in these animals. So a careful study was made of the children of women who had taken meclizine during the period of organ building in the embryo; there was no evidence at all of increased birth defects in these children. (If anything, the results were better than with no medication and uncontrolled vomiting.)

This illustrates very clearly the problem of screening new drugs for their effects on embryos. If mice and rats had been used as test animals for these two drugs, thalidomide would have been judged safe, and meclizine judged extremely dangerous, and the wrong drug would have been released for human use. It is true that thalidomide causes defects in rabbits and monkeys similar to those in humans. But the similar response in rabbits might not be true for some other drug, and monkeys are expensive and in too short supply for the mass trials needed in the testing of all new drugs. Perhaps a large breeding program to supply sufficient monkeys for testing purposes will be necessary in order to avoid another disaster similar to the thalidomide case.

A few other drugs, including steroid hormones, antibiotics,

and even large doses of vitamin A, have been known to cause occasional defects to the embryo when the pregnant woman was treated with one of them. Although the list of proven teratogenic drugs is small, it is now medical policy to use as few drugs as possible during early pregnancy, for maximum safety.

The effect of viruses

Viruses constitute another kind of agent known to cause wide-spread birth defects. Any virus causes infection by penetrating cells and taking over their synthetic machinery to make more viruses, thus interfering with cellular activities, or even killing the cell. If a virus infection is established in a pregnant woman, and if the virus is able to penetrate the placental barrier into the embryonic circulation during the sensitive period of organ building, it may cause a defect. As with radiation or drugs, viral damage during the first 2 weeks of pregnancy would probably cause spontaneous abortion rather than a birth defect. Luckily, not all viruses are capable of infecting the embryo and causing defects.

Rubella

The viral disease with apparently the greatest, and certainly the best understood, effect on embryos is **rubella**, often called German measles. It is usually a very mild disease in adults, with a measles-like rash and a fever for a few days. But when a pregnant woman has the disease during the first 3 months of pregnancy, the probability of birth defects is very high. Perhaps because the potency of the virus differs from one epidemic to another, there are many conflicting estimates of the risk. A summary of the data reported indicates that in cases of rubella contracted by pregnant women within the fourth week of embryonic development, about 50 percent of the babies born are affected. After that time the danger decreases steadily, to 20 percent at 10 weeks, and to no unusual risk at 18 weeks or thereafter.

The early embryo does not appear to have immune-type

defenses against viruses, and so it does not shake off an infection that has been contracted. It has proved possible to isolate virus from the affected organs of defective babies; the infection has been found to persist far beyond the few weeks it lasts in adults. Those types of embryonic cells that happen to be receptive to infection by rubella virus will probably grow more slowly and will fail to perform properly in forming organ structures.

Different effects at different times

The rubella virus can cause a variety of effects on the embryo. Of those babies that showed malformations from one epidemic, more than 80 percent had heart defects, 65 percent had disorders of the blood and its clotting mechanism, 60 percent had low birth weight and probable invisible defects, and 50 percent had eye defects. In other epidemics deafness was common, presumably because the virus involved was slightly different.

The time of infection influences not only the danger of defects but also which of the defects is likely to occur. Infections during the first 3 weeks of pregnancy probably cause spontaneous abortion. Infections from weeks 4 to 9 can cause defects of the heart; those from weeks 5 to 8 cause injury to the lens of the eye; infection from weeks 7 to 12 causes deafness from injury to the inner ear. This timetable correlates well with the periods of development. For example, the septa between the chambers of the heart are being formed during weeks 3 to 5; allowing for some delay in passage through the placenta, infection of cardiac cells could slow growth enough to cause septal defects. Later infection could interfere with valve development. Similar correlations can be seen for effects on the eye and ear.

It is extremely important that a woman not contract rubella during the first few weeks of pregnancy. Therapeutic abortion might be wise if the disease occurred in the second to the eighth week. Perhaps it will become possible to use amniocentesis to determine whether or not the fetus has the virus in its cells and is presumably damaged. It is, however, hoped that prevention of rubella-caused defects is possible. There is an ongoing program to immunize all girls with a harmless strain of rubella before they reach child-bearing age. It will not be known for several years whether or not this has been successful. The current program attempts to immunize all preschool children against rubella, with the hope of preventing epidemics.

Other viruses

Of the other common viral infections, there is most evidence that mumps causes an increase in both fetal death and congenital defects when it occurs in the pregnant woman. The likelihood of defects is much less than with rubella, however. Several other viruses are suspect, including some of the "cold" viruses, and hepatitis. There is good evidence that vaccination against smallpox during pregnancy can cause injury to the embryo. Probably it will eventually be shown that many viruses can cause occasional defects, and that taken all together, viruses are responsible for a large fraction of the defects that cannot yet be explained.

Can birth defects be prevented?

Since birth defects are brought about by such a variety of causes, there will be no one thing people can do to prevent them (Table 6-2). No one knows yet how important the various known teratogenic agents are in the total picture. There may be other agents that have so far not even been suspected. When the causative factors are discovered, at least some of the birth defects will be preventable.

There are a number of things already known that can reduce the incidence of birth defects substantially. They require, however, that people be aware of what they can do to lessen the chances of having a defective child and that they carry out these measures. Widespread immunization against viruses, especially rubella, can certainly decrease the number of defective births. Every woman, *before* she becomes pregnant, should be very sure that she is immune to

Table 6-2. A study of congenital defects

The following serious defects, sufficiently important to impair body function severely, were recorded in studies of all births (total: 58,686) in 1963 in the state of Iowa. Of these births, 777 were stillbirths, including 114 with recognized defects.

Part of body with defect	Frequency of defect per total number of births
Brain and spinal cord	1 per 326
Eye, ear, and nose	1 per 994
Face	1 per 326
Respiratory system	1 per 2000
Heart and circulation	1 per 252
Digestive system	1 per 435
Genitalia	1 per 1151
Urinary system	1 per 1140
General skeletal defects	1 per 994
Congenital hip dislocation	1 per 917
Club foot	1 per 524
Limbs reduced or absent	1 per 917
Skin, blood, or endocrine glands	1 per 524
Metabolic defects	1 per 917
Trisomy 21 (Down's syndrome)	1 per 1839
Whole-body malformations	1 per 3000

Note 1: Many defects, such as those of the heart and the metabolism, become more evident later in life

Note 2: One child out of every 48 had one or more major defects (similar to above).

One child out of every 28 had one or more minor defects (flat feet, extra finger or toe, absence of part of the external ear, or some similar condition).

One child out of every 23 had one or more insignificant defects (such as small birth mark, tag of tissue binding tip of tongue, etc.).

One child out of every 10 born in Iowa in 1963 had some defect, whether major, minor, or insignificant. The major defects were serious enough to handicap the person severely and prevent him from leading a "normal" life. The minor and insignificant defects were usually of nuisance value only and could be either easily corrected or totally ignored.

Source: Hay, S., 1971, Am. J. Epidemiol. *94*: 572–584.

this virus, either because she has already had the disease or through immunization to it.

Everyone—perhaps especially women—needs to be educated to the danger of taking drugs during pregnancy. Irradiation, especially fluoroscopy, should be avoided. In the future, we may learn that other factors—perhaps certain drugs or even certain foods—add to the risk of birth defects.

Genetic counseling is important for people whose families have any incidence of birth defects. Such counseling, combined with the possibility of amniocentesis and perhaps therapeutic abortion, can avoid many instances of defective children.

95 percent of births are normal

Zygotes enter embryonic development with a variety of different instructions on how to develop, no two sets of instructions alike or previously tested. Yet most of these instructions work very well indeed to produce individuals with impressive capabilities. The embryo may be subject to a number of insults and stresses during development, but it overcomes most of them without great injury. Right now, a child is born with a 95 percent chance or better of being reasonably healthy. There are so many things that could go wrong during embryonic development that the miracle is that most babies are healthy and that defects are so few. Hopefully, they can be made even less frequent in the future.

We are unlikely to be able to improve on the embryonic process for a great many years to come. What can be done to make birth defects even less likely is to reduce the frequency of harmful genes in the population and eliminate some of the agents that might injure the embryo. We can help make the successful miracle of embryonic development even more commonplace.

7 Pregnancy and birth

The course of pregnancy

The description of embryonic development in the previous chapter dealt with an unfamiliar process that is not a part of the conscious experience of most people. Therefore there is not much widespread, common information about it. In contrast, the progress of pregnancy (the same thing considered from the outside) is the very vivid experience of at least one woman in every family. Even those who have not observed a pregnancy of someone close to them have access to a great deal of information in magazines as well as gossip. The two aspects of the same process, embryonic development and pregnancy, are not usually considered at the same time. They are even dated differently. Embryonic development dates from fertilization; pregnancy is commonly timed in medical practice from the date of the last menstrual period. So the times given in this chapter should be adjusted by two weeks to make them conform to those in the previous chapter. The most common duration of pregnancy is 280 days—9 calendar months plus 7 days—after the last menstrual period.

Early signs of pregnancy

The first, and most important, sign that a woman has of pregnancy is failure of the menstrual flow to occur at the expected time. Other signs are similar to common symptoms of the last few days before a menstrual flow: breast enlargement, possibly edema, tenderness of the vaginal opening. In another week or so many pregnant women experience a profound tiredness, so that they can "barely drag themselves around." Presumably this is because the body is not yet adjusted to the hormone changes that are taking place. A little later there may be morning sickness—nausea, headaches, perhaps even violent vomiting. The cause of these symptoms is unknown; one suggestion is that they are caused by chorionic gonadotropin, which is in greatest concentration in the first few months of pregnancy. Morning sickness usually stops by the third or fourth month, and many women do not experience it at all. Interestingly enough, the occurrence and severity of morning sickness frequently differs from one pregnancy to another in the same woman.

Pregnancy tests

In situations where there is a reason or strong desire to find out immediately whether or not a woman is pregnant, there are fairly accurate tests that can be applied. Most of them are ways of detecting the gonadotropins of early pregnancy, which are excreted in the woman's urine. Injection of the pregnant woman's urine into a rabbit or frog will cause release of gametes (despite the fact that gametes are not being released by the woman herself). Recently it has become possible to make faster and less expensive chemical tests for the presence of these hormones in blood samples or urine samples, and the animal assays are used less often.

A very different kind of pregnancy test involves the injection of large amounts of progesterone into the woman who thinks she may be pregnant. If she is pregnant, the procedure causes no harm, since secretion of chorionic gonadotropin is not inhibited by progesterone. But if the woman is not pregnant, the additional dose of progesterone accentuates the feedback inhibition of pituitary luteinizing hormone, thus inducing the delayed menses.

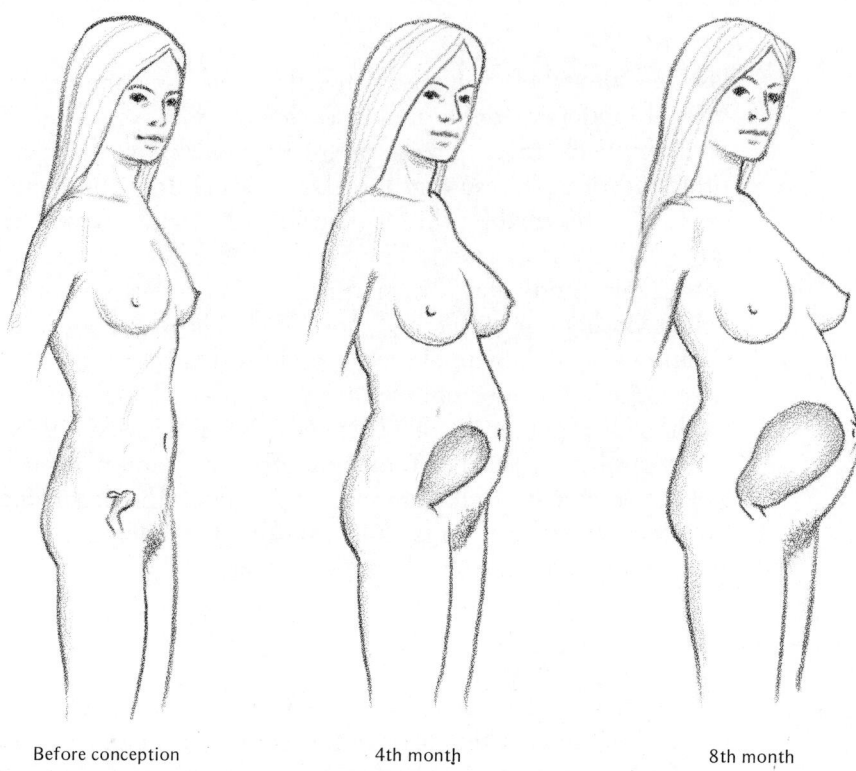

Before conception   4th month   8th month

Fig. 7–1: Body contours during pregnancy.

Changes in shape and weight

First pregnancies usually come to women who are young and slender, and they notice changes in their shapes very early. At first these are just the result of a slight water retention, but by week 12 there is a real thickening of the waist. By 16 or 20 weeks it is evident to others. In the last 3 months of pregnancy a woman must learn a whole new pattern of walking and balancing (Fig. 7-1).

The amount of weight gained during pregnancy is dependent, of course, on the woman's diet. She very often feels hungry enough to eat more than is advisable. The ideal weight gain is usually given at about 20 pounds, of which 7½ pounds represent the average weight of the baby and its membranes, 2 pounds the amniotic fluid, 1 pound the placenta,

1½ pounds the growth of the breasts, 2 pounds the growth of the uterus, and 6 pounds increased blood and tissue fluid. Several of these are not lost immediately at birth; it takes a few weeks for the uterus and blood volume to return to normal, and even longer for the breasts, if the baby is nursing.

Not everyone agrees on the importance of deviations from this figure. If the woman gains much more than this, it is obviously a nuisance to her to return to her desired normal weight. Some doctors feel that additional weight gain will show up in the baby, making labor more difficult. Others think that, within wide limits, the weight gain of the fetus is independent of maternal diet. A few doctors maintain that a maternal weight gain of considerably more than 20 pounds improves the health of the baby! Obviously no one really knows.

Fetal movements

During the third or fourth month the fetus gets large enough so that the movements of its arms and legs can be felt by the pregnant woman. The protrusion of a foot may even cause a temporary visible bump in the abdomen. Such movements increase in strength and prominence as pregnancy proceeds. They are quite erratic as to timing and may interfere with sleep. There is some evidence that physical activity by the woman may stimulate the fetus to move also. It has been suggested that the common observation by the mother of a new baby that the child is always awake needing attention when she is busiest in the evening preparing dinner, greeting her husband, dealing with the other children, and so on, may not be just her imagination. According to this suggestion, the late fetus, in the uterus, was habituated to activity at that time of day, and continues to be active then. Since birds are reported to learn to recognize the parental calls while still in the shell, this same kind of occurrence may be possible for humans.

Some of the fetal movements are not as easily identified as flexing of arms and legs. One of the more surprising ones is hiccups, which may occur in the fetus every so often in the last months and can be clearly felt by the pregnant woman.

Feeling pregnant

In the early part of pregnancy, the woman may have considerable discomfort, with morning sickness, lassitude, and general tenderness, although these conditions by no means apply to every woman. Later in pregnancy, there may be backache and clumsiness from the unaccustomed stance, as well as increased edema, frequently evidenced in ankle swelling. Not only do the hormones of pregnancy continue to mislead the kidneys into retaining too much salt and water, but the pressure of the older fetus on abdominal veins may cause considerable fluid retention in the legs and feet. Despite these complaints, a healthy woman in later pregnancy usually has a sense of great well-being, and a great deal of energy.

Health rules for the pregnant woman

Most of the health rules now advocated for pregnant women are the same as those applicable all the time. It is just that pregnancy is a stressful situation, physiologically, and so care is more important. The major criterion for any activity is whether or not it makes the woman herself feel well and does not put a special strain on her.

Prenatal medical care

The idea that a pregnant woman should visit a doctor regularly even when she feels well is an American invention of this century, which is spreading now to other countries. The number of spontaneous abortions, stillbirths, and maternal deaths has been greatly reduced in the last 50 years. The medical profession is inclined to give all the credit for this reduction to the fact that most women now have regular prenatal care. However, childbirth must certainly also have been made safer by the improvements in diet and general health that have occurred during that same period. Many women today have had an ideal diet available throughout their lives and have been spared most of the diseases that nearly everyone got a generation or two ago.

Regular prenatal care is not really necessary for the woman with an uncomplicated pregnancy. But that's the catch. If there is a serious complication, a woman is likely to wait longer than she should before seeking help, whereas a doctor would see warning signals at a regular checkup. Prenatal care is a kind of insurance, which one always hopes will be a waste of money. It's wise to have it. Even the healthy woman can benefit from assurance that the changes she thought were normal really are, and from advice on small matters. This is especially important during a first pregnancy. There is no lack of good advice available in books and magazine articles, or from friends and relatives. Everything one needs to know is available. Such advice is possibly mixed with a number of more-or-less dangerous pieces of misinformation, as well. A doctor's advice may include some unproven folklore and minor errors, but it is certainly the best source available, so the pregnant woman should, if at all possible, place herself under the regular care of a doctor.

Diets in pregnancy

General advice for pregnant women contains a great deal of useless folklore, both within and outside of medical practice. The same is true for nutritional advice. So any detailed counsel on diets for pregnant women is doubly suspect. Any faddish diet is certain to be based on very little foundation. At best it will be of no advantage and at worst, harmful.

The only clear criterion for diet in pregnancy is that of meeting the nutrient needs of the woman. There is an increase in blood supply, so extra iron is needed. The fetus is growing rapidly in the last few months, and vitamins, minerals, and proteins are therefore important. The easiest and safest thing to do is to eat a very ordinary diet, but with a vitamin and iron supplement.

The developing fetus has a stronger system for accumulating nutrients than does the pregnant woman. Pregnancy with a poor diet usually seems to lead to a healthy baby and an unhealthy mother. The mother's bones may be partially depleted of calcium; she may be anemic or have a vitamin deficiency. An adequate diet is necessary to preserve her health—the fetus will usually take care of itself.

It has been suggested recently that poor diet during pregnancy may interfere with the development of the fetal nervous system and cause poor performance in mental tasks later in life. There really is no reliable evidence about the effects on the fetus of long-term poor diets during pregnancy, because such diets are almost always associated with other aspects of poverty. There is convincing evidence, however, that a brief, severe period of starvation for a pregnant woman does not cause mental retardation in the baby born later. Near the end of World War II, parts of the Netherlands suffered a severe famine, brought about by a combination of the destructions of war and Nazi policy. This famine was limited to certain cities, and it began and ended abruptly. It has since been possible to obtain records of the performance on intelligence tests of people who were fetuses in the uterus during that period of famine. Their scores were compared with the scores of people exactly the same age from other Dutch cities that were not affected by the famine. There was no significant difference (those from the famine region were slightly higher). Birth rate after the famine was very low in the famine cities; many embryos had aborted because of the starvation. Those that did survive to be born were normal in intelligence and apparently in other ways as well.

Drugs and medicines

Since the thalidomide tragedy, doctors have become extremely conservative about prescribing any drugs for a pregnant woman, especially early in pregnancy. Some, of course, are necessary, such as insulin for a diabetic. A few are almost certainly safe. In the rare cases where a pregnant woman must be treated for syphilis, the antibiotic penetrates the embryo and cures it of infection, too. The general policy now is to avoid medicines unless the probable benefit to both woman and offspring is greater than the possible damage to the fetus.

The same advice holds for self-administered drugs. Nonprescription remedies should be used only in minimum dosages and not regularly. Perhaps the safest drug is alcohol; there is little evidence of fetal damage from small to moderate dosages. There is quite a bit of evidence, however, that tobacco causes a reduction in the size of babies born, thus indicating interference with growth. The tobacco interests claim that smokers tend to be people who would have small babies, anyway; that the reduced size is not an effect of the drug. Since they have not yet proven that contention, a wise woman will stop smoking during pregnancy. There are of course compelling health reasons for not smoking at all, quite apart from pregnancy. Concerning illegal drugs, there are inconclusive reports suggesting fetal damage from LSD and even from marijuana; a child of a heroin addict may be born with an addiction. Even a shred of prudence would indicate abstaining from such drugs during pregnancy.

Sleep and activity

The medical advice is always that the pregnant woman get more sleep than she did before her pregnancy. This will usually fit in with what the woman feels like anyway, since she is likely to be tired all the time in the early weeks. Even when she gets a burst of energy, as sometimes happens later in her pregnancy, she may tire easily. There is no clear evidence that insufficient sleep causes specific dire consequences, but getting ample rest seems wise.

There is a wide range of opinions among doctors as to what kinds of exercise a pregnant woman should undertake. Some doctors and many grandmothers are sure that some kinds of vigorous sports are likely to cause miscarriage. The bulk of current medical opinion is that there is no correlation, unless abortion is threatened according to other signs. A pregnant woman should remember that in the last few months she may have trouble keeping her balance, and that she stands the risk of spraining her wrist, for instance, if she stumbles. Most doctors now put few restrictions on activity, and they do not agree on which activities to restrict.

Sexual activity

Medical opinion in some past periods has discouraged or prohibited sexual intercourse during pregnancy. Now the prevailing medical opinion is that there is no problem during most of pregnancy. For comfort and practicality, many couples change to different positions for intercourse as pregnancy progresses. Most pregnant women are reported to feel a decrease in sexual desire in later pregnancy. Medical opinion is still divided on the advisability of sexual activity during the last 3 months.

Something may go wrong

There are still a number of complications that can arise during pregnancy, although the number of complications has decreased greatly in the affluent countries in the last two generations. Many of the problems that do occur can be treated effectively now. The possibility of a complication constitutes the principal reason for prenatal medical care. Some of the possible complications of pregnancy will be discussed in the next part of this chapter.

Spontaneous abortion

Estimates of how frequently pregnancies end in spontaneous abortion vary from 10 percent of the time to much higher figures. The most recent estimates take into account very early abortions and even failure to implant, and thus may reach over 50 percent. The real rate for known pregnancies that end in spontaneous abortions is probably close to the 10 percent estimate.

Many miscarriages—spontaneous abortions—are not tragedies, but quite the opposite. A large fraction are due to some serious congenital defect in the embryo. At least 20 percent of all spontaneously aborted embryos show abnormal chromosomes. Many more probably have other defects. It may be a good guess that spontaneous embryonic loss during the first 12 weeks is due to some embryonic defect and is desirable, even though disappointing at the time. Fetal loss after 12 weeks is more likely to be due to some uterine abnormality and should be prevented if possible. In cases of threatened abortion, many doctors prescribe bed rest and progesterone. Others feel that such measures are useless—that by the time there are signs of a miscarriage, it is too late to do anything.

Allergies during pregnancy

Women with allergies usually find them getting either much better or much worse during pregnancy. It is more common for the symptoms to diminish, but the effects may be opposite in pregnancies of even the same women. The same variable effect seems to be found with some types of arthritis related to allergy. It is known that steroid hormones have strong but complicated influences on allergy and arthritis. Presumably the effects during pregnancy are due to these hormones.

Rh incompatability

The best-known complication of pregnancy, Rh incompatability, is related to the immune system, as are allergies. The Rh blood type of a person is, like his ABO blood type, due to the presence of certain compounds on the surface of the red blood cell. There are actually many complications in the inheritance of the Rh factor, but in a simplified scheme applicable to most people, the presence of the Rh substance on the cell is determined by a dominant gene; the absence of the Rh substance is determined by its recessive allele. A person with either one or two dominant alleles will have the substance and will be called Rh positive. (His blood type will be written with a plus sign). The homozygous recessive person will be Rh negative (his blood type will be written with a minus sign). If the Rh negative person receives an injection of Rh positive blood, his system will build up antibodies against it. Rh negative blood in an Rh positive person, however, will not cause any reaction. When an Rh negative woman has an Rh positive husband, a fetus may be Rh positive. That is the only situation in which the common type of Rh incompatability can arise. Apparently, it happens sometimes, but not always, that late in pregnancy or at birth there are tears in the chorionic blood vessels, so that some fetal blood spills into the woman's circulation. It takes 2 weeks for antibody production to be very effective, and no damage is caused. In a second such pregnancy, the leakage of still more blood will cause still greater production of antibodies, and more quickly, because of previous immunization. Perhaps in the third pregnancy with an Rh positive fetus, the antibody concentration may be great enough so that enough antibody leaks into the fetal circulation to cause destruction of blood cells, kidney damage, and many secondary problems. The condition may be fatal at birth.

The worst of the Rh incompatability problems occurred before the cause was diagnosed in 1940, principally in women who had received transfusions of Rh positive blood some time before pregnancy. Now that such transfusions are never given, and with the trend to smaller families, the disease is much less common.

For some years it has been possible to withdraw (a little at a time) the injured blood of the newborn baby and to replace it with Rh negative blood. Such exchange transfusions have saved many lives, because any antibody still present in the baby's body will not react with the Rh negative cells. By the time the baby replaces these cells with its own Rh positive cells, the antibodies will have been largely lost.

It may now be possible to eliminate Rh problems altogether. It is thought that the greatest amount of leakage of fetal blood into the woman's circulation takes place late in pregnancy, perhaps even at and after delivery. In a new procedure, antibodies from sensitized women are injected into Rh negative women just after delivery of an Rh positive baby. These injected antibodies attack and destroy any fetal blood cells that have leaked through so that those cells are no longer available to cause the build-up of the capacity to produce such antibodies. Consequently, the danger in later pregnancies is no greater than in the first one.

Incompatabilities from the ABO system

It was puzzling to those studying the Rh problem that they rarely found evidence of similar problems with the ABO system. One answer seems to be that such incompatibility does cause some abortions, but such early ones that they are usually not noticed. There seem to be several other factors that account for the smaller problem with ABO incompatability. It has been found that problems with Rh incompatability occur less often when there is also an incompatability with the ABO system. Presumably this is because the naturally occurring anti-A or anti-B antibodies destroy the fetal cells before they can induce anti-Rh antibodies. (Thus the ABO antibodies act just like the therapeutically injected antibodies against the Rh substance.)

Toxemia of pregnancy

One of the most serious problems that prenatal care is designed to prevent is called **toxemia of pregnancy** (or preeclampsia). This is really just a collection of symptoms; it may actually be a collection of problems with different causes, rather than a single disease. The symptoms include severe edema, high blood pressure, and virtual kidney shutdown. They appear in some degree in about 7 percent of pregnancies. The major treatments used are restriction of salt intake to reduce edema and injection of drugs to increase the blood supply to the kidneys. Usually these measures are effective. The suggested causes include abnormal response to hormones (as is probably true with the common mild edema of pregnancy), or allergic reaction to the presence of the fetus.

Dangerous conditions during pregnancy

Several of the health problems that a woman may have will make it less likely that she can carry a fetus to term without injury to herself or the fetus. Pregnancy is stressful to the woman and tends to make many health problems worse. It has been known for a long time that if a woman has tuberculosis in an inactive stage, childbearing is likely to activate it into a serious problem. Now that TB is largely eliminated from affluent societies, it is rarely a problem in pregnancies. Careful medical management has now made it possible for most women with some heart problems to bear children safely, too.

Only in this century has it become possible for women with serious diabetes to have children; the disease was usually fatal to the fetus and often to the woman also. Women in a prediabetic condition, who have not previously developed symptoms, often first show symptoms during a pregnancy. (The incidence of diabetes increases with age.) Since insulin became available, however, it has been possible for a properly medicated woman to bear a live child. Nevertheless, diabetes still poses a serious problem in pregnancy. The child of a diabetic woman, of one in a prediabetic condition, or even of a diabetic woman treated with insulin develops a characteristic abnormality. Such babies are much larger than normal when born. They are puffily obese, with many signs of the condition over the entire body (Fig. 7-2). Presumably it is an extreme form of this same condition that is so often fatal to the fetus when the diabetic woman is not taking insulin. It is thought that most or all of the fetal symptoms are due to the abnormally high concentration of sugar in the circulation of the diabetic woman. This sugar crosses the placenta and causes the obesity and many other problems, including excessive fetal production of insulin. (Insulin, like other protein hormones, will not cross the placenta, and so the fetal insulin does not reduce the maternal diabetes.)

Making it possible for a diabetic woman to have a baby is not an unmixed blessing. Since the tendency toward diabetes is inherited, the child is likely to develop the disease.

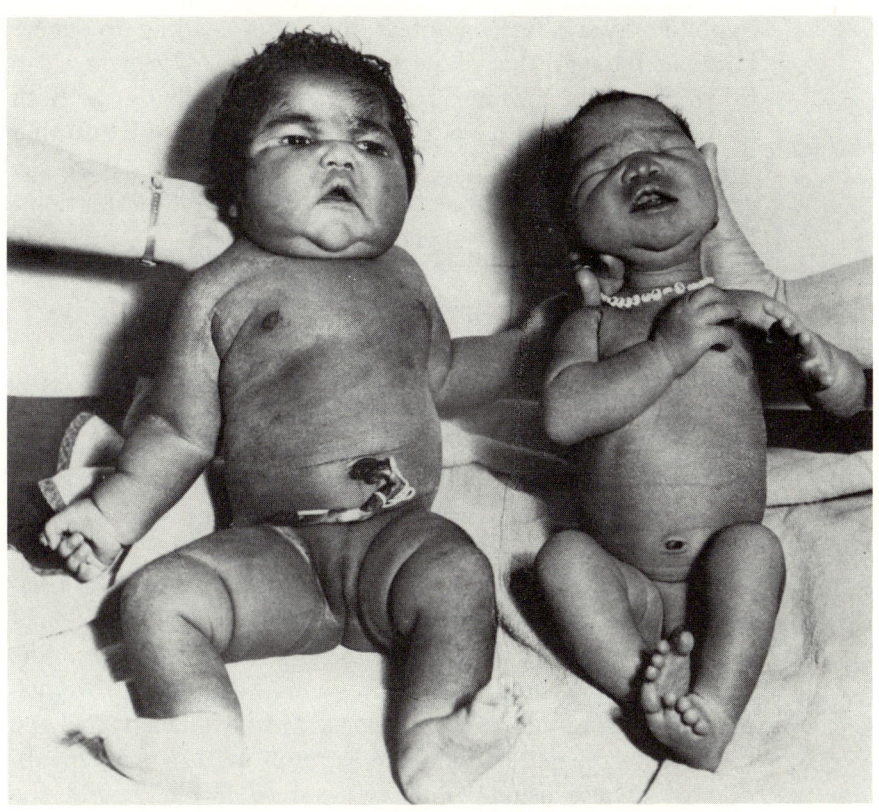

Fig. 7–2: Baby of a diabetic mother compared with a normal newborn. Note the fat, almost bloated appearance of the affected infant (left) in contrast with a normal infant (right). (Courtesy of W. P. U. Jackson, *Lancet*, 1955, Vol. 2, p. 625.)

It is observed that such children usually develop diabetes (if they do so at all) at a much younger age than the parent had. Since the onset of diabetes is hastened by excessive sugar in the diet, it might be argued that the excess sugar to which the fetus of a diabetic woman is exposed acts to hasten the development of the disease if the genetic factors for it have been passed on to the fetus.

The venereal diseases have long been a serious problem with children, especially before antibiotics were developed.

Syphilis, especially, can infect the fetus, leading to severe kinds of damage that show up in later life. Gonorrhea can infect the child during birth, often causing blindness. Both of these problems have diminished since antibiotics that would cure these diseases have become available. With the recent increases in venereal disease to epidemic proportions, this problem may easily become serious again.

Permanent effects of pregnancy

Despite the worries of many pregnant women, most of the changes that occur during pregnancy are reversed after birth; few are permanent. The most serious of the common long-term effects of childbearing is hemorrhoids. The same circulatory problems that cause foot swelling late in pregnancy act to distend the veins of the anal wall. The strong abdominal pressures during labor may distend them permanently, so that they protrude as hemorrhoids, or piles. Sometimes these are serious enough to require surgery later. The same pressure factors in late pregnancy cause some women to get varicose veins in their legs. Most women who have been pregnant get wrinkle-like "stretch marks" on their abdomens. These usually fade with time, or blend with the wrinkles everyone gets with age. Since our culture values the breast more as a sex attractant than as a nutritive gland, many women fear that pregnancy and especially nursing will "spoil their figures." This does not seem to be a valid fear, although most doctors advise wearing a special brassiere to prevent possible problems.

Timing of birth

No one knows why a baby is born exactly when it is, although animal experiments indicate that gestation time is more a property of the fetus than of the pregnant female. Among the many factors suggested as being responsible for the onset of the birth process are hormone signals from the fetus, aging of the placenta, and mechanical stimulation from the enlarged fetus. Perhaps the initiation of labor involves a combination of several factors working together. The most likely time for birth is 40 weeks after the last menstrual period; 95 percent of births occur between the thirty-fifth and forty-third weeks.

Early stages of labor

For a week or two previous to an actual birth, the uterus has usually been exhibiting waves of contractions that are not strong enough to be noticed most of the time. Finally, when birth is at hand, the contractions become strong and noticeable. The earliest contractions last about a minute each and are perhaps half an hour apart. Often called "labor pains," the contractions are compared by some women to menstrual cramps (which probably include uterine contractions, too). Some women experience such contractions as painful; others do not. The time that elapses from the first noticeable contractions of labor to the actual birth of a baby varies a great deal from one case to another. First babies are usually born in 7 to 10 hours, while later children usually emerge between 4 and 6 hours after the onset of labor. It involves very much of a guessing game to arrive at the hospital in plenty of time and still not spend many hours there to no avail.

As the contractions continue, they get somewhat longer individually (about a minute) and finally they may be less than a minute apart. The squeezing action of the contracting uterus changes the position of the baby, usually so that the head presses against the opening, the cervix (Fig. 7-3). This constant pressure forces the cervix slowly to open wider. The progress of opening (dilation of) the cervix is described by two measurements: the width of the cervical opening and the distance that the top of the baby's head has progressed through it.

Later stages of labor

Some hours after the onset of labor, it is common for the amnion (the "bag of waters") to break. This seems to speed up the progress of labor, and the amnion may be punctured

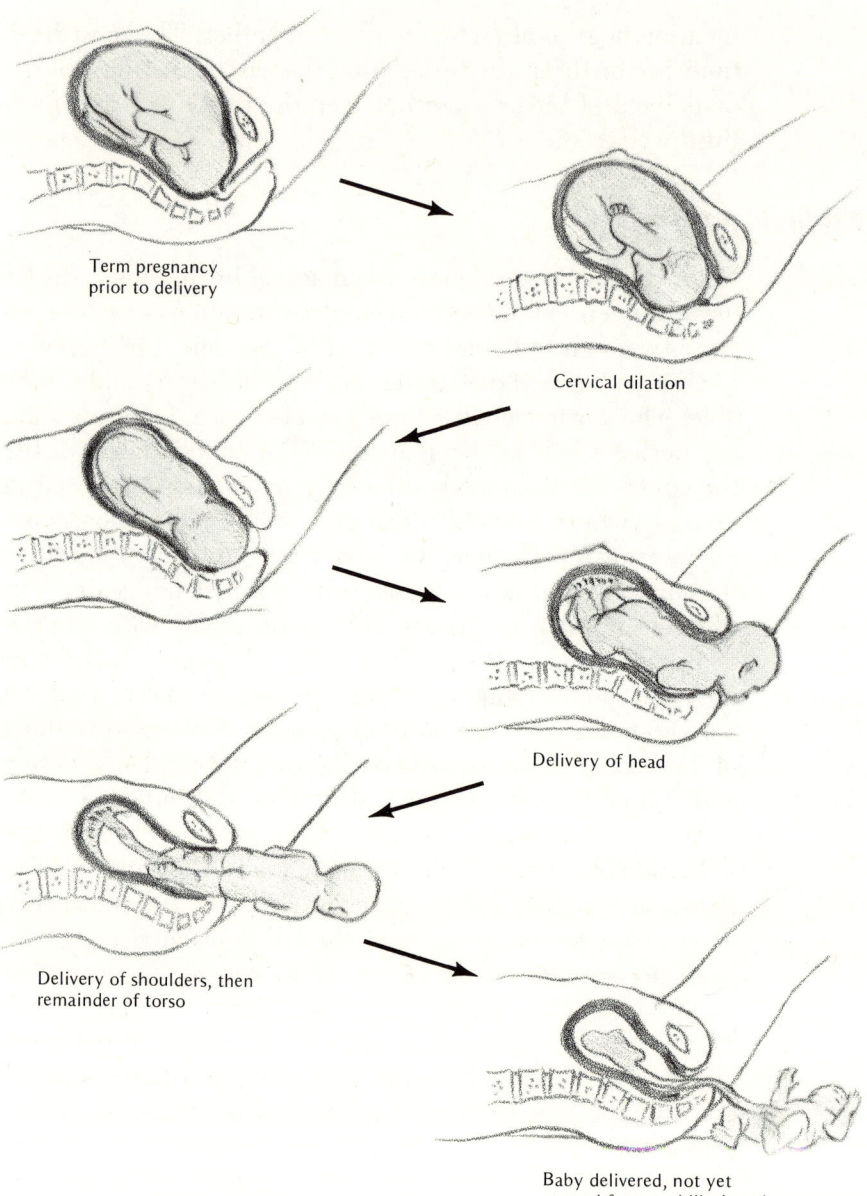

Fig. 7–3: The process of birth.

by the doctor for this purpose. When the cervix is sufficiently dilated, the baby moves through it into the birth canal (vagina). This part of labor is usually more rapid, since the head is the largest part of the baby and gives the most resistance to passage through the cervical canal. (The shoulders are of about the same diameter, but they are more flexible.) Since the bones of the skull are not completely ossified (thus the "soft spot" on the top of the baby's head), the head is capable of being squeezed into a more elongate shape, to pass through the cervix without injury to the baby. Newborn children, especially the firstborn, often have very strangely shaped heads for a few days.

When the head of the baby is about to emerge from the vagina, the doctor usually performs an **episiotomy**: The skin at the end of the vaginal opening is cut to prevent a jagged (and more painful) tear. The incision is sewed up right after the birth.

A few minutes after the baby has emerged, the uterine contractions usually begin again. They dislodge and expel the placenta and torn membranes in what is now called the afterbirth. There is some bleeding from the torn uterine wall, but the arrangement of the torn arteries between the muscle layers tends to close the vessels when the uterus is contracted.

Immediate care of the newborn

Immediately after the baby emerges, the mucus is cleared out of its mouth and nostrils with a syringe, and the baby is held by the feet to facilitate drainage of fluid from the airway (Fig. 7-4). Usually it immediately begins to breathe, and the passage of air in the first gasps causes a lusty cry. If the baby does not begin breathing right away, it is stimulated by cold water, rubbing with a towel, or even the proverbial spank. Artificial respiration is used if spontaneous breathing does not begin in the first few minutes. (A newborn infant can go uninjured for several minutes without breathing, considerably longer than an adult can.) The umbilical cord is cut and tied, but only after the lapse of a minute or so to

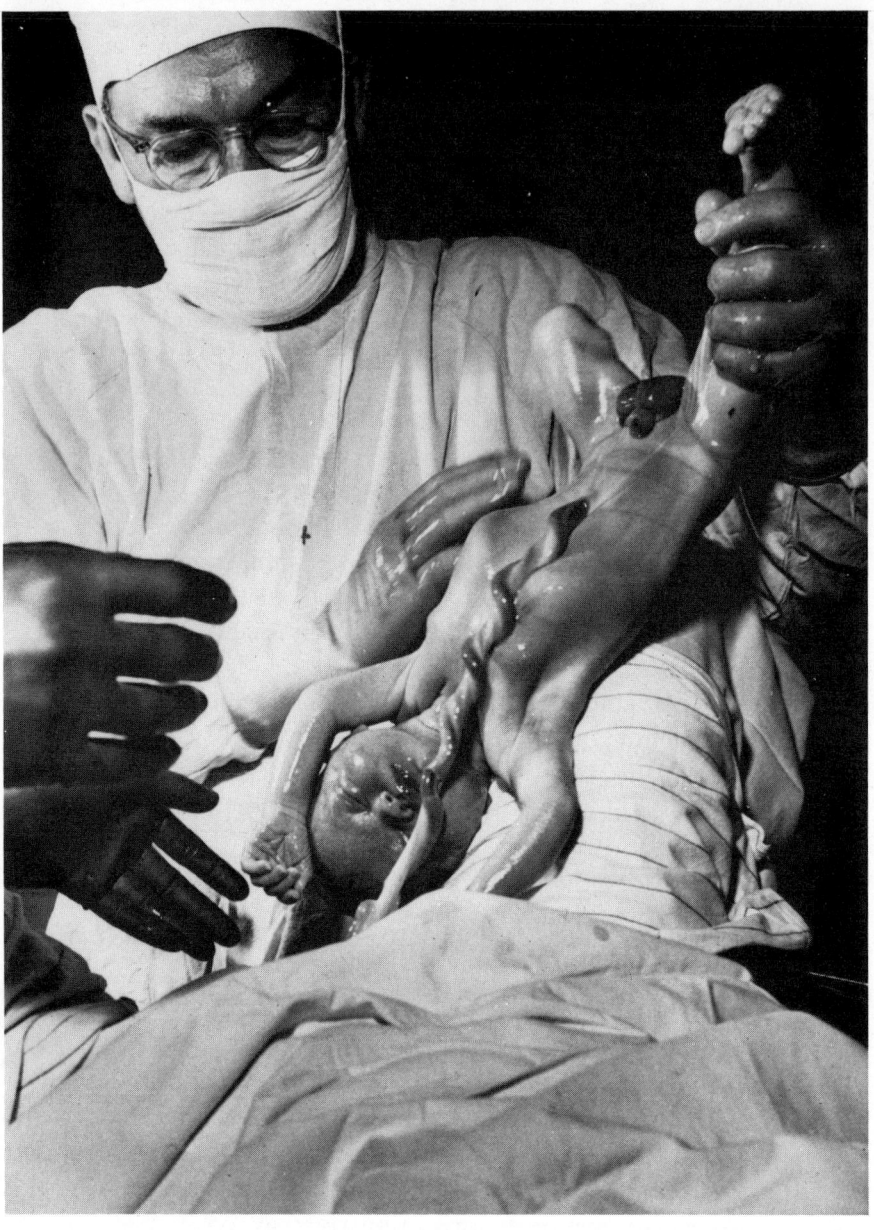

Fig. 7–4: A. The moment of birth. (Wayne Miller, from *The Family of Man* collection.)

Fig. 7–4: B. Cleaning mucus from the breathing passages. (George J. Coella, courtesy of Des Moines Register and Tribune.)

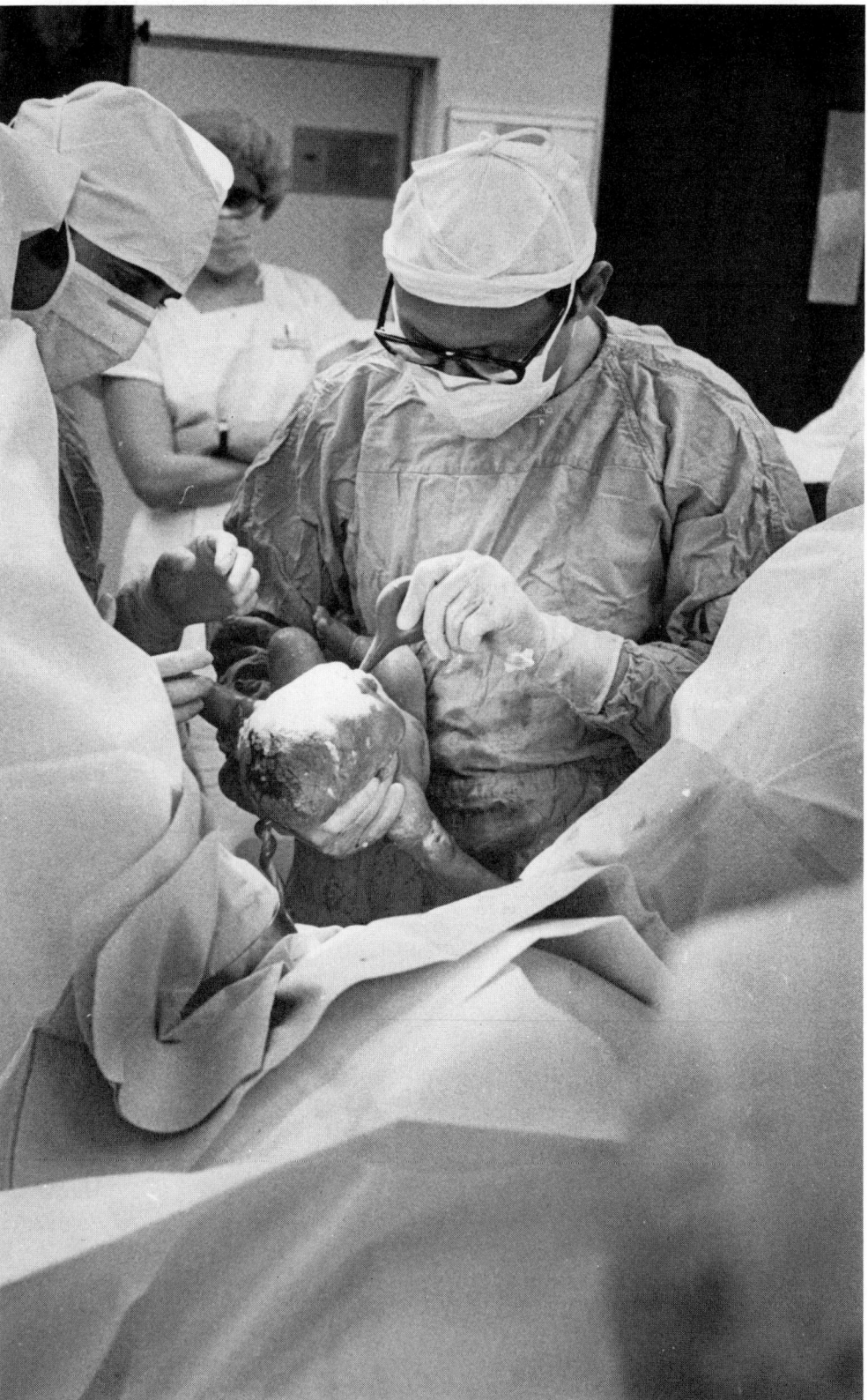

allow as much as possible of the blood to drain from the placenta into the infant's circulation.

Hazards of birth

In 1930, in the United States, the maternal death rate at delivery was 67 per 10,000 live births. Now it is less than 4 per 10,000 live births. This dramatic reduction in risk is due to a combination of factors: better general health, better prenatal care, the availability of antibiotics, the use of blood transfusions when necessary, better anesthetics, and better trained medical assistance. Another important factor may be the greatly increased use of Caesarean section. The removal of the infant by opening the abdomen and uterine wall is now so safe that it is done routinely if the opening in the pelvic bones seems too small for an easy delivery. It can be repeated in several pregnancies without complications.

The process of birth, especially in a difficult delivery, can be hazardous to the health of the baby. The use of delivery forceps can cause brain injury if unusual pressure must be applied. Any prolonged period of insufficient oxygen may have the same effect. Such oxygen deprivation may be caused by unusual pressure on the umbilical cord during a prolonged labor, premature separation of the placenta from the uterine wall, or by so much anesthetic as to greatly delay the onset of the infant's breathing. Serious injuries like cerebral palsy may result in about 1 out of 175 births. (This figure includes some developmental defects as well as birth injuries.) In addition, some workers feel that many learning disabilities can be traced to accidents and poor technique during delivery. Breech presentations (to be discussed in the next section) are associated with an increased incidence of congenital dislocation of the hip—apparently both from pressures during delivery and from the unusual position during late fetal growth.

Difficult deliveries

The most frequent problem of childbirth is a "breech delivery," when the buttocks of the infant, rather than the head, are forced against the cervix. It is sometimes possible to turn the baby around. If not, birth is rather difficult, and a little more dangerous than normal. Another problem is "placenta previa," in which the placenta happens to have formed over the opening of the cervix. Since the placenta must be dislodged for birth to occur, the baby is likely to die of oxygen deprivation before it emerges. Such a condition can now usually be detected early enough to do a Caesarean section instead. Still another problem that was once more serious is hemorrhage after delivery. This is often due to the incomplete expulsion of the placenta. If oxytocin is administered, it will usually stop the bleeding by causing uterine contraction. In any case, even large amounts of blood can now be replaced by transfusion. The same drug, oxytocin, can be used to induce labor when this would seem to be wise—for example, in a case of Rh incompatability, when the baby needs an immediate blood replacement.

Care of the mother after delivery

After a normal birth, the new mother usually needs very little special care once the placenta has been expelled and the episiotomy incision repaired. She will have to wear an absorbant pad for a week or so, since the return of the uterus to the resting state involves something like a very large menstrual flow. Couples are advised to abstain from sexual activity until after a medical checkup 6 weeks following delivery.

The breasts have a little milk in them at the time of delivery. The earliest secretion, called **colostrum**, is especially rich in antibodies that the newborn is able to absorb, and it is valuable for the baby to nurse at first, even if it will be principally bottle-fed. This first nursing is advantageous for the mother's health, too, since the stimulation of sucking causes the release of the hormone oxytocin from the posterior lobe of the pituitary gland. Oxytocin increases uterine contraction, which reduces bleeding and hastens the return of the uterus to its normal size. The breasts will not secrete large quantities of milk until a couple of days after delivery. They become enlarged, and the glandular elements grow, in the latter part of pregnancy under the influence of various placental and pituitary hormones; but the steroid hormones produced by the placenta inhibit actual secretion. With the placenta removed, the pituitary hormones stimulate the secretion of milk.

Many women feel unaccountably depressed, some greatly so, others only slightly, within the first few days after giving birth. This does not bear any necessary relationship to a woman's feelings about her baby but is usually hormonal in origin. The violent shifts in hormonal concentration simply have this unpleasant side effect on the emotional centers of the brain. Possibly it is due to the abrupt withdrawal of whatever hormones cause the feeling of well-being late in pregnancy. Usually, this "post-partum depression" clears up without further trouble. If not, medical advice should be sought.

Unsettled questions about childbirth

In most societies throughout history, helping with childbirth has been the function of the women of the community, often of an experienced midwife. The idea of birth controlled by male doctors is a European invention of the last century or two, intensified in recent American practice. Several European countries presently make considerable use of the trained nurse-midwife, a professional person with a good background of training. This job function has not been important in American society, but it may increase now that physician's assistants of all kinds are becoming accepted. A nurse-midwife would not be trained to deal with all possible emergencies—transfusions, Caesarean sections, other surgical interventions—but she could well become more skilled than an obstetrican of wider function at doing what was required at practically all births. If such specifically trained assistants worked in a hospital setting, where other help was available for possible emergencies, the general level of care would probably improve.

Most American babies are now born in the hospital, although that is a phenomenon of the last generation or two. For an uncomplicated birth, the home might be just as satisfactory. A few hospitals have so much trouble with problems such as antibiotic-resistant bacteria that a clean home would be preferable. In the rare emergency, however, only a well-equipped hospital provides the needed support.

Obstetric practice has changed dramatically within the last generation in the postdelivery care of mothers. In Victorian times, a woman who could afford help expected to stay in bed for at least 2 weeks after giving birth. This practice was due partly to the fact that many deliveries were difficult and any blood lost was not replaced, so the mother was often weak afterward. Primarily, it was just a bad medical habit, the result of unexamined folklore. Women were weaker after 2 weeks of inactivity than they would have been the day after giving birth. New mothers now take walks on the same day as the delivery with no ill effects.

The greatest current variability and disagreement in obstetrical practice centers around the use of drugs to relieve the discomforts of childbirth. A movement sometimes called "natural childbirth" maintains that most pain associated with childbirth comes from fear and muscular tension. This school of thought emphasizes the dangers to the newborn of the administration of drugs, and the emotional benefits to the woman to be gained from being conscious at her baby's birth. Recently the movement has become much less doctrinaire; it now emphasizes the value of education, of knowing what to expect in contrast with being frightened by the unknown. It is now merely suggested that the woman try to use as little anesthetic as possible, since more can be administered if needed. One can hardly argue with anything as reasonable as this, and "trained childbirth" is rapidly gaining acceptance in hospitals and among both doctors and expectant parents. Both husband and wife are urged to learn about the birth process in detail, and the woman is given a number of exercises to do and breathing techniques to use. The exercises are supposed to improve the health and muscle tone of the pregnant woman, and to make her able to speed delivery by using voluntary abdominal pressure. The breathing techniques, too, are supposed to aid in the birth process. There may be some legitimate doubt as to just how much direct

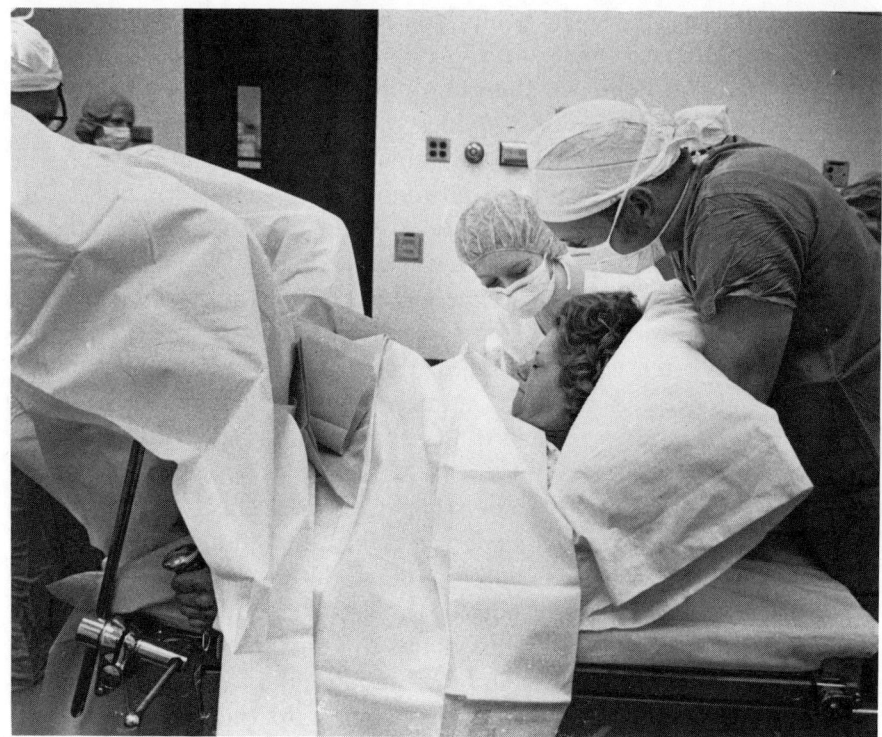

Fig. 7–5: Woman in labor using trained childbirth methods with her husband's aid. (George J. Coella, courtesy of Des Moines Register and Tribune.)

effect either the exercises or the breathing have; but they certainly provide the prospective parents with something to do when things seem to be out of their hands. Concentrating on a breathing pattern may take a woman's mind off the discomfort of labor contractions, as does having her husband there to "help" her (Fig. 7-5). Many people who have taken the courses of "trained childbirth" feel that the experience of giving birth has been made a "beautiful" one. The courses can certainly do no harm.

Physiological adjustments of the newborn: breathing

The newborn infant comes suddenly at birth into a very different kind of environment. The uterus had provided a completely controlled, nourishing environment that maintained constant temperature, salt/water balance, and nutrient concentrations. Even though a few of the systems of the late fetus were capable of performing their functions in maintaining these conditions, they were, for the most part, not required to operate. Suddenly, the sheltered environment disappears with the cutting of the umbilical cord. The most immediate physiological stress comes from the cutting off of the oxygen supply; the oxygen concentration in the baby's tissues goes down very rapidly. This provides a strong stimulus to several breathing centers in the brain, hitherto quiescent because of the constant oxygen supply. Under multiple nervous stimulation, the breathing muscles of the chest and diaphragm respond vigorously and pull in air in great gasps. (The fact that many pain killers and anesthetics tend to depress the breathing centers of the brain is what makes such drugs dangerous when inexpertly used during childbirth. An overdrugged infant may not breathe quickly after birth.) If the infant is slow in initiating a breathing cycle, the doctor may stimulate the skin in any of a number of ways that tend to bring forth a gasp reflex. If these measures are not sufficient within the first few minutes, mechanical respiration is used until the source of the problem is found.

Circulatory changes

The shift of the baby's oxygen supply from the placenta to the lungs requires a change in circulation and also provides the necessary forces to bring about the change (see Fig. 6-5). When the lungs fill with air in the first gasps, this opens the pulmonary vessels, which were previously folded and nearly shut off. The newly opened vessels provide much more space for blood to flow in the pulmonary circuit, and the pressure therefore falls in the pulmonary arteries. At the same time, the tying off (or the spontaneous constriction) of the umbilical vessels decreases the space through which blood in the systemic arteries can flow, and so causes arterial pressure to build up. Since now the pressure in the pulmonary arteries is less than in the systemic ones, the direction of blood flow through the shunts is reversed. Whereas before birth the blood flowed from the right atrium to the left atrium, it now flows from left to right. A valve-like flap of the partially complete septum between the atria is forced shut by flow in this direction, and the atrium now becomes effectively complete. In a few days or weeks in this condition, the flap grows permanently shut. The other shunt, the ductus arteriosus, had carried blood from the pulmonary artery to the systemic artery. The pressure change at birth reverses the direction of flow here, too. No one really knows just why, but flow in this direction induces the muscles in the wall of the ductus to clamp it shut. This, too, grows permanently shut later.

Food and warmth

Less immediate challenges to the health of the newborn, but challenges all the same, are the need for warmth and food. Only in the warmest climate is a human, without fur, subjected to no temperature stress. The newborn baby depends on additional aids, especially on being held close to the warmth of its mother's body. In a culture with central heating and warm baby blankets, this is not as obviously important as it has been for most newborn babies throughout human history. The physiological mechanisms to maintain temperature that adult humans have are not working very well during the first days or weeks of an infant's life. A newborn infant has, therefore, a compensating tolerance to withstand body temperatures that are so low that they would probably be fatal to an adult. The older infant, when appropriately clothed for the climate, regulates temperature well.

The infant does not really need to eat much for the first day or two; the amount of nourishment available in the secretion of colostrum is usually sufficient. As mentioned earlier, this source is probably as important for antibodies and other special ingredients as for ordinary nutrients. The ability of the young baby to digest other foods may be somewhat low, but the ability to use milk is high. The digestive secretions of babies contain the enzyme necessary to absorb and use milk sugar, or lactose. This enzyme is largely lacking in the digestive secretions of most adults. Only in populations (like those derived from Europe) in which raw milk is an important part of the adult diet can milk sugar be used by adults. In fact, large amounts of milk sugar support excessive bacterial fermentation with consequent digestive irritation when consumed by adults of most racial origins.

The ability of the infant kidney to regulate the concentration of salt in the body is probably less than in older children or adults. This may be more important than digestability in limiting the suitability of food for infant feeding (assuming that foods do not require chewing). Salty puréed ham, for instance, might not be tolerated in large amounts.

Characteristics of the newborn

The full-term baby is likely to be between 5 and 9 pounds, although the range seems to extend from 3.5 to 13.5 pounds. (These are values for American white populations; they are probably higher than for most other groups.) Whatever the color of the baby's eyes will be later, the newborn's eyes are always blue. Similarly, the skin pigmentation that a baby has inherited, especially the melanin, is poorly expressed at birth, perhaps because it requires some exposure to light to develop.

The books on infant care say that newborn babies spend practically all their time eating and sleeping, but many parents say that their babies haven't read the books! It is now being suggested by a few research workers that newborn babies should have more stimulation than our culture generally gives them in order to promote maximum mental development. Even a tiny baby should held, cuddled, and played with. It seems to be especially important that an infant be held while it is nursing, even if it is being bottle-fed rather than breast-fed. Beyond this important frequent physical contact, it can certainly do no harm (and may possibly be useful) to provide something more stimulating than a white wall for a baby to look at. A hanging mobile, colored pictures, a change of rooms—such things might stimulate even a small baby and feed the imagination early. A little bustle and activity by other members of the family may do more good than the sleep they may disturb. Infants in most cultures have been kept in close contact with other members of the family. (This usually includes physical contact, such as being strapped on the mother's back.) Researchers in early childhood development are becoming convinced that it is not good for a baby to be left constantly alone in a room and handled only at feeding time.

Neonatal mortality

In the affluent countries, a healthy baby has a very good chance of surviving to adulthood. It is in the reduction of early childhood mortality that the advances of Western medicine and public health measures have been most effective, and most obviously productive of human good. The mortality of infants during the first year of life is now from 30 to 35 per 1000 births in most technologically advanced countries. No one is really sure how much more this could be reduced. The death rate in the first year of life is far more than in any other year of life below the age of 70. Not all these infant deaths are tragic and preventable. At least 15 percent are due to significant congenital defects that cannot be effectively treated. More than half of the deaths of newborn infants are attributed to "prematurity."

A number of young infants die of what has come to be called "sudden crib death," in which a child that has seemed quite normal and healthy has been found suddenly dead in its crib. Such deaths used to be blamed on suffocation in the bed clothes, and bereaved parents were sometimes also made to feel guilty about having used a pillow in the crib or not having checked the baby often enough. Everyone who has studied the problem recently agrees that the cause is not suffocation and parental neglect, but they are not sure of the true explanation. The reasons for "sudden crib death" probably include unrecognized congenital defects and also sudden viral infection of the baby before its immune system has provided sufficient protection against such infections.

The premature infant

In American medical practice, a premature infant is arbitrarily defined as one weighing between 14 ounces and 5.5 pounds. (Below this range it is termed an abortion, since it is most unlikely to survive.) The true age of a fetus is difficult or impossible to determine, and the arbitrary range for premature infants may include healthy full-term infants of small ancestry. Other "premature" infants may be in that category because of some congenital defect that is not otherwise obvious. This may happen either because the defect slowed the growth of the fetus, resulting in a low birth weight in a normal period of gestation, or because the unhealthy condition of the fetus somehow triggers an early birth. In other premature births, of course, the problem may have been in the physiology of the uterus, and the baby may have been developing normally.

The survival of a newborn infant is directly related to its age at birth. The most common and "standard" age is 40 weeks; the survival rate for these babies is 99 percent. At a younger age the survival rate is less: at 36 weeks, 90 percent; at 31 weeks, 50 percent; at 27 weeks, 10 percent; at 23 weeks, practically never. This is a direct consequence of the poor capability of the fetus for independent physiological adjustment, and therefore also of the degree of sophistication of the supportive care available.

The care routinely available for premature infants in modern hospitals starts with careful temperature control. The diet is precisely formulated to be, presumably, suitable for the special needs of the premature infant. For the smaller infants, it may be supplied by feeding tube instead of bottle. Intravenous feeding is sometimes used. The breathing capabilities of the premature infants are often supplemented by supplying oxygen-rich air. For a number of years after this practice was begun, high concentrations of oxygen were used, as in ordinary oxygen therapy with adults. However, it was discovered that too much oxygen caused retinal damage and had become a common cause of premature-infant blindness. Care is now taken not to get the oxygen level above twice the normal concentration in air, and no retinal injury occurs.

Changing styles in infant care

Practically all aspects of the care of small children are subject to great variations from culture to culture, and from time to time in nontraditional cultures such as ours. Some of our current practices are clearly superior to certain others in promoting the health and well-being of the children. Other practices to which we adhere just as carefully may be unimportant or even harmful. Parents are quite properly reluctant or unwilling to risk the health and happiness of their children by subjecting them to experiments. Consequently, the process of discovering what actually works best goes slowly. Meanwhile, some theorists may convince the public that some new measure is best for children, and almost everyone will change to the new method, which may not actually be an improvement.

Food

The question of what to feed young babies has been a subject of debate in our culture in recent years. For most people in most of history, there was no argument possible: The mother's milk was the only food easily available. In cases when it was not sufficient, and when no other woman with more than sufficient milk for her own child was available, the child would starve to death. The development of dependable supplies of cow's milk, and of reasonably hygienic practices for keeping the milk and preparing it for infants, made it possible to eliminate breast feeding. In America and in some other countries, the breast-fed infant is in the distinct minority. (This trend probably has a complicated relationship with the importance of the breast in western sexual traditions.) There have been various movements and organizations throughout this time to promote the reacceptance of breast feeding. Some of the appeals are based on the health of the infant, the convenience of eliminating bottles, and the emotional closeness to be gained. As is the case in most crusading groups, the arguments given have often gone beyond the evidence. Plenty of perfectly healthy babies have been raised on some mixture based on cow's milk. New powdered formulas and safe water systems make bottle feeding very convenient. As long as parents hold the baby while giving the bottle, most of the same emotional closeness is experienced as in breast feeding, with the advantage that the father can get into the act, too. Breast feeding now has considerable popularity among the more educated mothers, and especially among those who are not employed. It costs nothing to try, and young mothers would be well advised not to reject it out of hand. It can't harm anything and may be good for a variety of reasons.

The introduction of foods other than milk into a baby's diet is supposed to be done according to careful rules given by the pediatrician, but not every doctor gives the same rules. There is probably no good set of data establishing just how it should be done, nor is there likely to be good evidence available very soon. Luckily, an older baby can stomach a great variety of diets without apparent harm.

Other details of care

In many cultures babies are swaddled in large amounts of cloth and changed only once a day. In our culture, babies aren't to be left wet more than a few minutes. European babies once were toilet-trained as early as possible. Since Freud claimed that such procedures caused life-long neuroses, American parents, at least, have largely abandoned rigid toilet training and are variably permissive on the subject. (People are still neurotic.) Obviously it isn't obvious which procedure is best.

A recent full circle in procedure has taken place in medical practice. Circumcision, long practiced by Jews and some other groups as a religious rite, seemed to be responsible for the low rate of cervical cancer among Jewish women. It was suggested that this was due to the greater cleanliness possible with a circumcised penis. (Circumcision is surgical removal of the foreskin, a fold of skin that partially covers the glans of the penis except when the penis is erect. The space under the fold can accumulate secretions and bacteria.) On the basis of these and other data, it became common practice some 20 years ago to circumcise almost all baby boys for hygienic reasons. A careful study within the past few years indicated that the freedom from cervical cancer among Jewish women was probably due to general cultural rules for cleanliness, rather than to circumcision of their husbands, since other circumcised but less fastidious groups had higher cervical cancer rates. Since routine circumcision of male infants results in an occasional mishap, the procedure is now being applied much less frequently.

Launched in a new direction

Any newborn mammal, in moving from the carefully controlled shelter of the womb into independent existence, goes through the distinct physiological adjustments described earlier. This is a shock even with the careful maternal care for their offspring that is common to all mammalian mothers. The human infant undergoes an even greater change in circumstance than other mammalian babies. The usual mammalian newborn exchanges the programmed development of embryonic and fetal periods for a postnatal period in which its inborn behavior patterns (reflexes, instincts) provide a built-in response to the situations in which it finds itself. The human infant has a much smaller stock of such inborn behavioral patterns and must learn by doing. So the human infant enters into a period of rapid behavioral change, in which learning of new behavior accompanies continued anatomical maturation of the brain. No one really knows what proportion of early behavioral change is ascribable to maturation and what proportion to learning. The creative process of childhood fascinates, delights, and frustrates parents. They are able to observe this fascinating process but also have the responsibility of contributing wisely to it.

8 From child to adult: growing individuality

The product of the process of human reproduction is an individual human being. Every person has some traits that are standard human ones, but he has many distinctive personal traits as well. Each person, unless he has been born with gross defects, has an immense range of capabilities, but his particular levels of capability can be very different from those of another individual. How can we describe the origin of both similarities and differences in the same process of reproduction? Many people seem to assume, without giving much thought to the matter, that the common human traits are reproduced before birth, and the individual traits after birth. But that is certainly too facile an answer. Both the shared humanness and the individual personality of each person are the products of at least two creative processes: embryonic development and childhood. Some of the elements that contribute to the development of individual personalities are described in the next sections.

Events before birth

Earlier chapters have described the intricate processes of gametogenesis and fertilization, which assure that nearly every zygote has a set of genetic instructions that is both complete and unique. During embryonic development, controlled interaction of these genetic instructions brings about the emergence of a unique individual. Some of the differences among individual sets of genes are minimized because other genes have a compensatory effect while other genetic differences among individuals assume exaggerated importance. Sometime, far in the future, it may be possible to know the complete set of genes of some zygote, but it will still not be possible to predict accurately and in detail the end product, since a laboratory understanding of a gene or enzyme does not tell us what it will do in a unique situation. The genes are participating in a creative process. Embryonic development can therefore never be completely predictable from a knowledge of the genes involved.

Once the parents of a newborn baby are satisfied that it is normal, they want to know its sex, and any noticeable characteristics. Many individual features are present at birth, whether or not they can be detected; other differences develop later.

The determination of sex

One of the profound differences among people, one that increases with advancing age, is the difference between the sexes. Sexual differences affect so many areas of life that it is not surprising that this is a subject of occasional confusion. Under ordinary circumstances there is a good correspondence among chromosomal sex, type of gonad, type of secondary sex characteristics, and social sex role. But any theory explaining how this all comes about must explain not only the normal correspondence among all these different aspects of a person but also the occasional lack of correspondence in exceptional individuals. No present theory does this completely and clearly. What follows is a simple scheme that will explain most but not all the known facts.

In the human, and in almost all mammals, the Y chromosome is the male determiner, while any combination of chromosomes without a Y chromosome leads to the development of general "femaleness," which may not always be complete.

(One primitive, egg-laying mammal, the Australian spiny anteater, the echidna, is an exception. It lacks a Y chromosome; the males have the XO condition.) It is not thought, however, that the Y chromosome acts directly to produce "maleness" in every cell in the male that is different from cells in the female. Rather, most of the differences that distinguish a male seem to be produced by the action of testosterone. Most of these male characteristics would have occurred in the presence of testosterone, even if the actual cells involved had been XX instead of XY. For instance, the cells of the lower part of the face, even if they are XX, will produce a male type of beard if treated with enough testosterone at the right times. Obviously, the chromosome type makes a difference in some cells, or else the testosterone would not be produced. The question is: In which cells does it make a difference? One sure answer to this question is, in the interstitial cells of the testis. Most, but probably not all, of what happens to produce male characteristics can be explained by assuming that these interstitial cells are the only cells in which the presence of the Y chromosome is important.

According to this simplest possible theory, the first event in sexual differentiation takes place when the embryo has developed for about 5 weeks in a uniform manner, without regard to sexual differences. This development includes the laying down of the tissues that are going to become the gonads, and the beginnings of the duct system for both sexes. All these are still in a primitive, unisexual stage. Gonads contain large cells that will later give rise to gametes, and several smaller types of cells.

The maternal and fetal circulations contain various gonadotropic hormones serving various functions, including the maintenance of secretion by the corpus luteum. At an embryonic age of about 5 weeks the small gonadal cells that contain XY become sensitive to these gonadotropins. They respond to stimulation by secreting testosterone, as they and the cells that arise from them will do in adult life. The similar cells in gonads of XX embryos, on the other hand, do not respond to the gonadotropins at this time, and so secrete no hormone.

The testosterone secreted by the gonad of the male embryo is carried to all parts of the embryonic circulation and causes profound effects on the development of tissues in several parts of the body. It causes the cells destined to be gametes to become spermatogonia rather than oogonia when they receive a general stimulation for development from some other hormone. As evidence for this effect of testosterone, if this hormone is injected into early female embryos of experimental mammals, the gonad will develop into a testis and carry on spermatogenesis. In the normal male embryo, the testosterone that is carried to other parts of the body influences the development of many tissues, especially the reproductive ducts and the external genitalia, turning them in the male direction. These male determining effects of testosterone before birth stretch over many months, including some actions that are still occurring at birth.

In the absence of testosterone (as in an XX embryo), the normal processes of development seem to produce oogonia from the undifferentiated gonial cells, as well as producing the female ducts and external genitalia. Perhaps what happens is that the estrogens and progesterone being produced by the placenta have a female-determining effect unless they are overridden by testosterone. A male (XY) embryo that has its gonads removed in the fetal period develops chiefly female characteristics. At some point in the development of the female, however, the normal set of XX chromosomes becomes important, since individuals with XO or XXX, while generally female, do not develop normally and are sterile.

The prenatally determined sex differences include at least a few differences in the central nervous system. The hypothalamus of the female, but not of the male, has an intrinsic tendency to produce gonadotropic hormones in a cyclic fashion, contributing to the monthly cycle. This effect on the brain can be reversed with hormone injections at birth; they do not have that effect in the adult. More important, in many experimental animals, and certainly in man as well, the male is more aggressive, and has certain other behavioral tendencies, because of a permanent effect of testosterone on the nervous system, acting late in fetal life. Baby boys are more active than girls even in the hospital nursery, before cultural effects have a chance to act. Some of the behavioral tendencies do not become visible until puberty, when the renewed secretion of sex hormones brings them about.

Most, but probably not all, of the various intersexes and sexual anomalies that are occasionally found can be explained by the assumption that the production of hormones was abnormal sometime during embryonic development, because of chromosomal abnormalities, developmental accidents, or external administration of hormones. When pregnant women have been given very large doses of progesterone to guard against miscarriage, female infants have sometimes shown partly masculinized genitalia. A dramatic natural experiment demonstrates this same kind of effect in cattle. A female who has had a male twin is always sterile; this is due to the fact that in cattle, unlike the situation in humans, fraternal twins share the same placenta and have a thorough mixture of their bloods. The early production of testosterone in the male twin prevents the development of the female's reproductive system, although the testosterone seems not to transform her into a functional male.

Hormonal changes at puberty

During childhood, the amounts of gonadotropins and of steroid sex hormones in the circulation are very low. In some experimental animals, the ovary is nonresponsive to gonadotropins before puberty (as it seems to be before birth in all mammals) but this is probably not true in the human. In both sexes, the levels of gonadotropins and of steroid sex hormones seem to be maintained by negative feedback mechanisms, which are set so that levels remain low. The gonadotropins are too dilute to allow gamete production, and the steroids are too dilute for much additional differentiation toward the characteristics of one sex or the other.

At about age 10 (in our society) a number of changes seem to occur that bring on the characteristics of puberty. Current theory identifies several changes that happen at once. The earliest observed changes in hormones include increases in both gonadotropins and steroids. These are ascribed to at least three primary changes.

1. Due to maturational events elsewhere in the nervous system, the hypothalamus starts to produce more of the LH (luteinizing hormone) releasing factor at age 10, causing a direct rise in LH production. It seems that the hypothalamus was ready to do this all along but was held back by inhibitory impulses from the rest of the brain. Destruction of the nervous connections between the hypothalamus and the other parts of the brain causes premature maturation.

2. Due to maturational changes in the gonads themselves, some of their cells become more sensitive to LH and produce more estrogen or progesterone per amount of LH than was the case previously. Since LH is also increasing, the steroid level goes up very fast, causing the noticeable external signs of puberty.

3. The hypothalamus, presumably because of its own maturation, becomes less sensitive to the negative feedback of steroid hormones on FSH (follicle stimulating hormone) production. Thus despite the rising level of steroids, the FSH production rises, causing production of gametes.

These three developments occur more or less simultaneously; the inevitable differences in timing between individuals could easily account for the variations in the time of onset, pattern, and difficulty of pubertal changes. The whole process is dependent on and influenced by nutritional level, genetics, and general health, as well as by other hormones, especially thyroid and adrenal cortex.

The behavioral differences between the sexes built into the brain by the action of sex hormones before birth are now further activated. At adolescence there is an awakening of sexual interest because of the new high concentrations of steroid sex hormones. The behaviors of the sexes diverge still more than they had during childhood because of the different effects of the two kinds of hormone. Human behavior is uniquely plastic, so a large amount of human behavioral difference between the sexes is undoubtedly of cultural origin. Since cultural influences tend to accentuate the differences that already exist, it is impossible to say how much of "masculinity" and "femininity" are hormonal and how much of social origin.

Today's children mature at an earlier age

The hormonal changes leading to sexual maturation occur fairly rapidly, many of them within a year or two. It takes much longer, of course, for these elevated hormone levels to act on the various growing tissues of the body to bring about the mature male or female condition. Thus the recognition of sexual maturation is arbitrary—any of several changes could be used as a marker. The word puberty refers to the hair that appears near the genitalia at maturity, but this, by its nature, occurs gradually. The single most dramatic event, most easily timed and remembered, is the menarche: the first menstrual discharge of the maturing girl. Sometimes used as an equivalent event in the boy is the first nocturnal emission of seminal fluid ("wet dream"), but this time is more irregular and may be unrecognized or unreported. Although the average time of actual hormone change seems to be much the same in boys as in girls, nocturnal emissions usually begin at a somewhat later age than does menstruation; this parallels the way some other maturational changes, such as growth in height, start and end a couple of years later in boys than in girls. Probably menarche is a good marker of the age of sexual maturation of youth in general. In the United States and western Europe the average age now is just under 13—a few years younger than it is in many less affluent regions of the world, but, surprisingly, similar to Hong Kong, among other places.

The present age of menarche has not always been so low but has moved to earlier ages at a surprising rate in recent history. Over the past 150 years, the average age at menarche for young people living in Western countries has moved to 1 year younger every 30 years (about every generation). Thus 120 years ago the average age was about 17. This change has come at the same time and approximately the same rate as a spurt in height that takes place earlier than menarche. Each generation, the average child reaches a certain height at a younger age. The graphs show that a certain height is reached 1 year earlier for every 40 years of history. There has also been an increase in final height reached. The average modern boy reaches his adult height at 17; his great-grandfather may have kept growing until he was 25. The final height of the modern boy is about 3 inches greater than that of his great-grandfather. Presumably, these simultaneous changes in growth pattern are related; they may all have the same cause or causes.

Of the various reasons advanced for the earlier age of maturation in recent history, by far the most plausible one is that the general diet of children has been improving rapidly in the last 150 years. Parents may now overdo the business of seeing to it that Baby has orange juice every day and that older children get a fruit, a vegetable, and a meat every day, but diet is certainly better than it was in many families in past generations. Both an increase in knowledge of nutrition and an economic improvement that has made it possible to improve diets have contributed to better eating habits. Tied in with this same progress has been an improvement in general health, especially in this century. Children's growth is not slowed or interrupted by major disease nearly as often as before. These explanations of earlier maturation and greater growth are supported by data that show that children of the poorer economic classes (but in the same climate and of the same racial stocks) may be a year or two slower to mature than those from more well-to-do homes.

While the diet and health explanations can be accepted as the major causes of these changes, several other explanations have been offered that are at least plausible. They may have contributed to the changes, too, but the evidence is weak for all of them. Several people have speculated that height differences, especially the increase in final height, may have been favored by the greater mobility of people in recent generations, so that the stunting effect of inbreeding is replaced by hybrid vigor. The earlier sexual maturation may have some relation to the finding that experimental animals mature earlier if exposed to light for longer periods of time than is normal—artificial light has lengthened the day dramatically in the last 100 years. It is even possible to imagine that greater social stimulation and sexual excitation in our culture in the last few generations has hastened maturation. As support of this supposition, it has been found that girls of the Oneida Community, introduced into frequent sexual intercourse before puberty, were found to have an average age at menarche that was a couple of years younger than girls outside the community.

Menopause

At the other end of the reproductive period things come again to a rather abrupt end for the woman. When she is about age 45, a woman's menstrual periods become irregular and then cease altogether, and this change is often accompanied by distressing disturbances. The blood distribution may be so out of adjustment that the woman breaks into cold sweats or has "hot flashes"—sudden flooding of the blood vessels of the skin. As a result of these discomforts and perhaps because of direct effects on the brain of hormone withdrawal, and sometimes certainly because of anxieties about role changes, the woman may become moody. These symptoms usually disappear eventually, but there are progressive and permanent changes due to the hormone withdrawal: some atrophy of the breasts and genitalia, aging of the skin, loss or thinning of hair, increased danger of heart attacks (up to the level of danger for men of that age), and softening of the bones. Because of these undesirable and sometimes dangerous changes, it is becoming more and more common to give continuing hormone therapy after menopause. Sex hormones are administered either cyclicly (as with birth control pills) or continuously. In either case, the level required for maintenance of structures is less than that used for contraception, so the side effects are less serious.

Only some of the reasons for menopause are understood. It is not a common phenomenon among animals; even in the domestic animals that die of old age (unlike most wild animals) such an abrupt end to the reproductive period is at least not generally recognized. Our preagricultural ancestors during most of human existence may not have been aware of menopause—they usually didn't live long enough. But the phenomenon is standard in our culture; our average life span is more than 20 years greater than the average age at menopause. The cause of menopause is certainly centered in the ovary. At menopause there are no more follicles capable of developing toward ovulation and releasing estrogen and progesterone. It is not a lack of pituitary gonadotropins; the levels of FSH and LH after menopause are greater than at any other period of life, because the hypothalamus and pituitary have been released from the feedback inhibition by the steroid sex hormones.

It is the reason for the disappearance of the ovarian follicles that is still unknown. The first explanation that might be considered—that they had been lost through ovulation—is clearly not the answer. The infant girl has 150,000 to 500,000 oocytes, and 34,000 of these are still present at puberty. Assuming 30 years of fertile life with no pregnancies and no contraceptive pills, there might be 390 ovulations. At each ovulation about ten oocytes come to nearly full size, and the nine that are not released degenerate. That accounts for a total of only 3900 follicles. The other 30,000 oocytes that were present at puberty have disappeared for some other reason, probably the same unknown process that destroyed before puberty most of the oocytes that had been present at birth. There is no clear reason why the oocytes disappear at age 45; they just do. From this discussion it is clear that prevention of ovulation by contraceptive pills cannot be assumed to have any important effect on loss of oocytes, and an earlier onset of ovulation may not be important either in determining the age of menopause. It seems that the general increase in health in recent years is causing a slightly older average at menopause, just by delaying follicle degeneration.

A man does not experience a similar abrupt change in his reproductive and hormonal situation. As a part of the general weakening and loss of body cells with age, the vigor and sexual activity of men tend to decrease from youth on. A gradual decrease in testosterone levels leads to a reduced secretion of the accessory glands, the prostate and seminal vesicle. Internal sex drive is partly due to accumulation of seminal volume, and therefore drive may decrease in older men. However, the readiness of a sexual partner is probably more important than these slight physical factors.

Human sexual behavior, its intensity and frequency as well as its nature, is probably more directly influenced by emotional patterns than by physiological changes. Decline in sexual activity with age in either sex is not always seen, nor is it usually necessary. When found, it is often due to illness, boredom, or negative expectations.

The genetic basis of individuality

With the knowledge that part of our individuality is of genetic origin, it is interesting to study human variation in greater detail. Each human zygote that is conceived is unique. There are so many different combinations of alleles at so many loci that it is highly unlikely that any two zygotes would get exactly the same set of genes. As the genes interact during the developmental process, two contrasting effects probably occur. Some potentially harmful genes (perhaps genes for ineffective enzymes) have less effect than might be predicted, because some compensating mechanism in the cells allows other genes and their enzymes to take over the missing function, just as a person can learn to overcome a slight handicap. On the other hand, a very slight difference from the ordinary in the effectiveness of some allele and the protein it specifies may happen to have such an effect at a key point during development that the whole pattern of embryonic growth is skewed in one direction. One might think here of the saying, as the twig is bent, so grows the tree. Each embryo develops somewhat differently from every other just because of its unique genotype, and also because it may respond individually to small developmental stresses.

After birth the environmental challenges to individuals are more varied than during embryonic life, and the range of response is also much greater. It is probably impossible to keep people from developing as individuals, although in a rigid, nonpermissive society neither the society nor the individuals may get much benefit from such individuality.

In considering the sources of biological individuality—human variation—in detail, it is necessary to discuss those traits that are inherited by combinations of many genes as well as the simpler cases. Which of these genes are likely actually to show up in parents and their offspring can be determined by knowing which human population one is discussing and which alleles are common in that population.

Quantitative inheritance

The most important traits by which normal humans differ (intelligence, physical abilities, creativity, health and stamina, general appearance) are typically influenced by genes carried at several different loci, all acting together. In some cases, the genes at each locus contribute something a little different from the contribution of the genes at other loci. In other instances, the contributions of genes at different loci cannot be distinguished; they seem simply to add together. It is this simplest type of quantitative inheritance by "multiple factors" or "multiple loci" that is most easily explained and understood and that will be described here. Actually, we do not know much about the inheritance of any human characteristic governed by multiple alleles. We do not know whether or not our conception of genes at several loci having identical additive effects really applies to any human characteristics; perhaps all cases of multiple loci are more complicated than this. About all that can be done in describing the inheritance of any quantitative trait is to make an estimate of the number of loci involved.

Inheritance of height

In a simplified picture of quantitative inheritance, it can be assumed that at each locus there is a choice between a contributing allele that adds to the amount of something and a noncontributing allele that adds nothing. We might postulate, for example, that the variation in body height within a population of people was due to multiple factors at four loci. Each contributing gene might be postulated to add 2 inches to a basic height of 5 feet (60 inches) for men. Each noncontributing gene would add nothing. Women would be proportionately smaller; we would add 1 inch for each foot of a woman's height to calculate what her height would have been as a male.

According to this imaginary scheme, a man 5 feet tall would have none of the eight possible contributing genes; he would be homozygous for the noncontributing allele at all four loci. A man of 70 inches—10 inches more than the minimum—would have $^{10}/_2 = 5$ contributing genes, which might be arranged among the four height loci in any combination of one or two per locus, for a total of five contributing genes. A man who was 6 feet, 4 inches tall (76 inches) would be homozygous for the contributing alleles at all four loci; he would have eight contributing genes (5 feet $+$ 8 $\times$ 2 inches). If environmental influences did not obscure the pattern too much, there should be nine size classes observable in the population. That is, a person might have zero, one, two, three, or up to eight contributing genes. Counting the number of quantitative classes represented by individuals in a population is one way of estimating the number of loci involved in the inheritance of the quantitative trait being studied.

The patterns by which it is possible to predict the appearance of offspring with regard to quantitative traits are somewhat more complicated than with single loci. In the simplified model given here of inheritance of height, we can see some of the complications. If a man with no contributing genes at any loci married a similarly short woman (he would be 5 feet tall, she 4 feet, 7 inches tall), all their children would be short (that is, the same heights as their parents). If two maximally tall people, each with eight contributing genes, married, all their children would also be tall (76 and 70 inches). If a maximally tall man (76 inches) married a maximally short woman 55 inches), each of his gametes would have four contributing genes (one for each locus) and each of her gametes would have no contributing genes at the height loci. All their children would then have four contributing genes and would be 68 and 62 inches tall. The same would hold true for the children of a maximally short man and a maximally tall woman; they would be 68 and 62 inches tall, according to their sex.

The analysis of inheritance in the families mentioned has been as simple as if there were a single locus with a sex influence acting in combination with the genetic one. If we try to predict the result of other crosses, however, quantitative inheritance begins to get much more complicated. The heights (and number of contributing genes) inherited by children of parents of intermediate heights depend on how the parents' genes are distributed among the loci. Suppose that the husband and wife each have four contributing genes. In one extreme situation, they may each be homozygous for contributing genes at two loci and homozygous for noncontributing genes at the other two loci. Thus each person would always produce gametes with two contributing genes each, and all children would have $2 + 2 = 4$ contributing genes, and would be the same heights as their parents. If the contributing genes of parents of intermediate height are differently distributed, the heights of the offspring may be much more variable. In an example of this kind, the two parents with four contributing genes each may be heterozygous at all four height loci. Since genes at different loci assort independently, any one gamete might carry anywhere from zero to four contributing genes. In such a situation, combination of two gametes would give offspring with anywhere from zero to eight contributing genes, the entire range possible. Thus it is impossible to predict whether the children of two people of intermediate height will be intermediate in height like their parents or will range from one extreme to the other. It is common (though not assured) that children of parents who represent the two extremes of a characteristic will exhibit characteristics somewhere beween those of their parents.

In the last hypothetical family discussed above, a family with two parents heterozygous at all four loci, it is possible to calculate the *probability* that any one child will have an extreme number of contributing genes (zero or eight). One such parent could produce gametes with four contributing genes, but the chances of including the contributing gene at all of the four loci is $(1/2)^4$ or $1/16$. The chance that the gamete with which it unites will also have four contributing genes is again $1/16$. The combination of the two probabilities is $1/256$. So the chances of two such heterozygous intermediate-height parents having a child of maximum height (or minimum height, for that matter) is $1/16 \times 1/16$, or $1/256$. This kind of information could, theoretically, be used to estimate the number of genes involved. If the genetic history of the parents indicated that they were heterozygous (that is, if their parents had been tall × short), the frequency of extreme children among such families would give an indication of the number of loci according to the following pattern:

Number of loci	Number of different phenotypes	Probability of extreme offspring
1	3	1/4
2	5	1/16
3	7	1/64
4	9	1/256
5	11	1/1024

In an entire population, many people may be heterozygous for all the loci affecting a quantitative trait. If so, the chart shown gives a rough indication of the frequency with which the most extreme phenotypes will appear. If only one locus is involved, a large fraction of the population, something like one-fourth, may be at one extreme. But if five loci are involved, only one in a thousand or so will have the most extreme quantitative condition.

Inheritance of skin color

The most famous of the attempts to analyze the inheritance of a quantitative human trait concerns skin color (amount of melanin) in black/white (African/European) crosses. An early worker studied the offspring of a number of such matches and concluded that there were two loci involved, because he thought he could distinguish five phenotypes. According to this scheme the African would have four contributing genes for melanin, the European none. More recent work has shown that the inheritance of melanin is indeed determined by multiple loci but that it is more complicated than at first suggested, with more like seven loci involved.

"Throwbacks"

Since historically our society maintained such decided handicaps for people with any degree of African ancestry, it was common for those whose pigmentation and general appearance made it possible to "pass for white" to do so. Thus the "white" segment of our society has a small but definite amount of genes of African origin. When the social disadvantages of "blackness" were even more cruel than they are now, there were many stories in popular mythology to the effect that two blond whites, one or both of whom unknowingly had a trace of African ancestry, had produced a black baby, with resulting social disaster for themselves as well as their child. These stories were reported in the popular press and were widely believed. A professional geneticist conducted a study a few years ago, tracing each such story and found all of them to be without foundation. (In one case, for instance, the child had been born with some congenital defect and was kept hidden—this giving rise to the rumor that there was a social defect, instead.) It would be predictable from genetic theory, anyway, that the folklore on "throwbacks" could not be true. A look at the probable mechanism of inheritance of such quantitative characteristics indicates that a child could not be darker than the sum of the "darkness" of both parents and that he would be even that dark only rarely. There is no indication that there is any dominance or recessiveness to "crop up" later in the inheritance of melanin (with the trivial exception of albinism). Anyway, Africanism is not some unitary genetic trait that one either does or does not have. African populations have more genes for melanin production than European populations, but the individuals appear on the same continuum and share the same genes. Folklore dies hard, however.

Gene frequencies: what are they?

For some characteristics, including many quantitative ones, it is practically impossible to try to predict what the offspring of any one family are likely to be, but one can say quite a good deal more about the distribution of characteristics among people in a whole population group. For any **genetic population** (a group of individuals of the same species that are more likely to mate among themselves than with other individuals of the same species), it is often possible to calculate measurements called **gene frequencies**. The gene frequency of a certain allele at a certain locus is equal to the probability that a gamete from the population, taken at random, will contain that allele rather than the other alleles for the locus. Another way to think of a gene frequency is to think of one somatic cell from each member of the population and calculate the proportion of all genes at the locus under consideration that are some particular allele.

Use of gene frequencies

The gene frequencies for alleles at a number of loci are fairly well known for many human populations. Some, like those of blood group genes, can be calculated directly from tests on many people. The frequencies of rarer genes have to be estimated indirectly. There are a number of practical applications to public policy that can be made from knowing about gene frequencies.

If certain simplifying assumptions are made (such as the assumption that people ignore the presence or absence of a certain allele when choosing a mate), is it possible to calculate the likelihood that certain combinations of genes will appear among the offspring. The frequency of any allele is the same number as the proportion of all the ova that will contain that allele, and also the proportion of all the sperm that will carry that allele. The fraction of the population that will be homozygous for the allele is therefore the proportion of the ova carrying it that are fertilized by sperm carrying the same allele. That proportion is, numerically, the square of the frequency of the allele. Conversely, if the proportion of the population that is homozygous for an allele is known, the frequency of the gene will be the square root of the frequency of the homozygous condition. Since recessive genes show their effects only when homozygous, it is easiest to calculate a frequency for recessive alleles.

Cystic fibrosis, a serious genetic disorder of the pancreas, is rare, occurring in about 0.0004 of the American population. The condition is due to a recessive gene for an ineffective enzyme; so the frequency of the gene must be the square root of $0.0004 = 0.02$. This means that 98 percent of all gametes will carry the normal gene at that locus, and 2 percent will carry the cystic fibrosis allele. It is possible to calculate the number of carriers for the gene (heterozygous individuals) by slightly different reasoning. The probability of children being conceived from an ovum with a normal allele and a sperm with a cystic fibrosis allele is $.98 \times .02 = .0196$. The probability of heterozygous individuals from ova carrying the cystic fibrosis gene and sperm carrying the normal one is also .0196. When these two figures are summed, the total proportion of the heterozygous carriers of cystic fibrosis in the United States turns out to be .0392. So about 4 percent of the American population carries the gene for cystic fibrosis, although only 4 in 10,000 people suffer from the disease. It was this kind of calculation that formed the basis for the estimates described in Chapter 5 that concluded that a screening program for detecting cystic fibrosis carriers would be practical. For rarer genes, such screening programs would not be justified.

Cousin marriage

The calculations just presented depend on the assumption that anyone is equally likely to marry anyone else. When people marry relatives, there is a fairly high probability that they carry the same harmful recessive genes. Most such harmful recessive genes are very rare; cystic fibrosis is unusually common. But there are so many rare harmful genes that it is estimated that each of us has, on the average, ten seriously harmful recessive genes. Usually, the person one marries has his harmful genes at different loci, and no harm befalls the offspring. First cousins, however, share half of their ancestry. Stillbirths or deaths of newborn babies (both often due to congenital defects) are about twice as likely in children of first-cousin marriages as in children whose parents are not related. The hazards with offspring of even more closely related people are even greater, but the hazards for children of second cousins are not measurably increased. Probably the nearly universal prohibitions against incest came partly from observations that such unions resulted in an unusually large proportion of defective children.

Population genetics

Gene frequencies of a fairly stabilized population may be changed by the migration into the region of individuals of different genotypes, by the random accidental loss of certain genotypes (in very small populations), and by differential fertility. Differential fertility, the most powerful influence in the long run, may be due to the early death of certain genotypes, to a greater number of offspring produced by certain genotypes, or to more successful raising of offspring to maturity by some types of parents than by others. Even a slight difference in fertility may lead to large changes in gene frequency over a number of generations. This is evolution, and may be observed going on right now in a number of present genetic populations, including some human ones.

Hemoglobin S

Human populations differ in frequencies of the various alleles at many loci. For instance, European populations have fewer contributing genes for melanin pigment than African populations do, and fewer contributing genes for carotene pigment than Asian populations. Genes for hair shape, blood type, and body proportion are obviously different in frequency from one population to another. In most cases we do not understand the origin of these differences. The origin of some genes affecting red blood cells is well understood, however.

In many tropical populations, the sickling gene and other genes with similar effect (thallasemia, G6PD deficiency) are present in the population in moderately high frequencies, because the heterozygous person is more likely to survive than either homozygote. A person homozygous for the sickling gene, for instance, may die from anemia before reaching reproductive age. In regions of the world where the mutant genes are common, a person homozygous for the normal gene is likely to die in childhood from falciparum malaria (the most dangerous type). Malaria is or used to be endemic in central Africa, the Mediterranean countries, India, and Indonesia, and these are the regions in which the genes for abnormal red blood cells are found in high frequency. If both selections were absolute—if neither homozygote could survive at all—only heterozygous individuals could contribute to each new generation, and the gene frequencies would be 0.5 for the normal gene and 0.5 for the sickling one. In the case of sickle cell in central Africa the malaria-caused mortality of those with normal hemoglobin (homozygous for the normal gene) was never 100 percent and the highest recorded frequency for the sickling gene was 0.31. In Americans of African ancestry, the present frequency of the sickling gene is lower than would be expected according to its frequency in the African regions where their ancestors came from, and making allowance for the 20 to 30 percent European ancestry estimated from other gene frequencies. Apparently, over the ten or so generations in North America, the lack of malaria has removed the selection in favor of heterozygosity for the sickling gene, and the early death of those homozygous for the sickling gene has reduced its frequency.

It is believed that in many situations in which two or more alleles are commonly found for a single locus among the members of the population, there must be some selective advantage for the heterozygous condition, as with the sickling gene. However, investigation into this possibility began only recently and we do not understand the distribution of any of these other sets of alleles very well.

Human races

A race, in any species of animal, is a population within a species that is partially isolated genetically from other populations, and that differs significantly from other populations with regard to gene frequency at several loci. (Races are often called "subspecies" by people studying wild plants and animals, while races of domestic animals are called "breeds," and races of domestic plants are called "varieties.") The estimate of how many races there are within any species (such as man) is an arbitrary one, since there is no general agreement about how isolated from each other two populations must be and how much genetic difference there should be for the populations to be called races. Thus some authorities would distinguish many races, others just a few. Once a population geneticist has decided on the levels of isolation and difference he will use for his criteria, the actual designation of populations as races is fairly objective. Races are, then, real divisions of a species.

The thing to remember that has significance in discussing social problems is that race is a designation of a population, not of an individual. Description of a race simply states that if we were to examine many members of population A (one race) we would be more likely to find certain alleles than if we examined many members of population B. Identification of a population as a race tells us nothing certain about any individual who is a member of the population and who is therefore said to be a member of the particular race. Within any one race there are many individuals who differ quite widely from the average of the population, and who have a combination of genes that would be more likely to be found within another population. A person's race does not tell much about him. Conversely, inspection of an individual does not make it possible to assign him, without occasional error, to a certain race. If his ancestors and parents are of a certain race, a population, so is he, no matter whom else he resembles. (This points up the tragic absurdity of the racial policy in the Union of South Africa, where people are assigned to races on the basis of phenotypic characteristics. There have been cases where parents light enough to be called "White" had children a little darker, who were then classified "Colored," were subject to legal restrictions, and were barred from living in the same sections of the city as their parents.)

The social practice of identifying people by race is based on the assumption that by noticing certain characteristics of a person we may assign him to a race, and that by knowing his race people will know other important things about him and can make decisions on that basis. But that assumption is absurd genetically. The various characteristics of any race are controlled by genes at many different loci on different chromosomes. Therefore, they are assorted and inherited independently. The presence of one racial characteristic tells nothing about whether or not another is present.

Racial discrimination is to be condemned not just because it is immoral and unfair (although it certainly is), but because it is bad for society, leading to foolish, wasteful decisions. Society needs to use the abilities of all its citizens. If it fails to employ people up to their full capacities, for baseless reasons, society itself is hurt, in addition to hurting the individuals who suffer from discrimination.

Race is a useless criterion on which to assign people to social roles, for two reasons. First, few of the known characteristics by which human races differ from one another have any importance in the modern world. A very lightly pigmented European might be a poor bet as an explorer in the desert, but a pith helmet and suntan lotion take care of his problem pretty well. A Congolese or a native of Yap in the Pacific might synthesize too little Vitamin D in the far North, but cod-liver oil solves that problem. The second reason that race is irrelevant is that even in cases where racial characteristics are important for some special task, individuals cannot be wisely chosen on that basis. Since spaceships are cramped, tall men may be unsuitable as astronauts. A Watusi who happened to be short would have no disability —his race would be irrelevant. Similarly, a Pygmy who was taller than his cousins might do all right in basketball. Race is rarely if ever a useful criterion for choice.

How did races come about?

There are amazingly few good ideas as to how human racial differences arose, except for the general assumption that races evolved in certain geographic areas to fit the special circumstances there. The development of high frequencies for abnormal hemoglobins is clearly due to the selective advantage of these genes in malarial regions. This is the only racial trait that is really understood. The most commonly noted racial difference—skin color—is still a mystery. A dark skin might be advantageous to a hunter stalking game in the forest (where the darkest peoples live), but that doesn't explain races with paler faces. The only plausible hypothesis to explain the lack of pigmentation in northern Europeans is that light skin admits the ultraviolet rays from the brief winter sunlight and allows survival by the manufacture of vitamin D. Dark skin, on the other hand, is supposed to protect tropical man from production of a toxic excess of the vitamin. That still does not explain the successful existence of Siberians and Eskimos, who live just as far north as Scandinavians but are fairly dark-skinned.

Racial intelligence

Dr. Arthur Jensen, who is a competent educational psychologist, has proposed that the American blacks, as a race, are of significantly lower intelligence than other American races. The first of the facts that led him to this conclusion was that black children and adults consistently scored lower on standardized intelligence (IQ) tests, on the average, than people of other races who live in somewhat similar circumstances. The second fact that Jensen uses to support his thesis is that a measurement called the "heritability" of intelligence is high, about 80 percent, when it has been measured in American and European white populations.

For any variable trait that can be measured quantitatively, it is possible to calculate this technical value called "heritability." The calculations are based on how nearly parents and children, and brothers and sisters, are alike compared to the variation in the general population. To the extent that the effects of heredity and environment can be untangled, the calculation of heritability probably gives as good an estimate as can be made of their relative importance, in a given population, living in some specified environment. However, if the population were put into an extremely uneven environment, the total variability of the population might be increased, and the fraction of it due to hereditary differences would be much less. Heritability is highest when environment is favorable and uniform. Recent data indicate lower values (as expected from theoretical considerations) for heritability of IQ among blacks than among whites. Even this tells nothing at all about heritability (relative contribution of genes and environment) in comparisons between populations.

Is Jensen's contention valid?

Both of the facts cited by Jensen to support his hypothesis have such serious uncertainties associated with them that they are not convincing to most geneticists. The effects of environment (both physical and social) and of inheritance on the intellectual development of American blacks are hopelessly intertwined. There is no way of untangling these effects by any statistical study until the environmental differences between races are completely eliminated. A valid statistical study of a thesis such as Jensen's requires that the two variables, heredity and environment, be tested independently. This is now impossible. Since the data of heritability are at least as suspect as the IQ values, there is no valid statistical information on racial differences, if any, in intelligence, nor is there likely to be any reliable information in the near future.

From general evolutionary principles, it is difficult to imagine one large branch of the human species with markedly lower intelligence than the rest of the species. There has been ample time in all populations for selection to provide some kind of an optimum intelligence. Evolutionary pressures must have been similar everywhere; there are no reasons to predict a difference. Jensen's thesis is basically unlikely, and the burden of proof is on him. No basis now exists for proof in either direction.

The importance of "racial intelligence"

The controversy described has grown out of proportion, and is really of only academic value, since social selection of people for certain jobs needs to be on the basis of individual abilities, not population averages. Even if Jensen were shown to be right, which is extremely unlikely, his thesis would make only a very few policy changes necessary. Because of the very great overlap known to exist in the distribution of IQ among races, race is, and will continue to be, a useless criterion for assigning individuals to schools, jobs, and social roles. It might make some difference in our expectations if we knew that there were average racial differences in important abilities. That would still not often be important, since we still have a long way to go until people are sorted out according to ability, no matter which of the estimates we use for that ability.

The most important present question about IQ tests and similar criteria for decisions is not whether what they measure is genetic. Rather, it is whether these tests are really helpful in giving data from which to make assignments of jobs and decide admission to schools. Or do they lead to selection of people with certain life styles, personal characteristics, and social classifications that are irrelevant for purposes of the evaluation? For example, if someone who does poorly on an IQ test really cannot hope to do well in a job, he should not get that job, whether his inability is genetic or environmental. (It is no kindness to a person to give him a job he cannot perform properly, and then expect him to do it.) But if the test measures only whether or not the applicant grew up in the same cultural style as the employer (or the person who constructed the test), and if someone who scores low on the test could learn to do the job fairly easily, the test is useless and may prevent the hiring of a useful employee.

Normal genetic variation

Not all inherited human traits are tragic defects or important quantitative variations. Most genetic variations that contribute to the uniqueness of each individual consist of alleles with very small effects, hardly noticeable. Some of these minor genetic variables are small differences in enzyme efficiency; others show up as small, unimportant structural variations.

It is probable that most or all genes are subject to mutation. A mutation is a rare error in the copying of the genetic instructions during cell division, causing one or more subunit in the DNA chain to be replaced by one of the three other possible subunits. The incorrect subunit (mutation) is faithfully copied in subsequent cell divisions. About a third of such subunit substitutions will have no effect on the protein produced. The rest of the mutations will cause one amino acid to be replaced by another, at a certain spot in the protein coded for by the gene. This change of a single amino acid (out of 200 or so in the whole protein) may have serious effects (as with sickle cell hemoglobin), but the change will more commonly be of little importance, since the protein is likely to function almost normally. Thus for any gene locus there can arise, by mutation, a large set of alleles (up to a hundred or more) differing from each other by only one or two DNA subunits, and most of them having only minor effects on the cell in which they are found.

Those alleles that code for normal proteins will be passed on to future generations, while those which contribute to ill health will be less likely to be transmitted. (Healthy people usually have more children.) Consequently, the frequency of any one allele in a population is determined by the balance among three influences: (1) how frequently other alleles mutate into this allele; (2) how frequently this allele mutates into other forms; and (3) how likely an individual with this allele is to survive and to have offspring.

Geneticists do not yet know the relative importance of these factors in determining gene frequencies. In those cases in which the genes being considered seem unimportant, the geneticists suggest that the frequency found is probably a balance between the "forward" and "back" mutation rates. When the frequency of an allele is fairly high despite an obvious harmful effect, they suspect that the gene must be advantageous under other circumstances. The well-established case of this kind, of course, is the sickling gene, which is often fatal in double dose, but which may be life saving in single dose in areas where malaria is a serious danger.

Most minor variations, on the other hand, have not been thoroughly studied, both because there is little incentive to study harmless, unnoticeable variations, and because such minor variations are difficult to measure or observe. Most good genetic information on minor variations concerns blood types (at many loci besides those for ABO and Rh groups), since these variations are easy to detect. The frequencies and pattern of inheritance for such blood groups is usually well known, but there is very little information on whether or not the groups are really important, and on what causes their frequencies to be as they are. We do not really know that any genetic variation is totally insignificant. Any differences may be important in special circumstances.

Interesting human genetic traits

A number of observable human traits have a simple inheritance and are interesting to notice but cause no trouble. The list that follows should be used with caution, since what seems to be a simple variation with a simple mode of inheritance may not always be inherited in that same simple way. For instance, baldness has been variously demonstrated, in different families, to be due to a sex-linked recessive gene, to an autosomal gene that is recessive in females but dominant in males, and to a Y-linked gene. The following traits are all dominant to their most common allele, so their inheritance is interesting to trace through several generations:

> ability to roll the tongue
> ability to taste the chemical phenylthiocarbamide
> astigmatism (a moderate visual defect)
> ear lobes free along the inner edge
> bass or soprano voice
> hair on the middle section of the finger
> tendency to fold the hands with the left thumb next to the body
> nonred hair
> "widow's peak" hairline
> eye fold of the Mongolian race
> high convex nose
> dimpled (cleft) chin
> freckles

Growth in childhood

Genetically determined individual traits are not always evident at birth. Often they become apparent only after a period of growth.

The growth that occurs during childhood is not just increase in size; it is partly a differential growth like that which occurs during embryonic development. Such developmental growth, which results in an adult shaped very differently from a baby, occurs in all parts of the body, from muscles to brain. The control of growth comes both from the genetic instructions acting within each cell and from the environmental experiences of the child. For example, the type and level of a child's physical activity are perhaps close in importance to the physical build of his ancestors in determining his own muscular development. Of greatest importance is the fact that the richness of mental experience interacts with genetic background in affecting the postnatal growth of the brain and of its internal connections.

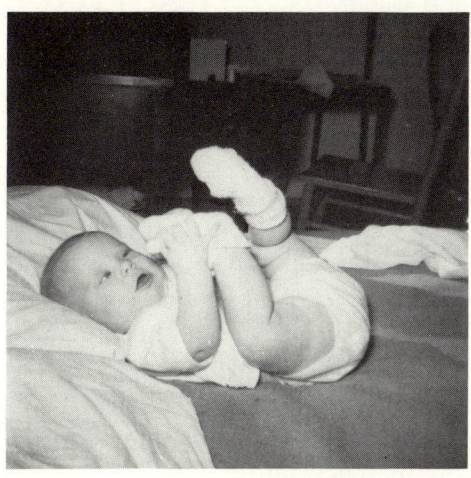

Fig. 8–1: An infant investigates her feet.

A child's discoveries

Parents have always been fascinated when they watched their children discovering the world and learning to live in it. Children's play is largely taken up with this process of discovery about operating within the surrounding world, about getting along with other people, and about what each child is capable of doing.

One kind of discovery that each person makes throughout his life is about himself. The baby's first noticeable discovery is of his feet, which he plays with and investigates in great detail (Fig. 8-1). The discovery, exercise, and development of capabilities need never cease. (Remember Grandma Moses?) Just as with developmental growth, self-discovery cannot be described as just "nature" or just "nurture." The question, "Did we inherit the ability to jump over a wall, or did we learn it?" is a futile one. We do not know how high we can jump, how fast we can run, or how much we can learn, until we try. Then the very trying helps to give us the abilities we discover in ourselves.

Discovering other people and how to get along with them occupies much of a child's energies. A baby has an automatic attachment to its mother from the first. The quality of this relationship then influences all of the relationships that follow. A young child's attitude toward other people, especially strangers, goes through several stages of easy acceptance alternating with suspicion. The process of building interpersonal relations continues as an important activity throughout life and is a source of great emotional meaning, both pleasant and unpleasant. The events of one relationship influence and alter the direction of future relationships. Whole sciences and professions deal with the understanding and furthering of such interpersonal interactions. The nature of a person's interaction with others depends in part on the individual's personality; the kind of interaction achieved in turn helps shape the personality.

Children are full of curiosity about the objects around them, and they devote much time and energy to discovering the nature of the world of objects. A baby picks up everything within reach and puts it in his mouth. From this he learns to correlate the appearance, the feel, and the taste of things. As he becomes more mobile, the older baby crawls, climbs, or toddles farther and farther afield, investigating new objects and wider vistas. By the time a child is learning to talk, he is obviously working at the gigantic problem of classifying the world and its contents, and of learning to describe a whole universe of objects with a relatively few words.

All of these discoveries are essential for the mental and emotional development of children. They need a chance to explore and discover as much as possible. No baby should be confined in a playpen longer than is required for his mother's sanity and his own safety. Babies should be supplied with a variety of harmless objects to clutch and bite, a variety of activities to engage in, and a chance to interact with people of various ages. As the baby's speech develops, parents who really listen can speed and improve the process by noticing and encouraging the flashes of insight that all babies have but which they should learn to seek. Each baby will grow to be even more of an individual with such a stimulating environment than with a bland, sparse one. At the same time, each individual can be expected to develop a standard competence to get along in the world as it is. This is because the results of all the discoveries are not arbitrary or accidental. The results, the discoveries, are largely predetermined by the very "nature of things." Part of this "nature" is the sensory and motor equipment of the child who is discovering the world.

Acculturation

A human child, left to itself, could never discover enough about how the universe operates to survive in it, even if protected and nurtured until biological maturity. Our human competence depends on our having received the benefits of discoveries made before our time, and having them passed on to us by parents and by the community. Human societies pass on to their children the set of discoveries and insights that the societies have found useful in coping with the universe. The significance of acculturation is probably greatest in supplementing our discoveries of the nonhuman world about us, but it is also important both in channeling human interactions and in encouraging specific kinds of personal understanding. Culture is all that distinguishes advanced peoples from primitive ones, and it is a large part of the difference between all mankind and the apes.

Individuality

The things that a child discovers, and what he remembers of all he is told, depend on what he is prepared individually to understand and remember. Each person develops his own pattern of understanding the world, and his own scheme for dealing with it. Just as a person has genes and experiences in common with everyone else, he also has a combination of genes and experiences that is unique. Even when people's outward behavior is standardized, their internal experience of what is happening is highly individual. Some of the unique differences from the "average" that are present at birth in any individual tend to be minimized by the process of growing up, while others increase in importance and shape the whole development of the personality.

How parents can make a difference

Parents have the responsibility of providing for their children all the assistance, both physical and emotional, that is needed for those children to grow up into adults that are both competent and individualized. Each baby is born an individual but with shared universal human characteristics. Parental care must meet both the common needs and the individual needs.

Several specific kinds of necessary parental support can be identified on the basis of what has already been said about human development. Children need physical closeness and expression of affection if they are going to develop a capacity for such expression themselves. They should be given the opportunity to make discoveries about themselves, about getting along with others, about the objects in the world. They need to be informed of the discoveries that have been made by others, both through their parents' answers to their questions and in school. Finally, they benefit from specific stimulation by their parents. One way that this can happen is for the parents to anticipate what their children will be and encourage them in this direction. For example, the mother of a tiny baby talks to it as though it could understand and answer; that is how babies learn to talk. Parents should ignore or discourage the kinds of behavior that are dangerous or unsuitable and encourage behavior that they deem appropriate.

Parental problems

In a traditional society, there is some degree of general agreement on how children should be raised. In a changing society such as ours, many styles of parenthood exist side by side, and the many models are confusing. Furthermore, what worked in a parent's own childhood on an isolated farm might not work at all in a crowded city. So parents have to figure out their own styles, and conscientious parents have considerable anxiety about whether or not they are doing the right thing.

The basic problem of parenthood is that children are unique individuals, and what will work with one may fail with another. Also, children change so rapidly that what worked last week may backfire today. The results of some course of action often are not apparent soon enough to make changes before damage is done.

One of the commonest problems of parents today concerns how much authority they should exercise. They want their children to learn to make the right decisions and to make these decisions by themselves. Parents have always attempted to modify the behavior of their children. If new psychological techniques make behavior modification really effective, the problem of how standardized we want our children's behavior to be will become even more important.

The styles of parenthood in advanced western societies probably are more variable than in any other culture in history. (Other societies with stronger class systems may have more divergence in child-rearing practices between classes than we have. But our society does not have just two or three contrasting styles; it has hundreds or thousands.) This extreme diversity of rearing practices is a mixed blessing. Where the challenges of a certain upbringing happen to just fit the individual needs and abilities of a youth, he will achieve the greatest possible success. But since rearing practices vary according to the parents rather than to the child, many children are raised in exactly the wrong way. If we had an "ideal" society in which all children were raised in the "ideal" way, we would probably have fewer disturbed people but fewer high achievers, too, because no one would be really challenged.

Human nature

Individual as each of us is, we are all human. Our shared humanness is quite obviously a result of the reproductive processes. The next and final chapter will summarize the technological advances that will soon change the nature of many aspects of reproduction. Will this, then, change human nature into something inhuman? To be prepared to assess this danger, we must consider what it means to be human, and how we are distinguished from other animals. Then we must consider what it is about human reproduction that makes us human. On this basis we will be able to consider the promise and threat of the new technology.

Humans as primates

Any biologist is impressed by what a large fraction of "human nature" is similar to the nature of other animals. It is possible to describe much about human beings by simply putting them in zoological categories. We are animals rather than plants, and so we move and respond to stimuli actively, and we get our food by eating other organisms rather than by manufacturing it in the sunlight. Since we are vertebrates, we are designed for forward motion and have our sense organs concentrated in the front; we have a centralized nervous system also concentrated at one end; we have a standard vertebrate set of organs and senses. Since we are mammals, we are warm and hairy, and we care for our young with "live birth," maternal milk, and parental supervision. In most additional details of our anatomies, we fall squarely in the group of mammals called Primates, with the monkeys and apes. As investigators learn more about our nearest primate relatives, the great apes, they are increasingly impressed by how much our behavior is a modification of standard higher primate behavior. It is hard to think of a human trait that is not found, at least in rudimentary form, in the chimpanzee or other apes.

Yet no one could really confuse a human with an ape. The same studies that have shown basic similarities have emphasized the distance from ape to man with regard to several very important qualities.

Human traits

The anatomical properties that we share with the higher primates include our versatile hand, with a thumb that can work opposite the other fingers, and a very large brain in relation to body size. Higher primate brains are competent to direct complicated hand movements. Higher primates have an unusual ability to respond to new problems with new behaviors. Primates have good ears and eyes, with the eyes adapted for depth perception. Anatomical traits that are distinctively human, on the other hand, and that appeared later than these others in our evolution, include an erect posture, with specialization for walking long distances, a general hairlessness (which may be an adaptation for better cooling at times of extreme exertion), and an exaggeration of the already large primate brain.

We think of our behavior as distinctively human as opposed to animal, but even here we share many traits with the chimps and presumably with other apes. Like us, the apes use hand holding and other forms of touching as modes of social contact (Fig. 8-2). Certain other characteristics of human interaction seem to be remarkably constant in all people, whether or not they are also shared with any other primates. Eibl-Eibesfeldt gives convincing evidence that the happy smile and the rage-contorted face appear in all young children. The smile is given even by children born deaf and blind, who could not possibly have learned it from others (Fig. 8-3). He has photographs of jungle savages, village maidens, and sophisticated city dwellers which show that all people share many ways of signaling to each other. They are usually not consciously aware of using these signals, which include an opening wide of the eyes in a flash of greeting, and a lowering of the eyes in casual flirting (Fig. 8-4). Eibl-Eibesfeldt argues convincingly that both sides of the interaction between mother and child are partially dependent on universal signals. Many experiments have shown that newborn babies respond more readily to objects resembling a human face than to any other kind of object and that babies smile very early. Eibl-Eibesfeldt contends that

Fig. 8–2: Physical contact is reassuring, even when it is only between two hands. (Hugo van Lawick, from Jane van Lawick-Goodall, *In the Shadow of Man*, 1971, Houghton Mifflin, Boston.)

Fig. 8–3: Deaf-blind 7-year-old girl laughing. When fully laughing she throws back her head, opens her mouth, and laughs audibly, although in a restrained manner. (Courtesy of I. Eibl-Eibesfeldt, *Ethology*, 1970, Holt, Rinehart and Winston, New York.)

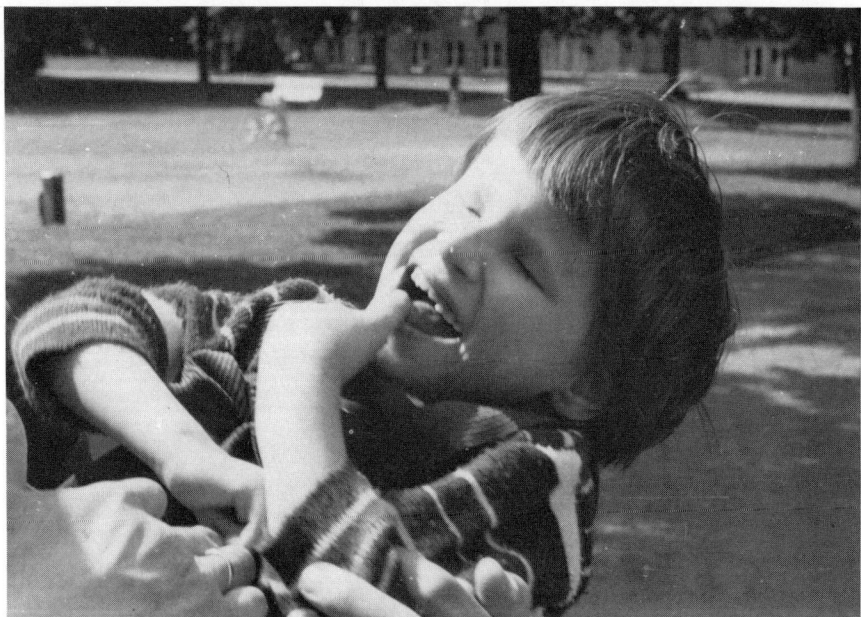

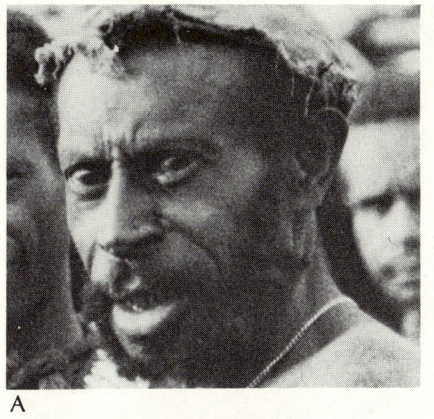

Fig. 8–4: Eyebrow flash during greeting seems to be a universal human expression. While this New Guinea tribesman is smiling (A), he briefly lifts his eyebrows (B) before completing the greeting (C). (Courtesy of I. Eibl-Eibesfeldt, *Ethology*, 1970, Holt, Rinehart and Winston, New York.)

we are adapted to respond more-or-less appropriately to the signals that a baby gives, and that this response is what has ensured enough infant care to keep the species going. He says that the infant cry that indicates need produces a strong desire in adults to solve the problem, while it is almost impossible not to respond affectionately to a baby's smile. The very shape ("cute") of the babies of all sorts of mammals evokes maternal reactions. (This reaction is exploited by advertisers and cartoon animators, who draw "superbabies" with extremely plump faces, large heads, and short legs.)

A behavioral trait that we share with other primates is usually called curiosity but might be expanded to include inventiveness and creativity. These are close to what we

Fig. 8–5: A chimpanzee carefully inserts a length of vine into an opening in a termite mound. The vine is prepared and used as a tool to draw out termites, which the chimpanzee eats. (Hugo van Lawick © National Geographic Society, from *The National Geographic*, Vol. 124, No. 2, p. 306.)

call intelligence and problem solving ability. In this whole category of behaviors, apes are amazingly like humans, but there is such a difference in degree that it is practically a difference in kind. The differences do not seem as large, though, as they did a few years ago. Until recently we could confidently identify a human as the only animal who could modify a natural object for use as a tool. But since wild chimpanzees were seen stripping leaves off a stick so as to use it in fishing termites out of a hole, that distinction has not seemed as distinct as it did (Fig. 8-5).

The most nearly exclusive human trait seems to be the use of spoken language for communication of information about the world around us. All animals respond in certain ways to body movements and sounds of other members of the same species. Sexual reproduction in all land animals, and other complicated interactions in all social animals, depend on such action and response. We have this kind of body communication, too. But such signals communicate to the other individual only "internal" information, relating usually to the emotional state of the communicator. Only humans seem able to communicate complicated information about other objects, which are out of sight in space and time. The process of speech by which this human communication is accomplished is a very complicated behavior indeed. It involves synchronizing breathing and specialized vocal cords to produce an immense variety of frequencies and sequences of sound. In contrast to the few stereotyped sound signals of other species of animals, human speech is minutely and directly related to complicated ideas and sets of information. The human brain has a sizable portion devoted to production and analysis of speech; nothing approaching the human speech center has been detected in any other animal brain, the ape included. It has recently become apparent that the inability of apes to talk lies not so much in their thinking ability as in their inability to vocalize thoughts. The pioneer experimenters in this field raised infant chimpanzees in the home, as if they were human children. The apes did very well in many complicated activities, but years of careful training produced the ability to name only two or three

Fig. 8–6: A. A chimpanzee has been taught to recognize and manipulate simple symbols and use them to express ideas and simple sentences, as described in "Teaching Language to an Ape," A. J. Premack and D. Premack, *Scientific American*, October 1972.

objects, such as the drinking cup. The little chimp had the vocal apparatus to make sounds something like human speech but not the mental resources to use the voicebox and lips in the right way. More recent experimenters, however, have used either hand signals (like the language of the deaf) or movable plastic shapes as media of communication. They have succeeded in teaching young chimpanzees to com-

Fig. 8–6: B. Sarah and her trainer. (Courtesy of David Premack, The University of California, Santa Barbara.)

municate surprisingly complicated sets of information, including what were essentially whole sentences (Fig. 8-6). Presumably our evolutionary ancestors, like the chimps, had something to say before they were able to say it.

It is really human culture that best distinguishes us from all other animals. Someone has said that a man is just "an ape with manners." But other animals do not develop cultures. All people do. A chimpanzee may learn to wear trousers, to eat with a spoon, to drink from a cup, and otherwise to ape his human friends. He will never be able to be a full part of human society, however, and he certainly will not contribute to its development. The question of why man alone is capable of culture is not easily answered, as we have seen. Human traits are primate traits, with certain ones emphasized and developed to the critical point where development and transmission of culture are possible. With culture, the bounds of human capability cannot be set.

Inheritance of humanness

The human zygote looks much like any other mammalian zygote. It does have a unique set of genes and chromosomes, but the number of chromosomes and their size are not especially extraordinary. Other primates have very similar (though not identical) sizes and shapes of chromosomes, and many animals have more, many have fewer, chromosomes than humans have. Our chromosomes are unique, but they do not have an unusual appearance. The genes in the chromosomes are less well known than the shapes of the chromosomes, but our uniqueness does not call attention to itself here, either. Our genes must code for about the same set of enzyme capabilities as do the genes of any other mammal. Many of our enzymes differ in a few amino acid positions from similar enzymes of other animals, but some are identical. Most of our genes are probably identical to those of our relatives the apes, a large proportion with those of all mammals, and many even with the genes of all vertebrates.

Some of the proteins of every human cell (zygote, embryonic, or adult) are different from those of other animals, and even somewhat different from those of other people. By using the immune systems of laboratory animals as a tool, one can distinguish human cells from the corresponding ones of animals, and often identify those of an individual person. But the working capabilities of these cells are not very distinctive. A liver cell does just about the same work, and carries out the same chemical processes, whether in one person or another, or even in a rabbit or a whale. The differences among individuals, and even the profound differences among species, can rarely be seen to be obviously related to the tiny differences in function between the corresponding cells of pairs of individuals or species. The most profound differences among men and between man and animals seem to be in the relative numbers and arrangements of the different kinds of cells, rather than in the major functions of the cells themselves. We could probably even use many of the organs of other animals to replace our ailing ones quite routinely, if it were not for the rather accidental fact that our immune systems destroy such foreign organs. *The genes that distinguish us from one another and even more profoundly from other species must act mainly during the development of the embryo and the child.* At those times, the effects of subtle differences among cells are intensified because of the interaction of different kinds of cells.

The embryonic processes that produce a human infant are altogether standard for primates and on the whole for mammals in general. The behavioral properties of the human embryo, the fetus, and perhaps even the newborn are not exceptional; they are probably duplicated by the corresponding stages of monkeys or apes. The organs produced by development are standard mammalian ones, but they do differ in relative size and shape. As previously stated, the most important of these differences seem to be in the upright stature (legs, feet, pelvic girdle), the hand, and especially the brain. It is the ordinary embryonic processes going on in a special way that produce the set of capabilities that confer human traits. Only a relatively few human genes may be needed to turn the trick.

It may help in understanding the influence of small changes in genes to consider here what profound effects on development may be brought about without even any difference in the genes present but just by a difference in their quantitative balance. The cells of a person with trisomy 21, Down's syndrome, are quite ordinary biochemically. All the genes are there for any single enzyme that might be needed. Trisomy would probably not be very serious for a single-celled organism. But in the subtle interactions among cells during embryonic development, the level of one gene product in one cell calls forth a certain pattern of gene expression in another cell; thus having one too many sets of one group of gene loci is usually fatal to the organism. Even trisomy 21 leads to multiple defects, including a serious mental deficiency.

The relatively few genes that are distinctively human must have their most important effects during embryonic development. By changing the growth rate of a limb bud or the differential spreading of a part of the neural tube, a few such genes can have a profound influence. They can make possible *human* development.

Social requirements for being human

For many kinds of animals, the inherited equipment and instincts are sufficient for a member of the species to function well. There is no need for interaction with other members of the species. A newly hatched sea turtle will probably never see either of its parents (and certainly wouldn't recognize them) and will rarely interact with other sea turtles until it is mature enough to mate. Yet it will be a perfectly competent sea turtle, a complete example of its species. This is less true of birds and still less true of mammals, but in most of these species the social requirements are not very extensive. Some birds know their songs instinctively; others learn them from parents. Rats raised in isolation are inefficient in mating and general social interaction but not totally ineffective.

Man is different. We would not recognize a "person" as really human, and we would certainly not admit him into our society, if he had grown up without any human interaction. A human infant probably would not survive such treatment; if it did, it would have the shape of a person, but its behavior would not be what we call human. A fully human being can be produced with human interaction even under great difficulties, as in the case of Helen Keller. However, unintentional experiments with poorly tended human infants have convinced most observers that "tender loving

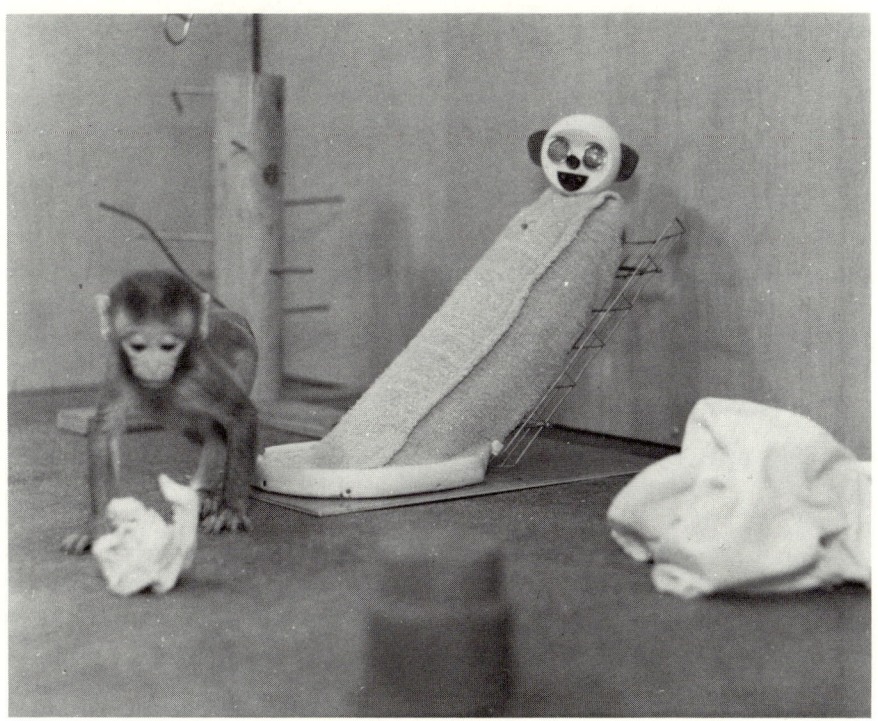

Fig. 8–7: A. Rhesus infant raised with a cloth surrogate mother displays security and exploratory behavior in a strange situation in the "mother's" presence.

care" is nearly as important as food and shelter for the very survival of babies. The remarkable studies by Harlow with infant monkeys have demonstrated that this is a primate trait intensified. The fascinating research by Harlow showed that infant monkeys had a primary need for clinging softness —they preferred a soft terry-cloth–covered model (quite crudely made) of a monkey mother to a wire model even when the wire one contained a feeding nipple. They obtained assurance from the soft pseudo-maternal comfort and were much more competent and willing to explore new situations than other monkeys reared without anything to cling to (Fig. 8-7). Infant monkeys reared in groups clung to each other

Fig. 8–7: B. Rhesus infant raised in isolation, when confronted with a new situation, retreats to a corner of its box, crouches, hides its eyes, and clutches itself. (Courtesy of H. F. Harlow and M. K. Harlow, *American Scientist,* 1966, Vol. 54, pp. 244–272.)

and developed into fairly competent adults. Those who had a cloth "mother" but no playmates grew into adults that would not mate, and if mated forcibly would not or could not care for their own infants. The descriptions of the monkeys suffering from various kinds of deprivation are suggestively similar to many human emotional ailments.

In most cultures other than our European one, young babies are carried nearly all the time, whether directly by the mother, strapped on a cradleboard, or by an older sibling. Ours is one of the few cultures that keeps tiny babies in cribs by themselves most of the time. Perhaps the "security blankets" or the stuffed toys that children love are their own soft mother-substitutes. Are the substitutes sufficient for their needs?

The process of acculturation

The human is born with few instincts and probably even rather few behavioral tendencies compared to any other animal, even including the apes. To become a "person" in any full sense of the word, the human infant must learn to operate within some particular culture, whether a widespread one or a new synthesis achieved by his parents. There is a fashion now among the young radicals, the social revolutionaries, to deplore the "cultural imprint" that each person has and to say that having to absorb a culture destroys freedom. But people would not really be people without having had a culture of some sort or another to grow up in, to respond to.

Do humans have instincts?

It is an article of faith among some social scientists that humans have no instincts (or perhaps only the sucking instinct). This statement can be true only if instinct is carefully defined as a very elaborate behavioral pattern, completely inborn. Even among most other mammals, some kind of developmental interaction is usually necessary for an instinct to become fully operative. Our relative lack of instinct, and greater dependence on acculturation, is simply a matter of degree. It would probably be more accurate to say that what we inherit are behavioral tendencies and likely responses to certain stimuli. The fact that infant humans and infant monkeys need both cuddling and social interaction for mental health shows that there is an inborn need or behavioral tendency for cuddling with something soft, and an inborn set of adaptive responses to social stimulation. The more doctrinaire of the anti-instinct people would contend that any behavioral regularities among all humans are either the result of accidental similarities of their cultures or the incidental result of bodily equipment. (For instance, humans hold hands because they have hands, whereas elephants couldn't, they might say.) This argument depends on a false distinction between brain and body. An inborn behavioral tendency means that there is a structural characteristic of the brain; this is as much an example of human structure as the shape of the hand. The hand, to be operative, must be hooked up to nerves that control its motion and convey sensations from it, and these connections are further joined to all higher brain centers. Much of human behavior is related to our inheritance of brain structures.

Nature vs. nurture

All the controversies about instincts are part of an old and largely fruitless argument about the relative importance of nature and nurture in determining one or another human trait and in shaping our whole personalities. The feudal nobility convinced themselves (and even many of their servants) that the privileged position of the upper classes was due to the intrinsic merit of its members, inherited biologically—"good blood." Modern social philosophers, impressed by the talent exhibited by the descendants of former serfs (including themselves), have tried to believe that all or nearly all the differences among people were due to differences in environment. In the extreme form of this opinion, a baby is a "blank slate" to be written on by experience. These two opposing absurdities have, in our lifetimes, been applied by some governments with a deliberate ruthlessness that destroyed individuals and social integrity. Hitler's racism was an extreme form of "guilt by inheritance," even though it went still further to ignore serious genetics and invent false racial differences. These were used as a pretense for depriving various disliked people of opportunity, possessions, position, or life, or even just to provide an emotional rallying cry to built a Reich of hatred. On the opposite side, the believers in the "blank slate" theory have not killed as many people on those grounds, but people have been killed. Lysenko, a Soviet biologist, convinced the Communist government of Russia that the accepted genetics was a capitalist fiction designed to enslave the masses and that better crop plants (and happier, more competent citizens) could be produced by providing them with the right environment. On the strength of this theory, the government exiled several geneticists to die in Siberia, and put Soviet genetics 30 years behind the rest of the world before the mistake was recognized.

The nature-nurture argument today may be a little more sane, but it is still vigorous. Researchers in animal behavior, who are biologists and stress evolution, explain most animal behavior and even quite a bit of human behavior by inheritance from ancestors. Psychologists, who are social scientists and believe in the effectiveness of education and the power of social circumstance, stress environment and experience in explaining human behavior and most animal behavior. The two groups may do the same experiments and get the same data, but they sometimes hardly talk politely to each other. Opinions in this area arouse strong emotions. Strong disagreement, confusion instead of understanding, and even hatred surround Jensen (who suggests on real but inconclusive evidence that there may be racial differences in intelligence) and Herrnstein (who suggests that differences in intelligence exist between social classes and that these differences are increasing).

What we have seen of the process of developing—that it is a creative process, always building on what is already present—demonstrates that both nature and nurture exist before birth, and their effects cannot be neatly divided. The same is true when we consider the further development of individuality during postnatal growth.

9 | *Revolution in reproduction: double-edged swords*

Science and technology change our lives. They make it easier to do what we had been doing before and sometimes make it possible to do things that were previously impossible. In addition to this effect on practical, everyday affairs, the new insights of science can change our whole way of thinking. When Galileo, Kepler, and Newton convinced people that the earth went around the sun and that the earth was not the center of the universe, they profoundly affected how important men felt themselves to be. Newton's discoveries in physics changed the belief that everything that happened was a whim of God to the idea that the universe was orderly —a kind of clockwork, with everything predetermined. Darwin's description and explanation of evolution placed man in the natural world, and made it necessary to rethink religious suppositions about the nature of man.

Most of the technological changes in the past have been derived from the physical sciences. Techniques of manufacturing, transportation, and communication have changed more in the last 200 years than in 4000 years before that. These changes have revolutionized our lives, even though they have dealt with the externals—the things we work with. New biological technology will touch our lives still more closely. Even what already exists—for example, the improvement in medical and public health practices—has outrun our conventional wisdom. The recent widespread reduction in premature death is making it necessary to reverse the age-old incentives for large families. The new technical possibilities for prolonging the lives of the dying are forcing us to reexamine our assumption that death is always to be avoided. The coming biological revolution, which will change the manner of biological inheritance and the circumstances of embryonic development and birth, will change even our perceptions of ourselves in ways that we cannot now fully predict. We will need the best thought and the wisest decisions we can produce to cope with the new possibilities and problems.

Technology is a double-edged sword. It creates new problems while it solves old ones. It causes problems because, as Garrett Hardin says, "You can't do just one thing." Any significant action, including the introduction of new technology, is sure to have several effects in addition to the one intended. Some of the unfortunate side effects of technology could not have been predicted and avoided, but others are the result of an improper use of new capabilities.

Technology as magic

Man has tended to use technology as a form of magic, trying to change the nature of things to fit our notions of what we want. We need, instead, to choose goals that make sense and that are appropriate in light of what can be learned about the world. The magician does little harm, because his spells and potions are powerless. A society that uses its powerful technology to reach poorly conceived goals can cause serious harm, as our dangerous worldwide pollution problem demonstrates all too well. We have tended to use technology (as magic) to avoid thought and effort—to keep from having to decide the right thing to do. Technology used arrogantly, as though it were magic, fails to bring lasting happiness. Appropriate goals and correct action are needed now more than ever before.

Goals and action

We are on the verge of a great flood of new technology in the realm of human reproduction. The new techniques could prevent many sorrows (such as many types of congenital deformity), but they could also cause new tragedy (perhaps another, more widespread, thalidomide case). These new possibilities in biological technology will cause frustration and bewilderment among people who cannot cope with the unfamiliar. In order to avoid tragedy and achieve increased happiness when using the new techniques, it will be necessary to choose goals thoughtfully and guide actions carefully toward those goals.

Increased scientific understanding of biological systems, and specifically of the reproductive process, is making possible the technology that will solve problems past generations have had in this area of their lives. The same kind of biological understanding can help us to cope with the new problems of the biological revolution. There are three important ways that science can assist. First, certain new problems can be solved with more new technology. Second, a scientific understanding of how biological systems work can aid us in predicting dangerous side effects of new technology. Third, better biological understanding can help us to decide what technology is wise and appropriate, and what applications should not be used.

The technological fix

Alvin Weinberg of the Oak Ridge National Laboratory has suggested that there are technological "fixes" or solutions to some, but not all, of our social problems. He has in mind, especially, new and cleaner sources of power to solve the fuel shortage and to make some presently inconvenient sources of raw materials available to bolster the diminishing supplies of metals and minerals.

We are dependent on technology to help solve the population problem, which is the result of older technology. As mentioned previously, the dramatic reduction in infant mortality in the last few generations, and the continuing reduction in premature death in general, represent some of the clearest possible examples of how improved technology has increased true human happiness. Meanwhile, this genuine improvement in the quality of human life is responsible for the population explosion, which threatens the health and well-being of all of us. New technologies of birth control now make it possible to achieve smaller families without as much personal effort as was previously necessary. Before about 1920 the only way to limit families was to abstain from sex before marriage and marry late, or to abstain from sex within marriage. Few people managed to do either very well. Condoms and diaphragms made birth control effective and acceptable for the first time. Even more acceptable are the newer IUD and the pill, which have largely replaced the older methods because they interfere less with lovemaking. It still requires some thought, and planning and preparation ahead of time, to limit families, and each couple must think about how many children they really want. The menace of the population explosion seems to be decreasing in the affluent countries, but overpopulation is probably the number one problem of the developing countries. The principal barrier to effective birth control in the poorer countries that need it most is not the expense or unavailability of contraceptive technology, but the fact that so many of the citizens still feel that large families are a virtue, and that the new methods are evil or dangerous. In the field of population limitation technology alone will not solve the problem; the will to right action is required, too.

The hazard of premature technology

As new kinds of technology are put into use, it requires all the biological sophistication that we possess to anticipate and avoid mistakes and dangerous side effects. A special problem comes when a new technology is first introduced. Some mistakes in the first application of new methods are probably unavoidable, but careful preliminary testing on experimental animals and carefully monitored trials on human subjects can minimize the dangers. With the benefit of hindsight everyone can now agree that thalidomide was introduced too quickly, without adequate trials. As discussed in Chapter 6, animal experiments cannot be counted on to prevent another tragedy such as that of thalidomide.

The ethics of experiments on humans

In the final stages of technical development, any treatment for humans must be tried out on humans. Since all new procedures may carry an unknown danger, there is always a risk to be considered. It is certainly ethical to subject a person to a risk to avoid a greater risk, giving him a chance to decide, when possible. Most people would agree that it is not ethical to subject some people to special risks for the benefit of other people, unless the risk takers are volunteers who are quite aware of what they are doing. (It may be observed that our society regularly violates this principal by drafting men to take mortal risks in military action for the protection of the rest of us.)

Widespread experiments with a new drug or medical procedure would seem to be ethical only during a limited period in the development of the new treatment. To be dependable, an experiment must be conducted with adequate controls, which means that a sizable group of experimental patients should not be given the treatment, to be sure that the treatment is really causing the effects that will be observed. To avoid imaginary symptoms, the control group must be comparable to those receiving the treatment. Preferably, neither they nor the doctors examining them should know, at the time, which patients are receiving the new treatment and which are not. For this purpose, control patients in such a test of a new drug are commonly given coded bottles of sugar pills, a "placebo," which look just like the real drug. Obviously, it is impossible for any of the people involved in such a controlled test to have chosen whether or not they wanted the new treatment.

Since patients involved in an adequate test of a new treatment cannot choose whether or not to receive it, such a test is ethical only when the people conducting the test are convinced that the probable benefit to the patients receiving the treatment balances its possible risks. One way of expressing this is that if members of their families were in need of such treatment, the investigators would find it hard to choose in which group they would like their relatives to be. If a new treatment seems too hazardous to meet this criterion, it is too soon to test it on humans. On the other hand, if a treatment is inadequately tested when first introduced and becomes the standard method, the opportunity for a valid test is lost, because it would not be right to deny the accepted treatment to a group of patients just because of lingering doubts about its effectiveness or safety. Only if there is considerable new evidence that a standard treatment is useless or dangerous is a controlled test acceptable.

Side effects

It is almost inevitable that any technology will have effects that were not intended, some of them undesirable. Even when some technique is suitable for use and should be employed, there are usually side effects. This is especially well exemplified by the steroid birth control pills, whose harmful or unpleasant side effects have been highly publicized. As discussed earlier, the hazards of the pill are not nearly as great as the hazards of not using it, and researchers have been trying to change the formulation of the medicine so as to reduce the undesirable side effects. Side effects of new technology may easily be more serious than they are with the pill, and national control agencies need to exercise their authority carefully to avoid unnecessary problems. With strict controls such as we have in this country, there is the alternative danger that useful new technologies will be greatly delayed because of safety rules, when their application would have saved more lives than would be lost during trials.

With new technologies in the field of human reproduction, such as amniocentesis, artificial insemination, and soon zygote adoption and choosing the sex of the child, the important side effects are not likely to be medical or physiological ones.

Rather, they will come from the fact that social institutions and personal feelings are not prepared to cope with the new practices. For example, the husband sometimes does not feel like the father of children conceived by artificial insemination. The availability of amniocentesis and therapeutic abortion precipitates a moral crisis for some families who face this risk of a detectable defect but who cannot easily accept abortion. Being able to choose the sex of a child may lead to a serious imbalance of the sexes when simple techniques become available to make the choice. The problems of expecting too much from zygote adoption have already been discussed, in Chapter 2.

Biological understanding

Wide use of the powerful new biological technology will require improved biological understanding not only on the part of scientists, but by the general public who will have to make many of the decisions. A scattering of facts mixed with a liberal portion of old wives' tales makes a poor basis even for standard decisions about personal and family matters. Such a poor understanding will be worse than useless for dealing with the new problems, both personal and social, that are sure to arise in the near future. Many people resist acquiring a deep understanding of their biological natures; they feel it is somehow "unnatural" to understand too much about themselves. They prefer to let things happen to them rather than to make deliberate decisions about the biological events in human lives. Yet, if citizens do not understand and make decisions about biological technologies and the resultant problems that will develop in the near future, someone else will make their decisions for them or the decisions will be made by default.

Science is constantly engaged in learning more about the nature of things. Biological systems have complicated and fascinating properties (like homeostasis and creativity) which cannot be understood without considerable effort. In the earlier chapters of this book we saw some of the places where biological understanding can aid us in making choices: A description of the processes of gametogenesis and fertilization helps in understanding how so many different kinds of contraceptives can work and suggests which kinds are likely to be suitable in various situations. A knowledge of the physiology of sexual response makes sexual problems less puzzling and suggests a few solutions. Studying the embryo's gradual acquisition of form and function provides background for difficult choices about abortion. And appreciating the interplay of inborn tendencies and individual learning during childhood gives a basis for deciding among different styles of parenthood. Finally, an overview of the nature of human variety adds perspective to the attempts to improve man's hereditary endowment.

Technology can change right to wrong

Whenever societies perceive that some specific kind of individual action is desirable, they establish sets of customs and institutions to encourage people to make the proper choice. If the original reason for the choice disappears, perhaps because of new technical powers, a period of unrest and moral uncertainty is likely to ensue.

Human fertility is a case in point. Early man, living by hunting and gathering, was always in danger of dying out. The mortality rate was probably greatest during childhood, but many women died in childbirth, and men were killed in hunting accidents. Without a steady stream of replacements, the band could dwindle away. If the family band got much smaller than neighboring tribes, it might be completely destroyed; an ample supply of warriors was necessary to fight off the enemy. All human groups seem to have developed fertility rites, with the intention of increasing the birthrate. Societies institutionalized marriage to reduce internal fighting and to ensure good parental care of the children. Couples with many children and grandchildren were honored; high fertility became a source of status. Later, when people settled down to till the soil, new incentives for fertility arose. A large family would ensure continuity in the operation of the farm and give economic security for the parents in their old age.

Hardship and poverty have been common human problems since before history began. Often this must have been due to local overpopulation, but the problem was rarely seen in that light until the time of the Industrial Revolution. Really widespread appreciation of the dangers of the population explosion has come only since 1960. Not all classes in all countries are truly aware of the extent of the problem even now. Fertility has dropped in the United States in the 60's and early 70's, but our culture still contains many of the old forces favoring fertility. Rice, a symbol of fertility, is still thrown at weddings, although usually without any symbolism intended. A couple is still considered decidedly odd if they decide not to have any children. If they have only one, they may be reminded that the "only child" is purported to suffer emotional difficulties. Groups that are trying to encourage good and responsible parenthood, whether in churches or as breast-feeding societies, assure women that motherhood is as high a vocation as they can have. This often comes through as a feeling that if mothering one child is good, having more is better. As with the glorification of motherhood, although the federal tax advantages awarded for having children are intended to aid in raising children already born, they have the effect of putting government on the side of population growth. Incentives to fertility are so imbedded in desirable parts of our culture that they obscure the pertinent fact—having many children was once the right thing to do but now is wrong. The technical changes that lowered the death rate so dramatically in the last two generations have actually changed right to wrong.

The sexual revolution

Societies have always had regulations about sexual conduct, primarily because sex meant children, and children needed both parents to support them. In our culture, premarital and extramarital sex have long been generally and strongly disapproved, despite numerous violations. To the primary reasons for having sexual regulations were added other incentives—the danger of venereal disease, the fear of social disgrace from pregnancy outside of marriage, the admonition that careless indulgence in sex leads to later problems, the belief that sexual promiscuity damages the personality.

Recently a host of new situations has shaken the consensus in our society about proper sexual activity. Not usually mentioned among these is the fact that the average age at puberty has become younger. Combined with an increasing level of education and a longer wait for financial independence, this has produced a period between the age of sexual maturity and the age of forming an independent family that is longer than ever before. Sexual restraint before marriage now has to be exercised for 7 or 8 years instead of 2 or 3. In spite of this fact, it is not clear that teen-agers are really involved in a great deal more sexual activity than their counterparts were a generation or two ago. It has been suggested that the increasing numbers of pregnancies in unmarried girls is partly a result of better health—spontaneous abortion is not as common as it used to be. Probably most important in changing sexual behavior is the fact that sex and conception are not necessarily linked, now that there are effective contraceptives. Venereal disease can be cured, too.

The result of all this is that many of the reasons for strict sexual prohibitions no longer have the force that they once did. The most important original reason—the need to assure two supporting parents for children—no longer exists. People who are careful and responsible in using contraceptives have very few unwanted pregnancies. Most come about because contraceptives are carelessly or ignorantly used, or not used at all. Many girls who feel guilty about premarital sex feel that they are planning something sinful if they make contraceptive preparations for sex, but that if they are "carried away" by the passions of the moment they need not blame themselves. This "head-in-the-sand" idea illustrates again that technology in itself is not enough. Choices and actions are required before technological advances can be effective.

For people who are confident of their competence to avoid both venereal disease and pregnancy, the real sexual situation has changed. Fear of these hazards no longer need enter into their decisions. Now a person must act strictly on the moral issues: Does unrestricted sexual behavior trivialize something which should be very important? Does such behavior harm the personality of the person engaging in it?

From the viewpoint of Christian doctrine, sexual correctness from fear of pregnancy, of disease, or of social disgrace is not true morality, in any case. If the person really wants to be promiscuous but refrains only for fear of the physical consequences, he is still being immoral. Therefore, it is incorrect from the Christian standpoint to say that sexual morality has decreased in recent years. Many people are still making decisions about sexual practices on a "moral" basis, but not so many decisions are based on fear of physical or social consequences.

Venereal disease

The two major venereal diseases have a fascinating relationship to the reproductive process and its technology. Gonorrhea and syphilis are two unrelated diseases; they have in common only the fact that they are spread by sexual intercourse.

Gonorrhea is caused by small round bacteria usually found attached in pairs and called *Neisseria gonorrhoeae*. Infection causes an inflammation of internal membranes, such as the lining of the urethra in both sexes and of the cervix in the female. Swelling of the tissues acts to nearly close the urethra, so that urination is painful; there is also a white, pus-like discharge. Infected males are usually painfully aware of their problem, but 30 percent of infected females have no noticeable symptoms. If those with the disease are untreated, they can infect all their sexual partners. The majority of untreated cases clear up in a few weeks without any permanent harm to the person, but in a sizeable minority of cases there may be further serious involvement, including sterility in either sex (from scarring of the ducts so that gametes cannot pass) and generalized infections elsewhere in the body, even leading to arthritis or heart disease. There is no lasting immunity, and a person may contract the infection several times. Gonorrhea is easily cured with penicillin, even though enough resistance has developed in the bacteria so that a larger dose is necessary now than was previously required. (For people who are hypersensitive to penicillin, there are other effective antibiotics available.) The gonorrhea organism is so easily killed by environmental exposure that it is practically impossible to get an infection by contact with drinking cups, toilet seats, or clothing. Intimate contact, principally sexual, is required for infection, since the bacteria do not infect the skin over most of the body. If a mother has an active case, her baby's eyes can be infected during the birth process, commonly leading to blindness. (The administration of silver nitrate or an antibiotic to the eyes of the newborn is now so common that the frequency of gonorrhea-caused blindness has been substantially reduced.)

Syphilis is caused by an unusual kind of bacteria, shaped like a corkscrew, called *Treponema pallidum*. It is a much more serious disease than gonorrhea but does not start out as threateningly. The first symptom of syphilis infection, occuring about 3 weeks after exposure, is called a **chancre** (pronounced "shanker"). It is a raised, button-like swelling, a centimeter or less in diameter. It is most commonly found near the end of the penis, or on the cervix. It will normally be noticeable to males but not to females. A chancre is usually not painful. The visible chancre disappears after a couple of weeks, but this is not the end of the disease. Six to eight weeks after infection the secondary stage begins; the bacteria spread throughout the body and cause more generalized symptoms. These may include flu-like aches and fever, a skin rash, and glandular swelling. Again after a few weeks, these symptoms also disappear, even without treatment. In about half the untreated cases, the immune system rids the body of the infection, with no further difficulty. But in the other half of untreated cases, the bacteria establish a long-term infection in some localized part of the body; it may be anywhere. It continues to stimulate the immune system of the body, and the infection does not again spread beyond this localization, nor can a new infection occur. Much later, after a lapse of from 2 to as many as 35 years, the most serious effects of syphilis, called the tertiary symptoms, occur. Antibodies produced by the immune system eventually have a destructive effect on the part of the body carrying the localized infection. The tissues first swell, many of the cells die, and the area ulcerates or wastes away. These dying patches of tissue are called **gummas**. The nature of the tertiary symptoms of syphilis depends entirely on the location of the gumma. If the area involved is small and not vital, the symptoms may be unimportant. A gumma in the heart or a large blood vessel can be quickly fatal. A gumma in a large nerve can cause partial paralysis; one in the brain can cause insanity. Parts of limbs or perhaps the side of the face may ulcerate or waste away. (A graphic account of the effects of syphilis can be found in *Jennie**, the biography of the mother of Winston Churchill. Her husband, Lord Randolph Churchill, suffered syphilis-caused insanity and finally died of the disease.)

Just as with gonorrhea, syphilis is easily cured with penicillin or other antibiotics. This cure must be effected in the primary or secondary stages. Treatment is sometimes helpful during the "silent period" before the tertiary symptoms start. But treatment during the tertiary period does no good, since these final symptoms are caused by an action of the body's own immune system, and killing the disease organisms can no longer help. Someone who has been cured of syphilis with early antibiotic treatment has increased resistance to the disease, but it is possible to become reinfected.

Syphilis bacteria are a little more resistant than those of gonorrhea, and there is a slight danger of transmittal through objects. Practically all infections, however, come from sexual contact. Unlike the gonorrhea organism, the bacteria of syphilis can be transmitted to the fetus in the uterus, from about the eighteenth week on. The effects on the fetus are disastrous. A variety of large and small defects may be caused, so that 25 percent of the fetuses die before birth, and another 25 percent shortly after birth. Of those that survive (with smaller congenital defects), 40 percent develop the various destructions of tertiary syphilis within a few years. Treatment of a pregnant woman with antibiotics will cure

* R. G. Martin, Prentice-Hall, Englewood Cliffs, N.J., 1968.

both her infection and that of the fetus, but this treatment must not be delayed.

The VD epidemic

When antibiotics came into common use in the 1950's and it was found that both gonorrhea and syphilis were very easily cured by them, there was a decrease in the incidence of these diseases. Public health officials had high hopes that the major venereal diseases could be eradicated, since they are not found in other animals, and there would be no way for a cured population to become reinfected. In the 60's the incidence of both diseases began to rise again, and by the early 70's it was as high as ever, with record frequency, especially of gonorrhea, among teen-agers.

Here again, technology was not enough by itself. Its benefits are lost if people do not take appropriate action as well. This is an old story, of course, for the venereal diseases. They are spread only by sexual contact; those who have contracted the diseases either before birth or at birth do not ordinarily transmit them to anyone else. So venereal disease has always depended on promiscuity; there would be no new infections, and the diseases would die out in just a few months if people behaved as conventions dictate: engage in sex only in marriage.

The failure of antibiotics to wipe out VD is not due to any loss of effectiveness of the treatment; the increase of resistance has been slight and adjustments to it are easily made. The reasons for the present epidemic are probably many, and all are related to people's behavior. The very fact that both venereal diseases can be treated without permanent damage makes many people careless. They do not realize that they may be infected without knowing it, or that the real dangers come after the first symptoms have disappeared, when they think they have recovered. A change in sexual behavior is probably even more important in explaining the epidemic. The amount of promiscuity may not be much different from what it was in the past, but the pattern has changed. Whereas once promiscuous young men frequented prostitutes or sought out girls of low economic or social status, today swinging sex is more likely to involve partners from the same social group. The public health measures against VD require that the recent sexual contacts of a person being treated also be located and given treatment, and boys are often reluctant to reveal the names of their friends. Some of the rise in incidence of VD may be apparent rather than real; there is less effort now to keep the diseases secret, and cases are more fully reported to the public health authorities than in the past. A final ironic reason for the epidemic is a side effect of technology. Before the 60's, the standard contraceptive, especially for casual encounters, was the condom, which protected against VD as well as against pregnancy. Now the pill and the IUD are the common contraceptives. Many people apparently think that the pill protects against VD, but it does not.

It is difficult to see any immediate decrease of the VD epidemic. There is no evidence that promiscuity is decreasing. Medical treatment of VD is already about as convenient and effective as could be hoped, but there is no early prospect of developing ways to immunize against the diseases. Detection of symptomless cases of gonorrhea is very difficult, with no easy method in sight. Syphilis infection can be detected with blood tests and probably with skin tests, but these are somewhat inconvenient for general use and do not work until a person has been spreading the infection for weeks. Technology alone cannot solve this problem; there must be a concerted effort by society if this very real and very dangerous situation is to be halted.

Sex education

The premise behind this book is that an understanding of the reproductive process helps promote wise decisions. There is no doubt that people need to be able to make good and informed choices about sexual behavior, family planning, childbearing and child rearing. Information about human reproduction, development, sexual function, and disease should be included as a regular part of the science and health programs at *all* levels of public education. Even if ignorant innocence is preferred by some parents, it is not one of the choices available. All children are exposed to floods of information and misinformation about sexual matters, and to strongly biased implications about sexual roles. Much of this comes from peers; other information and misinformation come through the public media. Solid, correct facts can be only helpful, not harmful. Facts are not enough, however. It is precisely because reproduction, growth, and sex are so tied up with emotions, life styles, and interpersonal attitudes that there are problems in knowing what to tell children about them. Sex is too important to be easy.

Sex education programs in the public schools are under attack from the radical right, who say that sex education is part of a Communist plot to destroy our nation. This attack is so mindless that it is difficult to fight, but it convinces some people. Many sane parents, however, are concerned that sex education deals only with the mechanics and does not do anything to help their children learn what is right. Since many programs describe various sexual behaviors without judgment or decisive advice, it seems to some parents that the programs imply that any behavior is as acceptable as any other. The "values" that would have to be taught in sex education if it were to go beyond mechanics and facts are so bound up with individual life styles and religious beliefs that any comprehensive treatment of standards of appropriate behavior would probably be unconstitutional. It would clearly be impossible to get any kind of agreement on what should be taught. The only admonitions that the schools can properly give are those that represent the clear concerns of the community: the health and welfare of its citizens. Today, that means preventing the birth of children to those who cannot properly care for them, either because they are too young, they are not married, or they already have as many children as they can care for. The dangers and prevention of venereal diseases need to be presented in detail. Admonitions to use contraception and practice VD prevention should be made strongly. Further than this, the schools probably cannot go. It would still be appropriate for a teacher to voice his own opinion on what is and is not appropriate sexual and social conduct, as long as he labels it as his own opinion; perhaps this is best done in response to questions. Detailed advice on what one should or should not do is the responsibility of parents and of philosophical and religious groups. Sex education is no threat to parents who are doing their duty.

A less fundamental problem of sex education, but sometimes a serious one, is the competence of the teachers. Elementary school teachers, especially, are unlikely to have sufficient background information to feel confident in answering questions about reproduction. (Books like this are in part designed to help meet this problem.) Teachers may have received poor education in the subject when they were children and find it emotionally difficult to handle. Careful selection of suitable teachers and training programs to improve their suitability can solve this problem.

Good news and bad news

The most exciting basic scientific research in the last 20 years has been in the field of basic cell biology, especially concerning genes and their actions. This improved understanding is bearing fruit now; more powerful technology and new developments related to human reproduction will come thick and fast in the next 50 years. Since technology can be a double-edged contribution to society, that is both good news and bad news. We can be glad that sorrows that have been unavoidable for all of human history will soon be prevented, and that wishes that have always been unmet will be granted. The bad news is that we will find that we would have been better off without having some of our desires fulfilled. In the process of avoiding sorrows we will confuse ourselves by changing the conditions of human existence and will precipitate an "identity crisis" for all of mankind by changing the definition of a person. Many of us are experiencing "future shock" right now from revolutions only in transportation, communication, and manufacturing. Will we be able to cope at all with equally profound changes in ourselves?

The biological revolution is not coming all at once. Some of it is already upon us, but some of the more drastic changes that have been predicted are many years away. I have given below my current guesses about when each of the listed developments related to human reproduction will become important. The promise and problems of each will then be discussed.

Probable initial date	*New development*
1975	Zygote adoption in humans
1980	Fail-safe IUD
1985	Choice of sex of offspring
1990	Significant program of eugenics
1995	Genetic surgery—replacing harmful genes
2000	Individualized medicine—tailored environments
2005	Corrective surgery on fetuses
2010	Artificial uterus
2015	Cloning—humans from single somatic cells
2020	Growing one's own replacement parts
2025	Genetic engineering—designing a better human
2050	Designing human-animal combinations

Zygote adoption

Transplantation of zygotes from one female to another has been done in many experimental animals and is now becoming commercially important in domestic animals (Fig. 9-1). It is used for the rapid genetic improvement of domestic livestock, greatly speeding up the process. For instance, a cow of superior qualities is treated with a fertility drug so that she produces many ova. She is artificially inseminated with sperm from an equally superior bull, and the zygotes are allowed to divide for a few days. Then a small abdominal incision is made to flush out the blastocysts from the uterus before they have time to implant. These superior zygotes can be, and are already being, carefully packed and transported even to another country, where farmers are trying to improve the

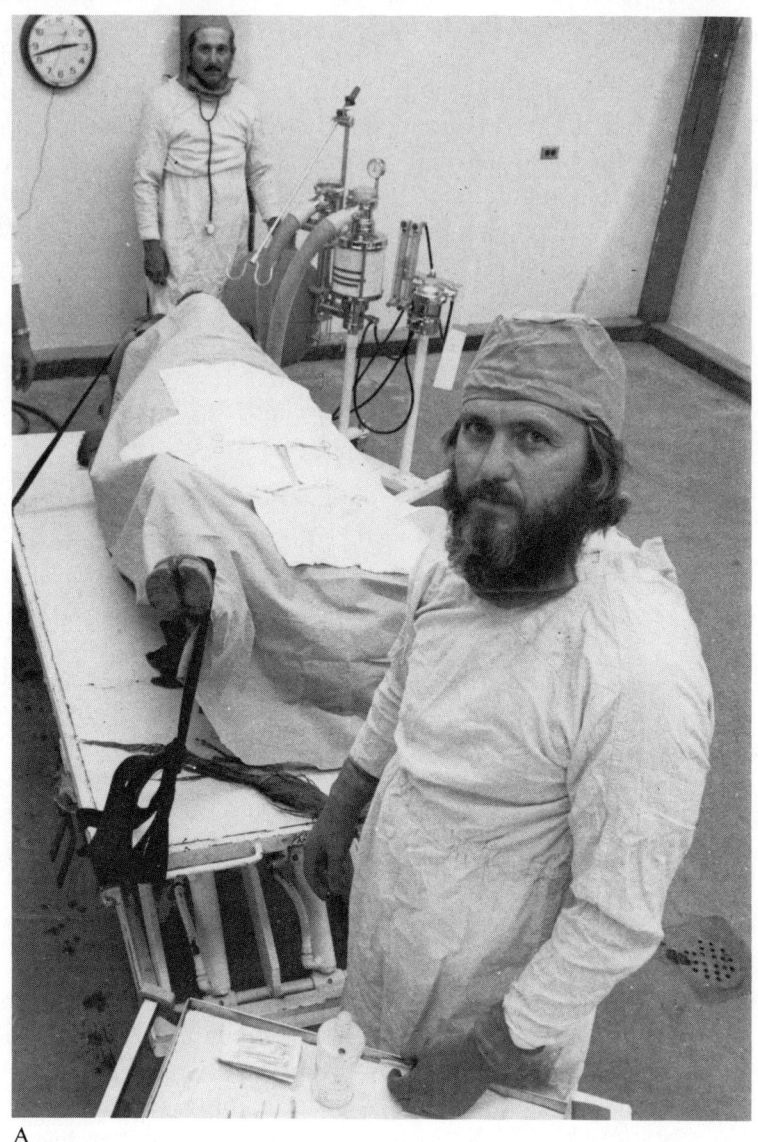

A

Fig. 9-1: Zygote adoption in cattle. A. Veterinarians prepare for the operation to remove the zygote from the donor cow. (Warren Taylor, courtesy of *Des Moines Register and Tribune*.) B. The uterus is exposed and the zygotes are flushed out of it with a syringe. (Photograph courtesy of International Cryo-Biological Services, Inc.) C. Cow with the calf "adopted" as a zygote. They have no genetic relationship. (Courtesy of Charles Vincent.)

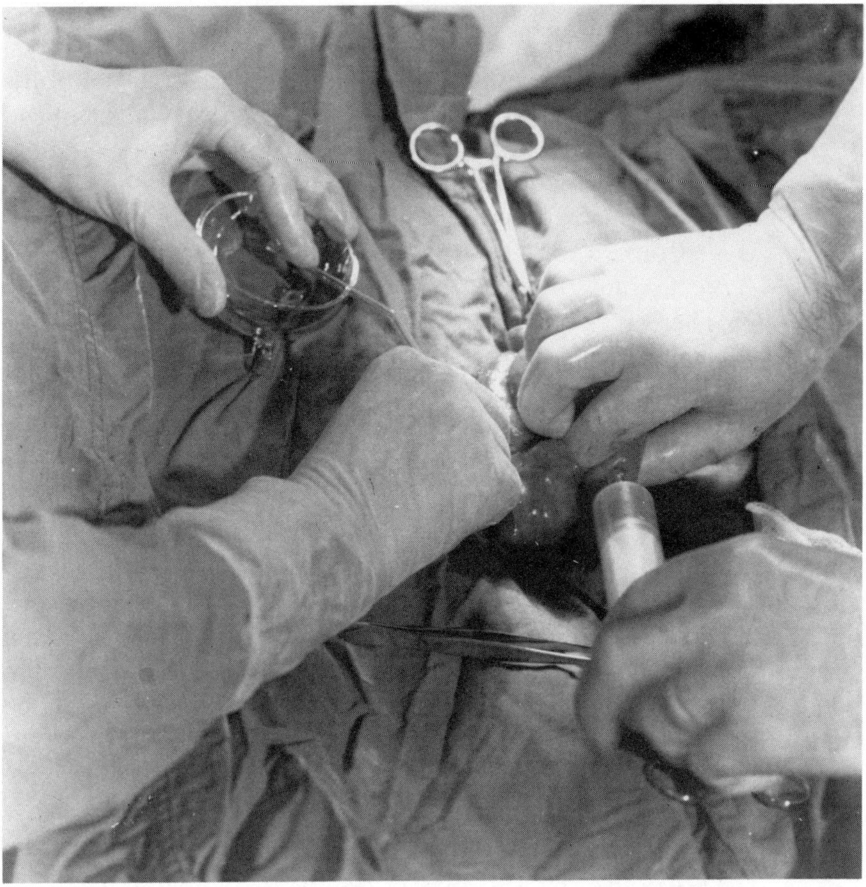

B

C

quality of their livestock. Each blastocyst is surgically transplanted into an undistinguished "nurse" cow that has been hormonally treated to be at the right stage for implantation. The embryo then develops normally. In one bovine generation a farmer's herd could thus be changed from cows of one breed or of undesired traits to purebred stock of whatever type he wanted.

There is no particular reason to suppose that transplantation of a zygote or blastocyst would be any more complicated biologically in humans than in cattle or rabbits. The only technical problem is one of safety. Some of the technical details will have to be determined by trial and error, and errors in human embryology are disastrous rather than just slightly expensive. I actually expect someone to try human zygote adoption soon, but it may not work until about 1975.

The first human applications of zygote adoption will not be for reasons of social genetic policy but because of personal motivations. Couples who desire a child but are infertile could usually be successful in adopting a zygote. Available adoptable babies may soon be almost entirely non-European (imported from Asia), and look unlike the adoptive parents. The parents for zygote adoption could be chosen so that the baby produced would be about as likely to resemble the adoptive parents as if it were their own genetic descendant. (Contributing parents would not even have to encounter one another if artificial insemination were used.) The adoptive mother of a zygote would have the same emotional experience of pregnancy and birth as if the baby came from her own ovum. The only practical problem would be one of persuading a couple to donate (sell?) a zygote. They wouldn't be emotionally attached to it, but the necessary operation would be unpleasant. In 5 or 10 years, however, it will probably be possible to flush out the zygote without an incision, and even to dispense with artificial insemination; ova could be fertilized in a test tube. Then, just as is now done with cattle, a contributing woman could take fertility drugs for multiple ovulation and meet the needs of ten or more adoptive parents in one month.

The recent American practice of adopting babies as totally accepted "own" children is something of which we can be very proud. I know of no other culture that has done this on a large scale. "Bastard" and even "stepchild" are still bad words in our language, attesting to how recent the change is. In practice, no one bothers about the high probability that an adoptive child was conceived out of wedlock; an adopted child is now the child of his adoptive parents. Our culture, perhaps alone among major cultures, has grasped the fact that parenthood is basically the commitment to the child and the care of it, not conception and birth.

The recent, unique, American virtue of true adoption has been clouded by the sad fact that the children adopted represent an emotional trauma to the biological mother, a problem pregnancy and then the need to give up a baby against all natural emotions. Zygote adoption will eliminate this sad side of adoption; the donation of a zygote is not a loss. Couples with genetic defects that they do not want to pass on, or those with fertility problems, will be able to have children happily and normally, without waiting for a foundling to become available. Zygote adoption for these purposes is really less radical socially than adoption of babies. It will be an undoubted human advance.

Any new power, no matter how useful, will be exercised by someone in some way that is unwise, dangerous, or even destructive. It is easy to imagine questionable applications of zygote adoption. One of these would be as a new form of leisure for the rich. The very wealthy have always been able to pay others to dial their telephones, hang up their clothes, and even raise their children. Zygote adoption will make it possible to go one step further. A rich couple who want their own genetic child without the bother of going through pregnancy and birth will be able to hire a "womb nanny" to carry their zygote and bear their child.

Less dramatic, but probably more important in the long run, will be the problem of exercising the new power to decide who should be the parents of your child. Some people, in speculating about the future, have envisioned "baby supermarkets" in which frozen zygotes would be identified by their probable characteristics, such as hair and eye color, general adult size, probable IQ range, possible special talents. Some parents have always caused emotional problems by expecting

"For crying out loud, Mildred, I wish you'd make up your mind!"

their children to share their own characteristics. If parents had the chance to pick out a child who was perfect by their standards, the dangers of unrealistic expectations would be increased.

If zygote adoption becomes common, there will be a sizable genetic effect on the population, since there are sure to be popular styles in what is considered a desirable parentage for the zygote. As will be discussed later, this will have both promise and dangers.

1980: Effortless contraceptives

Contraceptives have already changed an important part of the human condition by making it possible to plan confidently for sexual intercourse without the likelihood of producing children. For most current contraceptives, however, two kinds of choice of purposive behavior are required to avoid pregnancy. First, a decision must be made that children are not desired at some specific time. Second, something must be done regularly to implement that choice, either every day or before every act of sexual relations. The consequence is that the thoughtless, the ignorant, the irresponsible, the impulsive, and the immature tend to have children (whom they are not prepared to handle), while those most competent to cope with child rearing are those most likely to plan and limit their families.

It may be a case of wishful thinking to suppose that effective, effortless contraceptives will be available by 1980, since research and development in human medical applications must be slow to avoid danger. However, the steady improvement of the IUD is what makes such optimism possible. The mode of action of the IUD will probably be understood in just a few years, and it should then be possible to design a better model; development so far has been just by trial and error, without much understanding of the principle involved in the device. An IUD comes near the optimum for contraceptives if there can be improvement over the present models in effectiveness and freedom from side effects. Physicians' assistants could be trained to insert improved models without discomfort to the woman and for very slight expense. Only the single decision not to have children would be necessary; no daily or periodic responsibility would need to be exercised. A totally effective IUD that was free of discomfort would be, in effect, a simple kind of reversible sterilization. It might even prove feasible and popular to insert IUD's in the uteri of all girls before puberty, so that they would have to make a positive decision to have children. Family planning as practiced now is necessarily a negative activity. People must decide and plan *not* to have children. With an effective, removable IUD, the decision would be to *have* children. I think the effects on family life would be beneficial; children would always be desired and planned for, rather than, as they so often are now, being the tolerated side effects of sexual activity.

The development of a perfect IUD, coupled with an effective public health program that made the improved models easily available without cost to every woman, could mean that only wanted children would be born. According to interviews conducted in the early 1970's, American women desire somewhat fewer children than they have been having, and Americans would do no more than replace themselves if they had exactly as many children as they wanted. Really good contraceptives that did not depend on the exercise of responsibility would hasten the painful decision mentioned in Chapter 4: What about the virtuous groups who want many children, and who will teach these children that it is their religious duty to have many children in their turn? The requirement by the community that all families be limited in size will not be an easy one, but it will be necessary in the long run. Even a slow relative gain in population by a small minority group will eventually put that group in the majority.

1985: Choice of the sex of offspring

It is already legal and technologically possible to choose the sex of your next child. It requires simply amniocentesis, chromosome analysis, and selective abortion of whichever sex is undesired. This method will probably be tried within the early 70's. It is unlikely to become common, though, since amniocentesis is uncomfortable and expensive; the first months of pregnancy are uncomfortable, too, and most people would find such late abortion for so trivial a reason morally unacceptable. Several much less objectionable methods may soon be available. The interests of the livestock industry will probably assure the early development of a technology for separating sperm into X-bearing and Y-bearing fractions; this separation coupled with artificial insemination would work quite well. It also seems probable that within a few years there will be a way of harmlessly staining a zygote (during transplantation) to determine its sex; the choice of one sex rather than another from a pool of several zygotes would not be a serious moral problem for most people. Some enthusiasts already claim that the use of certain kinds of vaginal douches before intercourse will select for one sex or the other in the child conceived. It is easy enough to believe that this will really be true by the middle 80's.

As simple a power as being able to choose to have a son rather than a daughter might cause more problems than would be immediately imagined. First, no technology works just right every time. Suppose a couple that wanted a boy were assured that they could indeed choose a boy, but it then turned out to be a girl. Wouldn't they sue the doctor or technician who goofed? How would the girl feel about herself if she knew her father was suing because of having been saddled with her? When the techniques worked the problems might be even greater. There are fads in everything. One can imagine 10-year stretches with big runs on boys. Wouldn't they get lonely? The nearly equal sex ratio is a great convenience.

Human evolution

Evolution is a systematic change in gene frequencies. Evolutionary change is going on all the time, in every species. Human evolution may or may not be in directions we favor. The most obvious changes of human gene frequency in recent history have been caused by the differential fertility of whole groups of people: national, cultural, racial, and language groups. From the seventeenth to the beginning of the twentieth century, the advantage was to the advanced societies, especially those of western Europe. Europeans constituted 20 percent of the world's population in 1600; by 1900 the figure was 40 percent. This happened because Europeans had better health care and better nutrition, and because they expanded into the continents of North America and Australia. In the last 40 years, with the development of contraceptives and their use in the developed countries, and with the application in the less developed parts of the world of such public health measures as malaria control, the non-European groups have been rapidly increasing their share of the total world population.

Relaxation of selection

The international evolutionary trends have been of little basic biological importance, since there is no good evidence that racial differences are significant in the modern world. Not all human evolutionary trends are unimportant, however. Compared to a generation or two ago, there is now a great relaxation of selection against certain alleles at a variety of loci, because medical advances allow more people to survive. There is now much less selection against specific disorders such as diabetes or PKU, and selection has almost disappeared against susceptibility to smallpox, measles, diphtheria, and malaria. In addition there has probably been a small but real increase in the mutation rate because of radiation and chemicals, so harmful mutant genes are increasing in frequency.

Are we evolving in the right direction?

There is little evidence that evolutionary forces are acting at the present time in favor of socially desirable traits. The people who cope best with our modern society, and those we admire most, are not especially likely to have more children than other people. Perhaps there is a little selection against those people who cannot withstand the stresses of overpopulation, industrial speedup, pollution, and future shock; they get ulcers, heart attacks, and nervous breakdowns. But do these same people perhaps have more than their share of traits we value, too?

Many people have been afraid that we are evolving in the "wrong" direction; that we are becoming less intelligent as a species. People who score high on an IQ test come from smaller families, on the average, than people who score low on these tests. Contraceptives became available and popular first among the educated and well-to-do, and family size decreased in these groups. Since the educated may be expected to have more native intelligence than the dropouts, this seemed to be evidence that we were evolving toward less intelligence. The few test scores that were available comparing representative children of different generations did not show any measurable change in one generation, however. Further studies disclosed that the conclusion that there had been effectively a selection against intelligence had been premature. It seems that whereas the slightly dull people have more than their share of children, the very retarded have very few children, and the two effects just about balance out. With improvements in the ease and effectiveness of contraception, the class differential in births has already diminished, and it will probably disappear with further contraceptive improvement. In fact, we may expect to see a time when those who can afford more children will have more.

Eugenics

As the understanding of genetic inheritance has begun and grown in this century, there has grown with it the dream of **eugenics**. The hope of eugenics is that man can direct his own evolution, solving old problems and changing human nature for the better. Humans resist manipulation, though, and eugenics has never been really tried. Just as the idea of eugenics was gaining popularity in the 20's and 30's, it became badly discredited because of the atrocities committed by the Nazis in the name of eugenics and "racial purity." The Nazi programs were inconsistent, badly conceived, and antiscientific, and were generally the worst possible misrepresentation of the idea of eugenics. But the methods they used—compulsory sterilizations, incentive payments to encourage fertility of certain groups, and official matchmaking—were about the only kind of thing that could have been done then.

The necessary technology is now at hand or in sight to direct human evolution without Nazi-like violation of individual rights. At a time when most people limit their families, incentive payments would make a great difference by allowing selected parents to afford to have, raise, and educate the children they wished they could afford, anyway. With good contraceptives, those who should not pass on their genetic traits need not remain celibate; they can marry just like anyone else. If they desire children, adoption is now possible, and zygote adoption will soon make normal family life routine for them.

Zygote adoption and artificial insemination will make eugenics practical. H. J. Muller, a Nobel Prize winner in genetics, suggested that parents might want to adopt a zygote from a couple with superior attributes and thus promote eugenic selection by voluntary action. A decision by the community to regulate who could donate sperm or zygotes would not violate any basic human right, and it would be a powerful evolutionary force. These practices could easily become widely used before the turn of the twenty-first century.

Negative eugenics

Human genetic selection to eliminate undesirable traits is sometimes called "negative eugenics." Perhaps we need not continue treating the victims of genetic defects as they appear each generation, and perhaps it will someday be unnecessary to go through the stress of amniocentesis and therapeutic abortion to prevent the birth of an individual sufferer. Human society might, instead, decide to eliminate the undesirable genes from the population. The effectiveness of such a program of artificial selection would depend on the mode of inheritance of the traits we wished to eliminate, and on the technology we had available for the program. Both the necessary genetic knowledge and effective technology are becoming available, and a cautious program may not be long in starting.

Any serious defect that shows up early (such as some dominant gene having its effect during embryonic development) will already be strongly selected against in the natural course of events. Dominant lethal genes, to cite the extreme example, do not persist to cause trouble. There are a few dominant traits (such as Huntington's chorea, a serious deterioration of the nervous system) that do not show up until fairly late in life, after a major part of the reproductive period is past. A person whose father had suffered from Huntington's chorea would not know whether to have children or not; he has one chance in two of developing the disease himself. If the son does carry the gene, he in turn will pass it on to half of his children. New techniques, based on cellular biochemistry,

346

hopefully could detect such a trait in a young child of an afflicted person. If this child turned out to have the harmful gene, he could avoid having children, or adopt them in one way or another. Such a program could eliminate slow-acting dominant harmful genes in one generation.

The problems with the much more numerous undesirable genes that are recessive are much more difficult. Most people never know that they carry such a gene, even though we may average eight to ten of them apiece. The first problem with trying to eliminate undesirable recessive genes is that there are so many of them, but each individual harmful gene is very rare. In most cases, it will be some time before people heterozygous for such a gene can be identified. Even when tests become available for identifying people who are heterozygous for any of the 200 most common such genes, the expense of conducting all the tests on everybody would be enormous. Using techniques similar to those currently in use, the expense would generally be prohibitive. Only for the most common diseases (cystic fibrosis, sickle cell anemia) can a general screening program for the whole population be justified with present techniques. Screening for the other genes makes sense now only for relatives of afflicted homozygous persons; the relatives would have a high probability of carrying the gene. A major technical improvement, such as complete automation with miniaturization, could, however, change the whole picture. If a sophisticated machine could grind out 200 "yes" or "no" diagnoses from a single blood sample or tissue biopsy, mass screening could become feasible.

The second problem with genetic selection against deleterious recessive genes is that most of us have such genes. If maximal selective pressure were to be exerted against 200 harmful genes at once, only a few people could have children. Selection would have to be slower than that. (The growing practice of therapeutic abortion of homozygous sufferers detected in the uterus will have little if any genetic effect. The aborted pregnancy is usually replaced with another trial, in which the child is likely to be a heterozygous carrier for the gene. If the parents had gone ahead and had the homozygous recessive child, they would probably have been so overworked with its care that they would have had no more children. Since the victim of a serious genetic disease is usually unlikely to have children himself, the genetic selection is much the same whether or not the new techniques are used.) If, however, zygote adoption becomes common, and care is taken that the zygote donors carry none or only very few of the most common recessive harmful genes, selection will move fairly rapidly. It would be unwise to go to wholesale zygote adoption with just a few donors. For one thing, many good traits might be lost. For another, some previously very rare recessive genes not being detected in the standard surveys might become common if the selected donors contributed many zygotes each.

A third reason that negative eugenics must be conducted cautiously is even more serious. Geneticists are not sure that those recessive genes that are harmful in double dose really have no effect in single dose. They may even be helpful in the heterozygous state. That could be, after all, the reason that such genes are not even more rare than they actually are. The only clear examples thus far of genes that are helpful in single dose, dangerous in double dose, are the sickling gene and others affecting the red blood cell. They are desirable, even life saving, in regions troubled with malaria. Resistance to malaria is no longer important, and no harm will be done if these genes are eliminated from the population. But it is entirely possible that eliminating cystic fibrosis, PKU, or other undesirable recessive traits might, at the same time, eliminate some important valuable traits that, unknown to us, are caused by the heterozygous conditions. For all these reasons —recessive genes are difficult to detect, there are few people without undesirable recessive genes, and we are not quite sure that we want to get rid of all of them, anyway—it is prudent to go slow with eugenics. By the end of the century enough more questions will probably have been answered to make a eugenics program workable without tragic surprises.

Positive eugenics

Selection to promote the increase of good traits, positive eugenics, looks at the same problem as negative eugenics but from the other side. Why not increase the frequency of special human traits that we welcome? It is already certain that society could successfully promote the increase of any of a variety of attributes, simply by encouraging and helping people with such traits to have more children than the average person. With widespread zygote adoption, much-admired people could have several times as many offspring as they could ordinarily have. The fact that environment plays such a large role in the expression of such traits as intelligence does not prohibit selection. If all couples with IQ's above 140 averaged four children apiece, while other couples averaged only one, the IQ average for the population would be measurably higher in the next generation despite differences in environment. The setting up of social incentives to bring this about would not be very difficult, but before such a scheme was put into action, some cautions would have to be considered.

Does anyone know, and would everyone agree, what special human traits should be made more frequent? Should there be selection for beauty, size, strength, health, and intelligence? Would it help society to have more clever people, when society doesn't really make effective use of intelligent people at the present time? Maybe intelligence isn't what is needed most; perhaps there should be selection for patience, kindliness, and cooperativeness instead. It is very unlikely that people could agree on what traits were most desirable. There are perhaps also unforeseeable consequences of selecting for something even as presumably desirable as high intelligence. After all, there must have been strong selection in favor of intelligence in all cultures for millions of years of human existence. High intelligence has always been advantageous for solving problems, discovering new facts, and inventing and using tools. Despite the seeming advantages of high intelligence, human intelligence seems to have remained at much the same level for many thousands of

"Oh! Why in the world did we get in on that superbaby breeding experiment?"

years; there have always been some less intelligent people. Perhaps there is some disadvantage to high intelligence; we can do no more than guess at the possibilities. Possibly there is a greater difficulty at birth with a slightly larger brain; perhaps there is even a tendency for the most highly intelligent people to go drifting off in thought, neglecting duties and children. Evolution has already been quite efficient at selecting for desirable traits. It has generally been found by geneticists working with other kinds of animals that artificial selection for almost any trait will increase its frequency in the population. But if the selection is then discontinued, the average of the population will move back toward the original frequency found in the wild population. This happens because there is always some selective disadvantage in the extreme form of the trait that had been artificially selected for. We might easily find some unforeseen disadvantage if we succeeded in changing human nature. Eugenics, both positive and negative, constitutes a field in which our powers will keep increasing but in which the best counsel for the present is to proceed slowly.

1995: Genetic surgery

There has been a great deal of speculation recently that it is soon going to be possible to replace defective genes in affected individuals, correcting their individual problems and perhaps even correcting the genetic message that they would pass on to their offspring. If this worked, it could relieve a great many health problems and would provide a way of changing the gene frequencies of the population without restricting reproduction to genetically "correct" people. This speculation is based on the fact that it is already possible to change the genetic properties of bacteria by treating them with DNA from bacteria of a slightly different heredity. The bacteria so treated pick up genes from the DNA and incorporate them in their own chromosomes. The DNA may be in the form of chemical DNA crystals or it may be a special form of a harmless strain of virus (since viruses work by becoming associated with the genetic nucleic acids of the cell). Considerable progress has been made toward being able to bring about predictable genetic changes in bacterial cells. The bacterial cell is genetically much simpler than the cells of higher organisms, of course, and the application of the technique to man is distant. There are no clear and undisputed experiments that show that this kind of thing ever happens in any vertebrate animal, although there are a few experiments that suggest success. Even if these prove correct, the methods known so far are "shotgun methods," which may change many genes besides the one aimed at, and cause new problems.

Despite the extremely doubtful state of the art, genetic surgery has already been tried, unsuccessfully, on one patient afflicted with genetic disease. He was deliberately infected with a usually harmless virus that was thought to carry a healthy allele of the patient's mutant gene. Further research indicated that this experiment could not have worked because the genes weren't really alleles. Some of the experimenters in this research field are worried that this technique may be applied to humans before it is properly worked out (as was already done) and that some of the attempts will not only fail but cause direct harm. If purified DNA is used, there is a clear danger that some of the genes may be modified during the isolation procedure, bringing about other mutations in the person, some of which might cause cancer or new metabolic problems. Using viruses poses the same danger, especially if the viruses chosen for genetic reasons are not proven safe. It will probably take 20 years to work out the problems in this approach, if indeed it is even feasible to do so.

Individualized medicine—the year 2000

Whether or not "genetic surgery" finally proves effective, the same line of research will make it possible at least to identify people who can be helped by special treatment. This is already being done in the case of phenylketonurea, and in a similar disease in which milk sugar cannot be utilized, by simply eliminating the unusable foods from the diet. It will probably prove to be the fact that many or all of us have some unusual biochemical quirk, not really a defect, so that we would do better with a diet slightly different from the ordinary, a stronger than average dose of one kind of medicine, or a weaker dose of another. The optimum environment for each of us may be unique, and a whole new practice of supportive aid will arise to meet the needs. We may regard such a program, perhaps to become common at the beginning of the next century, either as a great blessing or a great curse. On the one hand, it would mean that everyone would be given the best possible conditions for his unique properties, whatever they might be. On the other hand, this would be a tremendous burden on society, and would also eliminate natural selection, since everyone would be living under optimum conditions. Would it be better to get perfect people to live in the world as it is, or to build perfect individual worlds for variable people? As long as it just meant better conditions to fit the slight quirks of those that we would now consider normal people, such a supportive program, if economically and practically possible, would be hard to criticize. Maintaining people for a lifetime on insulin, or on a kidney machine, are two early examples of individualized medicine. Where will it stop? If someone decided that additional effort should be expended to make it possible to support all manner of people with hitherto fatal genetic defects, humans would slowly become totally dependent on a perfectly functioning technology—prisoners of their machines.

2005: *Fetal surgery*

The danger of accepting too much human variability will also become more apparent when it becomes routine to inspect fetuses in the uterus and operate to correct any structural defect. If a complete repair can be made, this will produce nothing but good. Some conditions would certainly be correctable if treated before birth. But as with present surgery on accident victims and malformed infants, many instances of surgical repair will not solve all of the problems. What would have been otherwise a natural death will be prevented, without anything like normal human function being restored. The "repaired" person will then be a burden to himself and to others. There is no limit to the extent that some doctors might decide to go. It might be possible to start giving some kind of supportive therapy to embryos of any age that would have been naturally aborted. They could be kept in some kind of an **artificial uterus** forever. (This would perhaps be all too similar to the present practice of keeping "alive" the bodies of old people, such as victims of massive strokes, when the body is no longer anything but a mass of tissues, and human characteristics have long been gone.)

2015: *Cloning*

It has been possible since the 1920's to grow a snip of animal or human tissue in nutrient fluids for an indefinite period. Even a single cell can be induced to divide into a series of similar cells, called a **clone**. If each of these similar cells from one person could be treated in some special manner to become effectively a zygote, these zygotes would be as much alike genetically as a clone of bacteria derived by asexual cell division from a single ancestral cell. They would produce a "clone" of people, and the imagined technique of producing them is called cloning.

The technical problem that needs to be solved before cloning could be used to produce people is the technique of changing a regular cell into a zygote. Since no one knows just what the difference is between zygotes and other cells, we cannot know yet how to make the necessary changes. It is clear that the same genes are present in both cell types; it is just that different genes are active. Major progress in the 70's in discovering what chemical structures determine the pattern of gene activity for a zygote or for some particular kind of body cell will make a great contribution. It should be technically possible in a number of years to cause the change in somatic cells needed to make them zygotes. A breakthrough in the research might come much sooner than 2015, but work in the field during the last 15 years has principally demonstrated that the subject of differentiation is more complicated than previously supposed.

A different kind of cloning, much more difficult to accomplish and more susceptible to mishap, may be within our grasp sooner. It is now possible to transfer the nucleus from certain embryonic cells of a tadpole into a frog zygote, in place of its nucleus. With further progress, it may become possible to transplant the nucleus of a human skin cell into a human zygote. Since this technique will require such a high level of manual dexterity, and includes such danger of damage to the nucleus during transfer, it is unlikely to become important in human affairs.

At first consideration, cloning would seem to be a socially safer procedure than ordinary zygote adoption, since it would already be clear that the zygote that was identical genetically to a living person had no serious genetic defects. The problems of cloning would come from the new possibilities it offers. With cloning, we could readily obtain "carbon copies" of any much admired person who would contribute a bit of tissue. The first result might be disappointment. A person becomes great in any field, whether it be politics, scholarship, athletics, art, or public entertainment, from the creative combination of his unique genetic capabilities with his unique environmental experience. In a new environment, the same genes might produce a person who would have no outstanding success at all. For instance, an aggressive man who is careless of his own safety may become a war hero but be a complete misfit in peacetime. Children of prominent people already often suffer from the relationship. Parents who expected a genetic copy of some great person to achieve the same greatness might make life unbearable for the child. Some parents have always had unrealistic expectations for their children; the number of such parents would increase if they could pick a zygote with exactly the same genes as some admired person.

A second problem that will stem from the success of cloning will be a decrease in interesting and valuable human variability. Fads in types of children might produce crowds of thousands of genetically identical children in many families. In the 1930's many thousands of girls (and a few boys) were named "Shirley" after Shirley Temple; with cloning it would have been possible to have genetic duplicates of her, instead of just her name.

With the use of cloning certain types of people could be selected to be the genetic patterns for children who would be suited for special jobs. Whole basketball teams of Wilt Chamberlains could be produced, and there could be another group of little copies of the best jockey. Although the technique is illustrated most humorously by sports figures, a serious application of cloning might mean a decrease in the effort to adjust jobs to people; instead, mass-produced people would be sought to fill standard jobs. *Brave New World** was an imagined nightmare, but it may actually be on our horizon. Such possibilities need to be envisioned and seriously discussed before they are actually upon us.

2020: *Replacement parts*

A less dubious variant of cloning (but with its own problems) would be the production of replacement parts from a person's

---

* A. Huxley, Harper & Row, 1932.

"The Betty Friedans are really slow but here's an order for another 400 Raquel Welchs!"

own cells for his own lost organs or limbs. When organs are transplanted now, there are two problems: someone has to die or give up one of a pair of organs to provide the organ necessary for a transplant, and the transplanted organ may be destroyed by the recipient's immune system. Replacement organs producing by cloning would avoid these problems. If a person lost an arm in an accident, a technician would take a cell from elsewhere in the body, cause it to multiply into a mass of cells, and so stimulate these cells that they began to develop like a limb bud of an embyro. An artificial uterus would allow this to be done in the laboratory, where it could be watched and properly stimulated. Actually, an arm would be one of the hardest parts of the body to replace, since the muscles need stimulation and exercise to grow past the fetal stage. But if this could be somehow managed, a person could have a new arm attached that would cause no immunity problem and would not require that someone die in an accident. There is no way that people can grow new arms as soon as the year 2020, but the same sort of thing might be possible by then for organs in which shape and exercise were not so important. A healthy liver from which two-thirds of its mass has been removed will grow back to normal size within a month or so and will be perfectly functional. Someone whose liver was scarred from poisons might have a few healthy liver cells salvaged from his liver; these could then be induced to grow into a good, functional liver, complete with blood vessels. The "new" liver could be sewn into the right place and would grow to appropriate size and shape. It seems quite reasonable to expect development of this technique by 2020, with techniques for cloning of more difficult organs coming later. In fact, a start is already occurring. It is probably now possible to preserve, by freezing, some of the bone marrow of a person who is to receive a dangerous, marrow-destroying dose of radiation or anticancer drugs. After the treatment, the preserved bone marrow can be injected, and it will enter the bones, to resume the normal manufacture of blood cells.

With a ready supply of spare parts, derived from his own body and easily replaced, why should anyone grow old and die? Therein lies the problem with the technique, perhaps a more serious problem than any discussed so far. If people are to live forever, what will the nature of man become? Old people even now have trouble adjusting to a changing world. What if they lived many years longer? What would a society composed entirely of old people, however healthy, be like? Can you imagine a world without children? A world of immortal people with continually renewed bodies and continuously accumulating knowledge might be a good world, but it is not our world with humans as we know them. The philosophies and customs necessary to cope with mortality, and the practices of reproduction necessary to replace those

"Good morning, ma'am, what part do you need today?"
"Oh, the way I feel, I don't know which organ to replace first!"

who die, are a part of being human. Could we cope with a world where babies were unnecessary and undesirable? Physical immortality is probably more than any of us could stand. However, it may seem quite otherwise to our descendants, who will possibly have the chance to try it.

2025: Genetic engineering

It is gradually becoming more and more feasible to design a person. We do not know now what any specific combination of genes would produce during embryonic development, when subtle differences can have profound effects. Someday, perhaps by 2025, it will be possible to make some good guesses. Genetic selection and replacement of individual genes could contribute to an effort to design a new, superior type of person, but other methods are sure to appear. It will probably be possible, when the specific locations of genes on their chromosomes are known, to put together a zygote with one chromosome from one person, the second from another, and so on, to get a desired combination without generations of genetic selection. When new models of people can be designed, the question of what kind of people to choose will become more acute than ever. Our present characteristics were evolved in a culture of hunting and gathering. Perhaps we need to redesign man to fit his new, man-made environment. One obvious flaw in such a scheme is that our culture might change again, leaving us more ill fitted to it than ever. The outmoded characteristics that we might eliminate would probably only be certain of the obvious ones. Some hidden human characteristics that remained might fit very badly with other features of our new design, to the sorrow of everyone. The necessary techniques for genetic engineering will certainly be developed, since they will be useful in agriculture, speeding up what breeders have done for thousands of years. Once perfected there, the new techniques will work in people. Will the brave new people be a real improvement?

Someone will have to face the problem, perhaps in 2025, of deciding whether or not new models of humans should be designed and produced. Should genetic engineering be used to fit the nature of man to social needs, or should society be changed to fit the kind of people there already happen to be? Is there anything sacred about the gene frequencies our species happens to have in 1973 or 2025, or should we work toward improvement? Since we now dig ditches with machinery but need many skilled technical workers to run things, shouldn't we design for more brains and less brawn? Or should we adjust society to make good use of those people who would have dug ditches 50 years ago?

Obviously, the first thing to do is to improve our society's child-rearing practices and educational system so as to make the most of each person's potential. But as good as that sounds, it does not solve the problem. In fact, it runs up against the same kind of question. Should we change man to fit society, or change society to fit man as he is now? The parental practices and life styles of many individual families and some cultural groups in our society do not prepare their offspring for success in society as it is now constituted. For example, some Amish communities refuse to allow their children to go to school beyond the eighth grade, and resist having their children trained in science and modern technology. This kind of education is satisfactory if the child remains in the community, but it does not give him any competence for success if he moves away and becomes a part of modern society. Similarly, an American Indian youth taught only traditional skills and values would find it difficult to survive in most present-day situations. Both of these examples illustrate the dilemma raised by subcultures which have value and deserve to be preserved, but which sometimes handicap their children. Should such families and groups be forced to change their ways, or forced to give up their children, so that the children can have the experiences necessary to make them fit into the establishment? Or is society obligated to provide a living and a satisfying environment for children produced by families of any life style, even ways of living that most people believe destroy the children's chances of "success" in today's world? These are hard questions, but many of them need to be debated and even decided now, or the questions to be decided in the future will be even harder. Not thinking about a problem doesn't mean it will go away!

Combinations of man and animals: 2050

In the discussion of more and more radical possibilities, the questions and problems have gotten more and more serious.

In discussing genetic engineering, I tried to indicate that I feared the mistakes of such an enterprise at least as much as I hoped for good results. In the last possibility to be discussed, I see only threat, not hope, but not everyone agrees with me. Perhaps by the time it becomes possible, it will not look so threatening.

It has been proposed to design eventually not only an improved man, but some kind of creature with a combination of human and animal characteristics (and genes) that could perform certain tasks. One of the things suggested would be a monster with just enough human brainpower to do certain dull, repetitive tasks, but the necessary animal genes to have the right kind of body for the task. Not being human, such monsters would not require a minimum wage, educational opportunity or a vote, but just prudently humane treatment. A less horrifying idea, but one just as "far out," is that fully intelligent combinations would be able to establish civilizations in new environments. For example, some kind of porpoise people might live under the seas and establish an oceanic culture in which they would be at home. The idea has a bit of charm, but the horrible mistakes that could be produced in trying for such combinations would be hard to justify.

Although this particular idea sounds even more like science fiction than many of the others, techniques for this kind of thing are sure to be developed as far as possible, again for use in agriculture. Something of this sort has actually been done for thousands of years, ever since men began mating donkey males to mares to produce mules. The mule is an artificial species, created by man for its special virtues in agricultural use. Another example, the liger, a cross between a lioness and a Bengal tiger, recently died after being a star attraction at the Salt Lake City zoo for many years.

With new technology, crosses could probably be made between more distantly related species, although usually not with such good results. The first steps have already been taken toward the necessary techniques. In the first years of the 70's, two remarkable things have been done. Somatic cells of different species have been fused into single cells, with

"When they give us that human chromosome we'll be doing all the work and serving them instead of them serving us."

chromosomes of both operating to determine their characteristics. All kinds of cell combinations can live, although some of the chromosomes tend to be lost as the cells divide. The other development, even more recent than cell fusion, is the successful fusion of a blastocyst of a mouse with one of a rat into a large growing blastocyst. Mosaic animals and people have been known for some years to be formed from such fusion of two zygotes or blastocysts of a single species, but the fusion of two different species is new. Apparently the combined blastocyst has not yet been carried through embryonic development, but it will be in the future.

Neither one of these new developments means that man-animal monsters are likely to be produced in the near future. It seems quite unlikely that a cell, even a zygote, with mixed human and mouse chromosomes could develop through the embryonic stage; in humans even one extra chromosome is usually fatal. Similarly, the mixed blastocyst of rat and mouse may be found to develop easily into a new hybrid, but more distantly related combinations of species probably would not. Even if a half-human combination developed into a viable creature, this new form of being might not have any recognizable human traits. Whether or not to produce half-human beings is not a choice we have to make now, but such a choice may be coming. This is one new development I hope we don't get to for a long time; even 2050 is too soon.

Public discussion needed

The technological revolution that has been transforming our society for several hundred years has caused many problems, such as pollution, overpopulation, and urbanization, which we are just now beginning to try seriously to solve. The coming biological revolution, including fundamental changes in human reproduction, will cause staggering new problems. We need to be prepared for these problems to prevent them from becoming as serious as they will be if they are not faced.

This country, and every other country, needs to begin and continue a general public discussion of the biological revolution, how to profit from it and how to avoid creating new problems. All the topics mentioned in this chapter should be the subject of series of television programs, and of articles in newspapers and magazines. They should be discussed by citizens in all kinds of groups—churches, schools, League of Women Voters, men's service clubs, and special public discussion groups. Widespread participation in the discussion of these problems is even more important than for many other national issues, since many of the effects of new biological technology, both effects on families and effects on society, are the results of individual decisions. Discussions should first emphasize the problems that are already upon us: the population explosion, drugs and pollution, venereal disease, the care of the dying, the effect of birth control measures on the family, and management of abortion. New techniques that will soon appear need to be discussed too, including zygote adoption and choosing the sex of offspring. Finally, even distant developments need to be discussed briefly, so that people can start thinking about them now.

Some discussion groups, including many in churches, will want to emphasize the personal decisions that have to be made about the application of our new powers, such as when and why to use contraception, abortion, or zygote adoption, and the implications these have for the institutions of marriage and the family. Other groups, such as the League of Women Voters and other legislative discussion groups, will want to concentrate on proposed legislation for controlling the application of new procedures.

Ideally, each club member and each student in high school and college should obtain a copy of *Human Reproduction* as a source of information! Much good resource material is becoming increasingly available; a list of suggested readings is given at the end of the book.

Providing for research

Coping with new opportunities and problems will require the best understanding possible. Therefore basic research in all branches of fundamental biology, from DNA structure to chimpanzee behavior, should be steadily funded at some moderate level. (The current trend, which may introduce techniques before we are ready for them, is to fund applied research rather than basic research.)

Some kinds of research and development that are needed immediately, and are clearly in the national interest, need to be treated on a different scale. This group of priority items might include the development of a reliable test for infection with the venereal diseases, of a thoroughly reliable IUD as described in this chapter, and of rapid and inexpensive methods for detecting many more congenital defects early in development. Such projects should be funded until they are completed, at levels that assure that no worthy proposal goes unsupported, even if a little money is wasted on proposals that prove unproductive. Modern biological research is far too expensive to be carried on unsupported by any individual scientist, or by almost any school. It must be heavily supported by the federal government as well as by private foundations. However, such research is extremely inexpensive compared with the costs of armaments, highways, dam projects, and space programs, and especially compared with the costs of failing to solve the problems.

A legislative program

Earlier in this book, I suggested some changes in public policy to help deal with current problems. I feel that as we begin to deal with new techniques, new laws should be enacted to allow the use of techniques that will alleviate immediate suffering and unhappiness. However, techniques that may produce widespread social or genetic impact need to be avoided until the possible repercussions have been well studied. Here I will suggest a few possible laws to regulate new reproductive techniques of the near future, with the idea of initiating debate about their suitability. There are, of course, others which can and should be discussed as well. Out of such discussion could come really wise laws.

1. Medicines that would select the sex of a child at conception should be issued only to parents desiring both sexes of children (or with genetic reasons to desire only one sex).

2. Zygote adoption should initially be available only to those parents unable to conceive their own offspring, or with good genetic reasons for not doing so.

3. Donors of zygotes for adoption should not provide more than a total of five zygotes per donor couple, to avoid a reduction in general population diversity.

Because of the individual application of most reproductive techniques, laws cannot hope to control them absolutely. Laws can, however, prevent undesirable applications of new technology from being so frequent as to create large social problems.

The necessity of choice

These issues are all illustrations of the principle that when we gain new powers and new possibilities, we are forced to make new choices of how and even whether to use our new capabilities. If we decide not to use new methods, we are choosing the status quo, with all of its sorrows and imperfections. If we refuse to choose, we can be sure that someone will use the new techniques, but with no guidance or control. If we choose, as individuals or as a society, to employ new means to avoid old ills, we accept as well the responsibility of dealing with the possible unwanted results of our actions. We cannot avoid choice—even no choice is a choice. If we say that we refuse to "play God," we let the worst possible take place. (You might say that we hand things over to the Devil!)

Technology by itself just gives power, it does not solve problems. It requires the addition of correct action and wise control to make things go well. An understanding of the nature of man—where we came from as a species and how we got here individually—can help us decide which new proposals might lead to a better and more genuinely human life and which might lead to a denial of human dignity.

In our decision making, we are faced with changes that are possible but which perhaps should not be made. Just because something is possible does not mean it should be done. We need wisdom not only to know what *can* be changed, but what *should* be changed, as well. Will we choose wisely?

Glossary

abortion: death of an embryo; may be either spontaneous or intentional.
abdominal pregnancy: also called "ectopic pregnancy." Situation in which the embryo is deposited in the abdominal cavity, and grows for a time there.
acrosome: a cap-like structure on the head of the sperm; may contain enzymes that help the sperm penetrate the covering of the egg.
adenohypophysis: anterior lobe of the pituitary.
afterbirth: fetal membranes and placenta as expelled following birth.
albinism: genetic trait in which the person lacks the black pigment called melanin.
alleles: alternative genes (for one kind of trait) that may be found at one certain locus. For example, the genes for blue eye color and for brown eye color are alleles.
aminocentesis: diagnostic procedure in which a needle is inserted into the amniotic cavity and a sample of amniotic fluid is withdrawn into a syringe. Study of this fluid and of fetal cells found in it can indicate the sex of the fetus and the presence of certain abnormalities.
amnion: the innermost embryonic membrane. It surrounds the fetus.
amniotic fluid: found between the fetus and the surrounding membrane (the amnion); contains cast-off embryonic cells which can be examined for information about the embryo.
anaphase: part of cell division during which chromosomes separate and move to opposite ends of the cell.
anatomy: division of biology that deals with structures of organisms.
androgen: a male sex hormone, usually testosterone.
anomaly: a defect, such as a congenital anomaly of a newborn.

anterior lobe of pituitary: endocrine gland producing hormones that stimulate several other endocrine glands. It is stimulated by hormones generated from the brain.

antidiuretic hormone: produced by the posterior lobe of the pituitary; regulates the amount and concentration of the urine.

aorta: also called "dorsal aorta;" the large artery opening directly from the heart and giving rise to smaller arteries that lead to all parts of the body.

aortic arches: multiple blood vessels in the early embryo that connect the heart with the aorta. Most of them disappear in the course of embryonic development.

aphrodisiac: something that increases sexual desire. It is usually a drug.

artificial insemination: transfer of sperm without copulation; common in raising of domestic animals and used occasionally in human medicine. The sperm is inserted into the vagina with a syringe.

artificial uterus: a proposed technical device that would be able to support the development of a human embryo.

atrium: chamber of the heart that receives blood from the veins and discharges it into the ventricle. (The atria are sometimes called auricles.)

autosomes: chromosomes other than the sex chromosomes.

bag of waters: popular term for the amnion, which breaks during childbirth, releasing the amniotic fluid.

balanced translocation: genetic condition in which an individual has both a translocation (an extra copy of a chromosome segment) and a deletion (the lack of that same segment where it would normally be) so that there is a normal set of genetic instructions. However, the children of such an individual may have a genetic disease.

biological revolution: profound changes in the events of human life resulting from technological changes affecting human biology.

biopsy: sample of an organ from a living person; used for diagnosis. The familiar blood sample is a kind of biopsy.

birth canal: the vagina.

birth control: contraception; also used to mean planned number and spacing of pregnancies.

bivalent: see *tetrad*.

blastocyst: early embryo, in the form of a hollow sphere of cells.

blood type: description of a person's blood according to the combination of special chemical molecules (antigens) on the surface of the red blood cells, especially those substances designated A, B, and O.

breech delivery: childbirth in which the buttocks are the first part of the baby to emerge. Such deliveries are more difficult than the normal ones.

Caesarean section: delivery of a baby by an operation to open the abdomen and wall of the uterus; performed when the baby cannot pass through the birth canal.

capacitation: effect of vaginal fluids on sperm, causing them to increase their capacity to fertilize ova.

cardinal veins: prominent but temporary veins in the embryo.

centromere: the point on the chromosome that is last to divide during mitosis. The chromosome is attached by the centromere to the spindle.

chancre: the swelling or sore that is the first symptom of syphilis

chimera: originally a mythical animal with characteristics of several species; now used to refer to a cell, embryo, or adult formed from the cells of two different species. Man-animal chimeras have been proposed as possibilities for the future.

chorion: the outer most embryonic membranes, developed from the trophoblast. Part of the chorion will give rise to the placenta.

chorionic gonadotropin: a hormone produced by the placenta (derived from the chorion), causing sustained secretion by the corpus luteum, and having other effects on the ovary.

chromosome: structure that contains the genetic instructions; visible during cell division.

circumcision: surgical removal of the prepuce, a fold of skin that partially covers the glans of the penis when the penis is not erect. If the prepuce is not removed, secretions accumulate under it requiring frequent cleaning.

clitoris: a small protuberance just above the vaginal opening. It is derived from the same embryonic structure that gives rise to the glans of the penis, and like it, is an organ of sexual sensation.

cloning: producing many cells from one original cell by ordinary cell division. Also, the proposed production of many individuals from such cells; all would be genetically identical.

coitus: sexual intercourse.

colostrum: the first secretion of the mammary glands after birth.

condom: a rubber sheath placed over the erect penis to prevent conception; also used to prevent transmission of venereal disease.

congenital: present at birth.

consanguinous marriage: marriage between relatives.

contraceptive jelly: a special medicated preparation placed in the vagina immediately before intercourse to prevent conception; acts either by destroying sperm or by mechanically preventing their passage.

copulation: sexual coupling; the series of activities by which semen is deposited in the vagina.

cornea: the outermost layer of the transparent part of the eye.

corpora cavernosa: distensible blood sinuses in the penis that fill and stiffen to cause erection.

corpus albicans: scarlike white spot remaining on the surface of the ovary after the corpus luteum has ceased to secrete.

corpus luteum: a structure in the ovary derived from a ruptured follicle and producing the hormone progesterone.

corticosteroid: any of several steroid hormones produced by the cortical portion of the adrenal gland. These hormones are similar to the sex hormones.

creative process: actions of an organized system by which it becomes more competent in resisting stresses.

crossing over: a process during early meiosis in which homologous (synapsed) chromosomes exchange segments.

cystic fibrosis: genetic enzyme deficiency causing abnormal mucous secretions; leads to early death or lifelong invalidism.

cytoplasm: the portion of the cell outside the nucleus; contains most of the metabolic machinery.

D & C: *(See dilatation and curettage).*

deletion: abnormal chromosome condition in which a segment of a chromosome is missing.

delivery: the successful completion of childbirth.

depression after childbirth: see *post-partum depression.*

diabetes: a condition characterized by excessive production of urine. It involves a serious deficiency in the body's ability to use sugar because the hormone insulin is not produced by the pancreas.

diaphragm: a rubber membrane placed over the cervix to prevent conception.

differential growth: important embryonic process; the cells in one portion of the embryo grow faster than those in another part, drastically changing shape of the whole embryo.

differentiation: process during embryonic development by which cells descended from the one original zygote change into about a hundred different kinds of cells.

dilatation and curettage: a medical procedure in which the opening to the uterus is stretched (dilatation), and the lining of the uterus is scraped (curettage); used to remove unexpelled tissues after a spontaneous abortion, and as a method of intentional abortion; commonly referred to as D & C.

diploid: having two of each type of chromosome; the usual condition of the ordinary body cell.

dominant gene or allele: an allele that gives the same effect in the phenotype whether that allele is present in single dose (heterozygous) or in double dose (homozygous).

double blind experiment: testing procedure in which neither the patient nor the doctor checking his response knows whether the patient received the drug being tested or a placebo; used to prevent patients or doctors from imagining positive or negative results just because they expect them.

ectoderm: one of the three layers of cells in the earliest embryo; will give rise to the amnion, skin, and nervous system.

endoderm: one of the three layers of cells in the earliest embryo; will give rise to the yolk sac and to the lining of the intestinal tract.

entoderm: endoderm.

egg: the ovum, or the female gamete.

ejaculation: involuntary muscular activity of the male's reproductive tract, expelling the semen.

ejaculatory duct: part of the urethra; the duct from the prostate to the tip of the penis, involved in the discharge of semen.

embryo: the early stages of a new individual.

endocrine: a hormone, or having to do with hormones.

endocrine gland: a gland that expels its product directly into the blood stream rather than through a duct; a gland that produces hormones.

epididymis: coiled tube next to the testis in which sperm are stored and nourished.

epigenesis: philosophical idea that the events at any one time in embryonic development depend on previous events, but are not totally determined by them.

episiotomy: an incision to enlarge the vaginal opening in order to prevent tearing during childbirth. The incision is repaired after delivery.

erection: stiffening of the penis by engorgement of its tissues with blood; similar stiffening of other body parts.

estradiole: the most important of the estrogens.

estriole: one of the female sex hormones.

estrogen: any of a group of female sex hormones acting to promote estrus in many mammals.

estrone: a derivative of the female sex hormones.

estrus: in most mammals, the limited period of time during which the female will accept the mating approach of the male.

eugenics: planned improvements in human inherited traits.

eugenics, negative: improvement of human heredity by eliminating unfavorable traits.

eugenics, positive: improvement of human heredity by increasing the frequency of favorable traits.

extraembryonic membranes: the chorion, amnion, and yolk sac. They are of embryonic origin, but function only during embryonic life, and are therefore not considered a part of the embryo proper.

FSH: see *follicle stimulating hormone*.

Fallopian tube: oviduct.
family planning: planned spacing of pregnancies; may or may not involve choosing a small family.
favism: one form of a group of diseases called "glucose-6-phosphate dehydrogenase deficiency" (or G6PD deficiency); causes fragility of the red blood cells, and may give some protection against malaria.
fertility drug: any chemical that promotes conception, generally by stimulating ovulation; usually a hormone.
fertilization: combination of a sperm with an ovum to form a zygote.
fetal surgery: surgical repair of a fetus in the uterus.
fetishism: sexual aberration in which the person is sexually stimulated by some inanimate object.
fetus: a late embryo, especially after the second month.
fields: regions in an early embryo which are already partially committed to become a certain organ, despite the lack of any visible sign.
follicle: cluster of ovarian cells containing the oocyte that is developing into an ovum. Some of the follicle cells produce the female sex hormones.
follicle stimulating hormone: pituitary hormone that stimulates maturation of gametes; known as FSH.
forceps, obstetric: a kind of tongs for grasping the baby's head during the birth process without injuring it.
foregut: the anterior part of the embryonic gut.
foreplay: activities between sexual partners that initiate and increase sexual arousal.
fraternal twins: individuals conceived and developing at the same time in the same uterus, but from two separate ova fertilized by two separate sperm.
frigidity: inability of a woman to achieve a normal sexual arousal; specifically, the inability of a woman ever to reach climax during intercourse.
fronto-nasal process: structure growing down from the top of the embryo's head and giving rise to much of the face.
G6PD deficiency: (see *favism*).
gamete: a special cell, with half the usual number of chromosomes; combines with another gamete in sexual reproduction. Gametes are sperm or egg.
gene: hereditary instructions for one characteristic.
gene frequency: in any certain population, the proportion for each allele at a certain locus.
genital swellings: paired elevated regions in the early embryo, developing into the scrotal wall or the labia majora, depending on the sex of the embryo.
genital tubercle: a median protrusion in the early embryo that becomes the glans (tip) of the penis or the clitoris.

genitalia, external: the visible organs associated with sexual intercourse, including penis and scrotum in the male, and vaginal opening, clitoris, and labia in the female.
genetics: the study of biological inheritance.
genetic engineering: the proposed design and production of a vastly superior type of human by skillfully combining parts of cells of people with selected characteristics.
genetic population: a group of individuals of the same species that are more likely to mate with each other than with other groups of individuals. A genetic population usually, but not always, occupies a certain geographical area.
genetic surgery: the proposed possibility of replacing an undesirable gene in a living person with a desirable one.
genome: the whole set of genetic instructions of a cell or organism.
genotype: the actual set of genes that an individual has, stated according to how many loci are being studied.
gestation: embryonic development.
gill arches: embryonic structures in the throat region of the embryo. They tend to develop into supporting structures containing bone or cartilage.
gill pouches: embryonic structures consisting of expansions of the body wall between adjacent gill arches. Some develop into glands.
glans: the terminal portion of the penis. It is smooth, somewhat larger than the shank of the penis, and well supplied with nerve endings.
gonads: the primary sex organs, testes in male, ovaries in female.
gonadotropin: a hormone that stimulates growth or secretion by the gonads.
gonorrhea: a very common venereal disease causing inflammation of membranes inside the sexual tracts, and discharge of pus; may sometimes cause infertility and other serious damage.
gumma: a localized breakdown of tissue because of the late effects of syphilis.
gut: embryonic structure that will give rise to the digestive tube and associated organs.
gynecology: medical specialty in disorders of the female reproductive organs.
haploid: having only one chromosome of each type. Gametes are haploid.
heat: (see *estrus*).
hemoglobin: the red protein pigment that carries oxygen in the blood.
Hemoglobin S: the abnormal type of hemoglobin found in cases of sickle-cell anemia.
hemophilia: inherited tendency to bleed too easily due to improper clotting of blood; famous for its prevalence in the royal families of Europe.

heritability: a term in population genetics that is a rough measure of the fraction of variation of a certain trait caused by genetic differences rather than environmental ones.
heterosexuality: sexual attraction by members of opposite sexes (as opposed to homosexuality).
heterozygous: having two different alleles occupying the two examples of a certain locus.
hindgut: the posterior part of the embryonic gut.
homeostasis: maintenance by a cell or organism of an internally steady condition; for instance, our body temperatures are the same summer and winter.
homologous chromosomes: chromosomes that come together (synapse) during meiosis. They are usually of the same size and shape and carry genes for the same kinds of characteristics.
homosexuality: sexual attraction toward a member of the same sex: carrying on sexual activities with a member of the same sex.
homozygous: having all (usually two) of the examples of one locus filled with the same allele.
hormone: special chemical produced in one organ and having an effect on another organ.
Hutterites: a religious group (conservative Protestants, related theologically to Baptists) who live in colonies in the Dakotas, Montana, and Canada, where they own their economic assets in common. They have the highest growth rate by birth of any group studied, since they ascribe a positive virtue to large families and are prosperous enough to support them.
hymen: a thin fold of skin that partially covers the vaginal opening in many women who have not had sexual intercourse. It is used as an indicator of virginity, although not always reliably.
hypophyseal portal system: blood vessel carrying the releasing hormones from the hypothalamus to the pituitary.
hypophysis: (see the *pituitary gland*).
hypothalamus: a part of the brain concerned with regulation of several internal organs, and especially concerned with influences on the pituitary gland.
hysterectomy: an operation to remove the uterus.
hysterotomy: an operation to remove a fetus from the uterus; a type of induced abortion.
IUD: (see *intrauterine device*).
identical twins: individuals that developed from a single zygote.
implantation: active process in which the early embryo (blastocyst) imbeds itself in the uterine wall.
impotence: generally, lack of power; specifically, the failure of a man's penis to remain erect during intercourse, so that his efforts to complete intercourse fail.
incest: sexual union between close relatives.

independent assortment: the genetic observation that the genes found at different loci are distributed to the offspring independently; the combination of alleles that an individual receives at one locus does not predict the combination of genes he receives at another locus.

individualized medicine: medical support procedures tailored to the needs of the individual.

induction: embryonic process in which a cell of one type influences a cell of another type to change into a cell of still a third type.

inguinal canal: the passageway through which the testes descend in early life into the scrotum; a common site for hernia (protrusion of a loop of intestine.)

inner cell mass: the inner part of the blastocyst. It will give rise to the embryo proper.

instincts: inborn behavioral traits, particularly complicated ones.

intercourse, sexual: also called sexual union, copulation, coitus, making love, sleeping with, carnal knowledge; a set of activities usually culminating in the insertion of the penis into the vagina and ejaculation of semen.

interstitial cell: small cell type found in the testis; produces testosterone, the male sex hormone.

intrauterine device: a contraceptive device, usually a specially bent wire or plastic rod, placed permanently in the uterus; commonly called IUD.

intromission: the insertion of the penis into the vagina.

karyotype: an arrangement of photographs of the chromosomes according to size, so that individual chromosomes may be identified, and abnormalities may be detected.

Klinefelter's syndrome: condition caused by the XXY condition; a sterile, feminized male.

LH: (see *luteinizing hormone*).

labia majora: the pair of raised swellings surrounding the labia minora and the vaginal opening.

labia minora: the pair of skin folds immediately on each side of the vaginal opening.

labor: the strong uterine contractions that expel the baby from the uterus to the outside.

labor pains: the sensation of uterine contractions during childbirth.

lens: the variable part of the transparent portion of the eye. Changes in the curvature of the lens can change the focus of the eye.

lesion: an injury to a tissue or organ caused by disease.

limb buds: projections from the trunk of the early embyro that will elongate into arms and legs.

linkage: the slight tendency for genes at loci of the same chromosome *not* to assort independently.

locus: from the Latin for "place;" a location on a certain kind of chromosome, containing information for one kind of hereditary trait, such as eye color.
lubrication of vagina: secretion of fluid by vaginal wall that facilitates intercourse.
luteinizing hormone: pituitary hormone that stimulates hormone production in the gonads; generally referred to as LH.
Lyon hypothesis: the idea that in any certain cell of a female mammal, only one of the X chromosomes is active; different cells of the same female individual may have the genes of different X chromosomes expressed.
masochism: sexual aberration in which the person is sexually stimulated by the infliction of pain on himself or herself.
masturbation: sexual self-stimulation using the hand or some other object as a substitute for the penis or vagina of a partner.
maxillary process: an embryonic structure, derived from a gill arch, that grows in from the side of the face to form the upper jaw.
median nasal process: structure projecting down into the face region and growing to form the nose and the middle of the upper jaw.
meiosis: the special type of cell division that gives rise to gametes, each with half the number of chromosomes found in the original cells.
melanin: the standard black pigment found in all mammals and in some other kinds of organisms. Humans vary considerably in the amount of melanin they have, but all except albinos have some.
menarche: the first period of a girl's menstrual cycle.
menopause: the ceasing of a woman's menstrual cycles; the end of the reproductive period.
menstrual cycle: the monthly buildup, breakdown, and discharge of the uterine lining, brought about by a complicated interaction of hormones.
menstrual phase: portion of the menstrual cycle during which the uterine lining is being broken down and discharged.
menstruation: the monthly discharge of a mixture of blood and fragments of the uterine lining.
mesoderm: the middle (and last formed) of the three layers of cells making up the earliest embryonic structures; will give rise to most of the bulk of the body, including muscle, bone, and urinary system.
metaphase: part of cell division during which the chromosomes are lined up and ready to move apart; the nuclear membrane has disappeared.
migration: important embryonic process in which a whole group of cells streams to another part of the embryo, altering relationships.

minipill: popular name for small doses of synthetic hormones that prevent pregnancy while allowing ovulation; has very few side effects.
miscarriage: spontaneous abortion, especially when it is early in pregnancy.
mitosis: ordinary cell division, giving rise to two cells, each with the same set of genes and chromosomes that the original cell had.
monoploid: (see *haploid*).
monosomy: condition in which an individual has only one chromosome of a certain type; compatible with life only when it involves the X chromosome.
morning-after pill: a preparation of synthetic sex hormones that prevents pregnancy when given after sexual intercourse.
mortality: death; sometimes the statistical probability of death, as in numbers per thousand.
mosaic condition: a rare condition in which an individual has developed from the combination of two zygotes, so that some of the cells have one genetic constitution and some have another.
Mullerian ducts: paired ducts in the early embryo that will become the oviducts in the female.
multipara: medical term for a woman who has borne more than one child.
mutation: a permanent change in the genetic instructions of a cell; may be accidental and random.
natural childbirth: (see *trained childbirth*).
Nature-Nurture controversy: the argument over the relative importance of heredity and environment in determining human characteristics.
neonatal: referring to the newborn child, or to the period immediately after birth.
neural crests: clusters of cells derived from the neural tube that develop into peripheral nerve cells.
neural tube: hollow derivative of ectoderm that will develop into the nervous system.
neurohypophysis: (see *posterior lobe of pituitary*).
nocturnal emission: ejaculation of semen during sleep, often accompanied by a sexual dream.
nondisjunction: abnormal event in gamete formation in which a pair of synapsed homologous chromosomes fails to separate, so that gametes get either two or none of that kind of chromosome.
notochord: an early embryonic structure that is related to the development of the vertebral column.
nullipara: medical term for a woman who has not borne a child.
obstetrics: medical specialty concerned with pregnancy and birth.

once-a-month-pill: a contraceptive medication that requires administration only once a month. Most of these act by causing abortion of the just-started embryo.

Oneida Community: a religious colony of the last century that practiced not only common ownership of property, but complete sexual sharing. It would now be called "group sex," "open marriage," or a "free-love commune."

oogenesis: the meiotic divisions and cellular transformations that give rise to the ovum. Oogenesis is stretched out from the time of the female's birth to the release of ova.

oocyte: a cell in the ovary that may divide to become an ovum.

oogonium: the cell of the ovary that begins the process leading to mature ova; all of these start their transformation at a period near the birth of most female mammals.

oral contraceptives: a chemical, usually a synthetic sex hormone, administered by mouth for the prevention of conception; popularly called "the Pill."

orgasm: in either sex, as a culminating effect of sexual stimulation, intense and involuntary contractions in the genital tracts, accompanied by other movements and by intense pleasure.

ossification: hardening of softer tissues into bone.

ovary: a female gonad, producing eggs (ova) and sex hormones.

oviduct: the tube carying the ovum from the ovary to the uterus.

ovulation: the release of the ovum from the ovary.

ovum: the egg; the gamete produced by a female.

oxytocin: hormone released from the posterior lobe of the pituitary that causes muscular contractions in breasts and the female genital tract.

paraplegic: having the spinal cord severed so that the lower part of the body cannot be voluntarily moved.

parthenogenesis: reproduction by the development of unfertilized eggs.

parturition: medical term for childbirth.

pelvis: the portion of the body trunk below the waist; the hip region.

phallus: Greek term for the penis.

phenotype: the actual discernable characteristics that an individual has.

phenylalanine: an amino acid (one of the building-blocks of protein). Some of it is transformed by body cells into black pigment, into adrenalin, or into other kinds of amino acids.

phenylketonuria: a genetic defect in metabolism, so that dietary phenylalanine cannot be used for many normal functions. Because of the buildup of abnormal products, this causes mental deficiency in affected infants.

physiology: study of functions of the body or of its parts.

Pill, the: (see *oral contraceptives*).
pituitary gland: endocrine gland under the brain that provides a link between the nervous system and the endocrine system.
placebo: a harmless, noneffective preparation designed to resemble a medicine or treatment and intended to make the patient believe he has received an effective treatment. Used for patients with imaginary ailments and in programs to test the effects of new medicines.
placenta: organ formed of embryonic tissues joined to maternal tissues that allows the exchange of food, wastes, and oxygen between the two circulations.
polar bodies: the small, nonfunctional cells produced by unequal division of the oocytes; contains the chromosomes to be discarded.
poles of the cell: locations in the dividing cell toward which the separated chromosomes move.
population explosion: recent rapid growth of the human world population, so that the number of people is doubling in little more than a generation.
population genetics: study of the distribution of genetic traits in a population, and of the change in time of this distribution.
post partum: medical term for the period after childbirth.
post-partum depression: a temporary depression that women often feel in the first few days after giving birth. It is hormonally caused.
posterior lobe of pituitary: organ concerned with the release of two hormones produced in the brain.
pre-eclampsia: (see *toxemia of pregnancy*).
preformation: the philosophical idea that the gamete or zygote already contains the structure that will appear in the adult, but in a preformed miniature state.
premature infant: defined in medical practice as a newborn weighing less than 5.5 pounds.
prepuce: fold of skin that normally covers most of the glans of the nonerect penis.
primary germ layers: the three layers of cells (endoderm, mesoderm, and ectoderm) in the earliest embryo; will give rise to all later embryonic structures.
primate: any of the group of animals including monkeys, apes, and man.
primigravida: medical term for a woman who is pregnant for the first time.
primipara: medical term for a woman giving birth to her first child.
progesterone: the special female sex hormone produced by the corpus luteum and acting to maintain the health of the uterine lining.

proliferative phase: the portion of the menstrual cycle during which the uterine lining is thickening. This phase terminates at ovulation.
promiscuity: sexual activity with many partners.
prophase: early part of cell division when chromosomes are becoming visible.
prostate gland: located at the junction of the vas deferens ducts and the urethra; contributes secretion to the seminal fluid.
puberty: the time of sexual maturation.
pubic region: the lower abdomen; the part around the genitalia that is covered with hair in mature people.
quantitative inheritance: the genetic transmission of the information leading to variations in height, weight, intelligence, or any other trait that may occur in various amounts.
race: a genetic population noticeably different from some other populations.
radiation, ionizing: penetrating types of radiation from x-ray machines or from radioactive isotopes that can cause profound chemical change and therefore genetic mutation, cancer, birth defects, or early death, depending on the dose and the circumstances.
recapitulation: theory in embryology that the sequence and appearance of embryonic structures is a reflection of evolutionary history, that the embryo goes through the life forms of many of the ancestors of the animal developing.
recessive gene or allele: an allele that gives a visible effect in the phenotype only if it is present in double dose (homozygous).
recombination: the genetic observation that offspring may have different combinations of alleles at a certain locus than the parents had.
releasing factors: hormones produced by the hypothalamus of the brain, that stimulate the pituitary to release specific hormones such as FSH or LH.
retina: the portion of the eye that sends out nerve impulses when struck by light.
rhythm method: a method of attempting to prevent pregnancy by avoiding sexual intercourse at the time when ovulation is expected.
rubella: three-day or German measles; a mild disease that may cause serious defects in an embryo carried by an infected woman.
sadism: sexual aberration in which the person gets sexual stimulation by inflicting pain on someone else.
safe period: the time during the menstrual cycle when ovulation is unlikely, and during which conception is also unlikely.
saline infusion: a recent technique for inducing abortions comparatively late in pregnancy. Strong salt solution is injected into the amniotic cavity, killing the embryo, which is then expelled.

scrotum: the external pouch of skin containing the testes.
secretory phase: period of the menstrual cycle after ovulation has occurred, when the glands of the uterine wall are secreting actively.
segmentation: anatomical plan found in many lower animals in which much of the body is made up of several nearly identical segments.
segregation: process in gamete formation by which the two genes for the same kind of characteristic become distributed into separate gametes.
semen: mixture of sperm and glandular fluids expelled from the penis during copulation.
seminal vesicle: a gland adjacent to each sperm duct that contributes a secretion to the fluid that is ejaculated; once thought to store sperm.
septal defect: birth defect affecting the heart, in which the wall separating two heart chambers has an opening through it.
sex chromatin: a stainable spot in the nucleus of certain cells indicating the presence of more than one X chromosome.
sex chromosome: an X or Y chromosome. An individual with two X chromosomes is female; with one X and one Y chromosome, male.
sex hormones: hormones produced by the gonads of either sex which act to cause the visible differences between the sexes and to promote the events leading to sexual reproduction.
sex-linked genes: those carried on the X chromosome.
sexual control center: a nerve center in the brain that integrates the effects of sexual stimulation and inhibition so that sexual arousal does or does not occur.
sexual revolution: recent rapid changes in frequencies of various sexual practices and in social attitudes toward them.
sickle-cell anemia: an inherited disorder, in which the hemoglobin differs slightly from normal and the red blood cells are easily destroyed. Victims usually die young.
sickle-cell trait: inherited condition in which half the hemoglobin is abnormal. The trait rarely causes any trouble, but two parents with the trait may have children who suffer from sickle-cell anemia.
somites: clumps of mesoderm cells in the embryo which develop into bone and muscle.
somatic cells: ordinary cells of the body, as opposed to the special cells that give rise to gametes.
Spaceship Earth: a phrase employed to remind us that space and resources on the earth are limited, just as though we were traveling on a giant spaceship.
sperm or **spermatozoan:** the gamete produced by the male.

spermatid: the product of meiotic division in the testis; loses most of its cytoplasm to become a mature sperm.

spermatocyte: cell of the testis; part of the developmental sequence from spermatogonium to sperm. It is the spermatocytes that undergo meiotic division.

spermatogenesis: the processes of meiotic divisions and cellular transformations that give rise to sperm.

spermatogonia: cells of the testis which either divide to form more spermatogonia or begin the series of steps leading to sperm production.

spermatozoa: (see *sperm*).

spindle: a temporary structure to which chromosomes are attached during cell division. The chromosomes are transported to opposite ends of the cell by shape changes of the spindle.

spinal cord: the part of the central nervous system extending down within the backbone.

spinal sexual center: set of connections in the spinal cord responsible for integrating the nervous activity of sexual climax.

spreading of layers: important embryonic process in which cells of a certain region move in a layer over a surface or between other layers.

sterilization: permanent elimination of fertility, by preventing the release of gametes.

suction curettage: (see *vacuum aspiration*).

sudden crib death: unexplained deaths of apparently healthy infants during sleep. These are sometimes erroneously blamed on suffocation by bedcovers.

synapsis: the unique process during meiosis by which homologous chromosomes come together and become temporarily attached.

syphilis: very serious venereal disease with minor early symptoms, but severe late effects.

taxonomy: division of biology that deals with classification of organisms.

technological fix: a proposed technical advance that is expected to cure some social problem.

telophase: final part of cell division during which the chromosomes fade from sight and the nuclear membrane reforms.

teratogenic agent: any influence that can cause a birth defect.

tertiary syphilis: the late, serious, effects of syphilis, including some which are debilitating or fatal.

testis: a male gonad, producing sperm and sex hormones.

tetrad: a bundle of four chromosome strands during gamete formation; made up of two synapsed chromosomes, each split in two.

thalidomide: a mild sedative that was found to cause failure of limb development in embryos that were in the uterus when the drug was used by pregnant women.

throwback: popular term for the appearance of an offspring that resembles a fairly remote ancestor without intermediate generations having shown the noticeable characteristics; roughly equivalent to what a geneticist calls "segregation."

toxemia of pregnancy: also called "pre-eclampsia"; a collection of symptoms which may not have a common cause. The symptoms include edema, high blood pressure, and kidney malfunction. Possible causes include hormonal imbalance and allergic reaction to the fetus.

trained childbirth: the practice of giving prospective parents information about what to expect at birth, and training in procedures that are designed to reduce or eliminate the need for anesthetics during childbirth; related to natural childbirth.

translocation: abnormal chromosome condition in which one kind of chromosome or a part of it is permanently attached to another kind of chromosome so that the cell has an extra dose of that segment. Translocations are at least as frequent as the trisomies.

trisomy: a condition in which an individual has three of one kind of chromosome, instead of the normal two. This generally occurs in just two of the 23 kinds of chromosomes. Trisomies may be fatal before birth.

trisomy 21: Down's syndrome, mongolism; condition caused by having three chromosomes designated as chromosome 21 in each cell. It is an important cause of severe mental deficiency.

trophoblast: the outer layer of cells of the blastocyst; important at the time of implantation in the uterus. The trophoblast functions principally in the absorption of food; it later becomes the chorion.

tubal ligation: female sterilization by tying and cutting the uterine tubes (oviducts, fallopian tubes).

tumescence: swelling with blood; when applied to sexual organs, nearly synonomous with erection.

Turner's syndrome: condition caused by the XO condition; a sterile female.

umbilical cord: bundle of blood vessels joining the fetus and the placenta.

urethra: the duct from the bladder to the outside. In the male, the outer portion of the duct also carries semen.

urogenital folds: paired structures in the early embryo which become the shank of the penis in the male, the labia minora in the female.

urogenital groove: a median slit in the early embryo that fuses shut in the male, but is retained as the vaginal opening in the female.

urogenital sinus: space in the early embryo derived from the hindgut that will later become the urinary bladder.

uterine tube: (see *oviduct*).

uterus: hollow organ in which the embryo develops. Embryologically it is an enlargement of the fused oviducts.

VD: see *venereal disease*.

vacuum aspiration: new procedure for induced abortion in which the uterine contents are sucked out through a tube.

vagina: the tube connecting the uterus to the exterior; it receives the penis and the sperm at intercourse, and is the birth canal for exit of a baby from the uterus.

vaginismus: muscular spasm of vagina that prevents intercourse.

vas deferens: tube carrying sperm and fluid secretions from the testis to the urethra.

vasectomy: male sterilization, by cutting and tying the vas deferens.

venereal disease: any disease spread by sexual intercourse, especially syphilis and gonorrhea.

ventricle: chamber of the heart that receives blood from the atrium and pumps it into the arteries.

wet dream: popular term for a nocturnal emisison of semen.

withdrawal: a contraceptive practice in which the penis is withdrawn from the vagina just before ejaculation of semen.

Wolffian duct: one of a pair of tubes in the early embryo, which in the male will become the vas deferens and associated structures.

womb: common term for the uterus.

X chromosome: one of the two sex chromosomes. There are two in the female, one in the male.

Y chromosome: one of the sex chromosomes, present only in the male.

yolk sac: a temporary embryonic membrane, which in the bird or reptile serves to absorb foodstuffs from the yolk, but in the mammal simply gives rise to the gut.

zygote: cell formed by the union of a sperm with an egg; it becomes the embryo.

zygote adoption: also called embryo transplantation or ova transfer. The transfer of a zygote from the uterus of one female to that of another, where it develops normally, as though it were her genetic offspring. Already done with domestic animals and predicted for humans in the future.

Suggested readings

Chapter 1

Sadleir, R. M., *The Reproduction of Vertebrates*. New York, Academic Press, Inc., 1973. Compares reproductive patterns of animals, including man.

Chapter 2

Carter, C. O., *Human Heredity*. Baltimore, Penguin Books, Inc., 1962. A somewhat technical book.

Winchester, A. M., *Human Genetics*. Columbus, Ohio, Charles E. Merrill Publishing Company, 1971. An excellent small introductory book.

Several more advanced texts in human genetics are available.

Chapter 3

Guyton, A. C., *Basic Human Physiology*. Philadelphia, W. B. Saunders Company, 1971. Probably the most readable college text on the subject.

Masters, W. H., and V. E. Johnson, *Human Sexual Response*, 1966, and *Human Sexual Inadequacy*, 1970. Boston, Little, Brown and Company. Very technical. Much of the information is summarized in the following book by McCary.

McCary, J. L., *Human Sexuality*. New York, D. Van Nostrand Company, 1973. A wide-ranging elementary college text.

Morris, D., *The Naked Ape*, 1967, and *The Human Zoo*, 1969. New York, McGraw-Hill, Inc. (Dell, paper). A zoologist's view of human behavior and misbehavior, respectively.

Odell, W. D., and D. L. Moyer, *Physiology of Reproduction*. St. Louis, The C. V. Mosby Company, 1971. A slim but very technical book.

Chapter 4

Ehrlich, P. R., *The Population Bomb*. New York, Ballantine Books, Inc., 1968. A widely read, popular book that may occasionally overstate the case.

Guttmacher, A. F., *Birth Control and Love*. New York, The Macmillan Company, 1969 (Bantam, paper). One of the most complete of the many good books and pamphlets on the subject.

Hardin, G., ed., *Population, Evolution, and Birth Control*. San Francisco, W. H. Freeman, 1969. A collection of readings, including several essays by Hardin.

The Population Bulletin. Washington, D.C., Population Reference Bureau, Inc. A small, informative, and readable magazine on the population problem, costing $8 a year.

Chapter 5

Who Shall Live? Man's Control Over Birth and Death. New York, Hill & Wang, 1970. Discussion of the ethics of abortion and of other problems posed by modern medicine. Commissioned by the American Friends Service Committee.

(The following books listed for other chapters are equally applicable to this one: Chapter 2, Winchester, Carter; Chapter 4, Hardin; Chapter 6, Allen.)

Chapter 6

Allen, F. D. *Essentials of Human Embryology*. New York, Oxford University Press, 1969. A clearly written book, but it assumes an advanced vocabulary.

Saxén, L., and J. Rapola, *Congenital Defects*. New York, Holt, Rinehart and Winston, Inc., 1969. Technical but clear review of recent research.

Tanner, J. M., G. R. Taylor, and Editors of Life, *Life Science Library: Growth*. New York, Time, Inc., Book Division. An incomparable series of photographs of human development.

Volpe, E. P., *Human Heredity and Birth Defects*. New York, Bobbs-Merrill Company, 1971. A small book designed for the layman.

Chapter 7

Guttmacher, A. F., *Pregnancy and Birth*. New York, The Viking Press, Inc., 1962 (Signet, paper). Probably the best popular book on the subject. Written from the medical point of view.

Chapter 8

Eibl-Eibesfeldt, I., *Ethology: The Biology of Behavior*. New York, Holt, Rinehart and Winston, Inc., 1970. Large but readable book, with a chapter on human behavior.

Harlow, H. F., J. L. McGaugh, and R. F. Thompson, *Psychology*. San Francisco, Albion, 1971. Elementary text that draws heavily on Harlow's work with developmentally deprived monkeys.

King, J. C., *The Biology of Race*. New York, Harcourt Brace Jovanovich, Inc., 1971. Vocabulary moderately easy.

Van Lawick-Goodall, Jane, *In the Shadow of Man*. Boston, Houghton Mifflin Company, 1971. Fascinating popular account of the author's remarkable pioneering observations of wild champanzees, and her conclusions.

Chapter 9

Handler, P., ed., *Biology and the Future of Man*. New York, Oxford University Press, 1970. A large volume on the future of biology and its applications, with reproductive changes included in the discussion.

Hardin, G., *Exploring New Ethics for Survival: Voyage of the Spaceship Beagle*. New York, The Viking Press, Inc., 1972. Painful decisions that will be required soon, especially about population.

Hubbard, C. W., *Family Planning Education*. St. Louis, The C. V. Mosby Company, 1973. Principally on contraception, but contains a chapter on venereal disease.

Huxley, A., *Brave New World*. New York, Harper & Row, Publishers, 1932 (Bantam, paper). Still the most frightening of the antiutopias depicted in science fiction, Huxley's brave new world depends on technological control of human reproduction.

Rorvik, D. M., *Brave New Baby*. Garden City, Doubleday & Company, Inc., 1971. A journalistic account of many of the dubious technological improvements that are coming in the reproductive revolution.

Toffler, A., *Future Shock*. New York, Random House, Inc., 1970. On coping with revolutionary changes in our lives.

Index

aberrations of sexual behavior, 106
abortion, 171
    spontaneous, 244
acculturation, 299
adoption, 338
albinism, 162
alleles, 35, 159
allergies during pregnancy, 245
amniocentesis, 152
amnion, 146
anal intercourse, 107
anterior lobe of pituitary, 75
aortic arches, 194
aphrodisiacs, 96
arms and legs, formation of, 201
artificial uterus, 356

babies, *see* infants
baby supermarkets, 339
behavior, sex hormones affect, 74
biological revolution, Chapter 9
birth, 250
    difficulties of, 256
birth defects, 220
    prevention of, 233
blastocyst, 118
blood group inheritance, 44
brain, development of, 206
breast feeding, 266

capacitation, 114
centromere, 29
chancre, 329
chemicals that cause birth defects, 228
childbirth, 250
chorion, 144

chorionic gonadotropin, 150
chromosome, 27
circumcision, 267
cleft palate, cause, 199, 226
cloning, 23, 353
congenital defects, 234
compulsory pregnancy, 174
condom, 128
contraception, methods, 120, 138
corpus luteum, 64, 71
cousin marriage, 286
creative process, development as a, 218
crossing over, 48
cystic fibrosis, 170

D & C, *see* dilatation and curettage
depression after birth, *see* post-partum depression
determination of sex, 49, 202, 270
development of individual organs, Chapter 6
developmental defects, 220
diabetes during pregnancy, 248
diaphragm, 128
differential growth, 224
differentiation, 168, 222
difficulties of childbirth, 256
digestive system, development of, 197
dilatation and curettage, 171
dominant allele or gene, 39
double blind experiment, 122, 322
Down's syndrome, 52
ductus arteriosus, 195

ectoderm, 145
embryonic development, 175, 179

embryonic processes, 213
embryonic stages, 188
endoderm, 144
epididymis, 113
epigenesis, 186
episiotomy, 253, 257
estradiole, 71
estrogens, 73
eugenics, 345
evolution, 160, 287, 343
    of human sexuality, 111
experimentation on humans, ethics of, 322
external genitalia, development of, 204
extra-embryonic membranes, 147
eye, development of, 208

face, development of, 199
family planning, 141
favism, 164
fertility drugs, 119
fertilization, 116
fetal movements, 239
fetal surgery, 353
fetishism, 107
fields, for embryonic organs, 224
follicle stimulating hormone, 75, 120
foregut, 197
frequencies of genes, 285
FSH, *see* follicle stimulating hormone

G6PD deficiency, *see* favism
gametes, 16
genes, 27, 156
    action, 157, 221
    frequency, 285
genetic engineering, 359
genetic populations, 285
genetic screening, 167
genetic surgery, 351
genetic traits, 295
genetics, population, 287
    quantitative, 280, 285
genital system, development of, 202
genotype, 38
gill arches, 199
gill pouches, 199
gonad, 59
gonadotropic hormones, 75
gonorrhea, 328
growth in childhood, 296
gumma, 330
gut of the embryo, 197

heart, development of, 193
    origin of defects, 226

height, inheritance of, 280
hemoglobin S, 287
hemophilia, 163
heritability, 291
heterozygous, 36
hindgut, 197
homeostasis, 78
homologous chromosomes, 28
homosexuality, 107
homozygous, 36
hormone, changes at puberty, 273
    effects on behavior, 74
    interactions, 78
    interactions with nerves, 82
    reproductive effects of, 75
human nature, 301
inheritance of, 310
Hutterites, 142

immortality, technological, 357
implantation, 118
incest, 286
independent assortment, 46
individuality of humans, origin, 279
individualized medicine, 352
induction, 222
infants, care of, 266
infertility, 118
inheritance, Chapters 2, 5, 8
    quantitative, 280
inheritance ratios, 43
inner cell mass, 144
instincts in humans, 314
intercourse, frequency of, 102
    physiology of, 87, 100
intrauterine device, 135, 341
IUD, *see* intrauterine device

Jensen, Arthur, 291

karyotype, 49, 134
kidney, formation of, 201
Klinefelter's syndrome, 58

labor during childbirth, 251
legislation proposed, 142, 365
LH, *see* lutenizing hormone
linkage, 48
locus, 28
lutenizing hormone, 77
Lyon hypothesis, 56

male accessory glands, 113
male contraceptive pill, 126
man-animal combinations, 360

masochism, 107
masturbation, 106
maturation, 189
meiosis, 33–37
menarch, 275
menopause, 277
menstrual cycle, 78
menstruation, 78
mesoderm, 146
midwives, 259
minipills, 125
mitosis, 28–32
mongolism, 52
monosomy, 51
morning after pill, 137
mortality of the newborn, 264
mosaic inheritance, 184
mouth-genital contacts, 107
Mullerian duct, 202
mutations, 60, 159

natural childbirth, 259
Nature and nurture controversy, 316
nerve-hormone interactions, 82
nervous system, development of, 206
neural tube, 190
newborn, care of, 253
    characteristics of, 261, 263
    mortality of, 264
nondisjunction, 51
normal sexual activity, 101
notochord, 190
nuclear transplantation, 21

oogenesis, 64
once-a-month pill, 137
Oneida Community, 108, 277
oral contraceptives, 120
organ building, 189
ovary, 59, 64
overpopulation, 140
ovulation, 68
ovum, 17
oxytocin, 85

parents, importance of, 300
parthenogenesis, 8, 23
penis, development of, 204
phenotype, 38
phenylketonuria, 161
physiology of sexual intercourse, 87, 100
Pill, the, 120
pituitary gland, 75
placebo, 122, 322
placenta, 118, 147

playing God, 366
polar bodies, 66
population explosion, 138, 142
population genetics, 285
post-partum depression, 258
posterior lobe of pituitary, 75, 84
predictions of the future, 335
pregnancy, Chapter 7
    activities during, 243
    allergies during, 245
    complications of, 244
    compulsory, 177
    diabetes during, 248
    diets for, 241
    duration of, 236, 250
    early signs of, 236
    health rules for, 240
    sexual activity during, 244
    toxemia of, 247
    use of drugs during, 242
pregnancy tests, 237
premature infants, 265
prenatal medical care, 240
preformation, 182, 210
primary germ layers, 146
primates, 302
promiscuity, 106
prostate gland, 113
puberty, 273
pubic region, 71

races, 289
racial discrimination, 289
racial intelligence, 291
radiation as a cause of birth defects, 227
recapitulation, 211
recombination, 37
releasing factors, 83
replacement parts for people, 355
Rh incompatability, 245
rhythm method, 126
rubella, 226, 231

sadism, 107
safe period, 126
saline infusion, 172
scrotum, development of, 204
segregation, 25, 37
seminal vesicle, 113
sex, choice for offspring, 342
    determination of, 49, 202, 270
sex chromatin, 55
sex chromosomes, 49
    abnormalities of, 58, 155
sex education, 332

sex hormones, 70
sex-linked genes, 49
sexual aberrations, 106
sexual activity, 86
    during pregnancy, 244
sexual control center, 91
sexual inadequacy, 91, 103
sexual instincts, 95
sexual intercourse, 87, 100
sexual maturity, 275
sexual reproduction, importance, 7
sexual revolution, 326
sickle cell anemia, 165, 288
side effects, oral contraceptives, 122
    technological, 323
skeleton, formation, 201
skin color, inheritance of, 283
speech, human and chimpanzee, 307
sperm, 18
spermatogenesis, 61
spinal sexual center, 96
sterilization, 131
suction curettage, *see* vacuum aspiration
sudden crib death, 264
synapsis, 34
syphilis, 329

technological fixes, 320
technology, side effects of, 323
teratogenic agents, 228
testis, 59, 61

testosterone, 73
thalidomide, 228
toilet training, 267
tool-using, 307
toxemia of pregnancy, 247
trained childbirth, 259
trophoblast, 144
translocation, 53, 59
trisomy, 51
tubal ligation, 131
Turner's syndrome, 58
twins, 184

umbilical cord, 152
urinary system, formation of, 201
urogenital sinus, 202

vacuum aspiration, 171
vasectomy, 131
venereal disease, 328
    epidemic, 331
venereal diseases during pregnancy, 249
viruses as causes of birth defects, 231

withdrawal, 131
Woffian duct, 202

yolk sac, 146

zygote, 16, 116, 174
zygote adoption, 335